ANALYSIS OF OCEANIC WATERS AND SEDIMENTS

ANALYSIS OF OCEANIC WATERS AND SEDIMENTS

THOMAS ROY CROMPTON

CRC Press
Taylor & Francis Group
Boca Raton London New York

CRC Press is an imprint of the
Taylor & Francis Group, an **informa** business

CRC Press
Taylor & Francis Group
6000 Broken Sound Parkway NW, Suite 300
Boca Raton, FL 33487-2742

First issued in paperback 2022

© 2016 by Taylor & Francis Group, LLC
CRC Press is an imprint of Taylor & Francis Group, an Informa business

No claim to original U.S. Government works

ISBN-13: 978-1-498-70152-5 (hbk)
ISBN-13: 978-1-03-234009-8 (pbk)
DOI: 10.1201/b19088

Visit the Taylor & Francis Web site at
http://www.taylorandfrancis.com

and the CRC Press Web site at
http://www.crcpress.com

Contents

Preface

This book covers all aspects of the analysis of seawater using both the classical and the most advanced newly introduced physical techniques.

Until fairly recently, the analysis of seawater was limited to a number of major constituents such as chloride and alkalinity.

It was generally agreed that any determination of trace metals and metalloids carried out on seawater prior to about 1975 was questionable, principally due to the adverse effects of contamination during sampling, which were then little understood and led to artificially high results. It is only during the past few years that methods of adequate sensitivity have become available for true ultra-trace metal determinations in water.

Similar comments apply in the case of organics in seawater as it has now become possible to resolve the complex mixtures of organics in seawater and achieve the required very low detection limits. Only since the advent of sample preconcentration and mass spectrometry coupled with gas chromatography and high-performance liquid chromatography, and derivatisation and conversion of original sample constituents into a form suitable for chromatography, has this become possible.

Fortunately, our interest in microconstituents in seawater from both an environmental and nutrient balance point of view has coincided with the availability of advanced instrumentation, which is capable of meeting the analytical needs.

Chapter 1 discusses an important aspect of seawater analysis, namely sampling. If the sample is not taken correctly, the final result is invalidated, no matter how sophisticated the final analytical procedure. Recent important work on sampling is discussed in detail.

Chapter 2 deals with the method of preservation and integration of samples between the moment the samples are taken to the analysis for which they are being referred.

Chapter 3 discusses the application of techniques such as atomic absorption spectrometry with and without a furnace and Zeeman background correction, inductively coupled plasma mass spectrometry, X-ray fluorescence spectrometry, neutron activation analysis, voltammetric techniques, and more. In the first part of the chapter, the determination of 69 metals and metalloid elements is discussed singly in alphabetical order, then as groups of elements, as the newer techniques often cover the range of elements. Finally, a section on metal preconcentration techniques is provided. By concentrating all the metals present in a large volume of a sample into a few millilitres, dramatic improvements in detection limits can be achieved. Covered in Chapter 4, this enables the techniques to be applicable at the low basal concentrations at which many metals exist in seawater.

Methods for the determination of organic compounds and dissolved organic matter in seawater are examined in Chapter 5; and in coastal and estuary waters in Chapter 7. These include organic compounds such as antibiotics and pharmaceuticals, which are now being found in increasing amounts.

The determination of organometallic compounds in seawater is discussed in Chapter 6. The presence of metals (Chapter 8) and organic compounds in marine sediment (Chapter 9) provides an indication of the time dependence of their concentration over large time spans (Chapter 8). A concentration of sediments is found in estuarine and oceanic sediments, and thus, sediment analysis provides a means of tracking organics from their source through the ecosystem.

Another consideration is that fish, particularly bottom-feeders and mustacea, pick up contaminants when the sediment enters their gills and the contamination of these creatures has a definite toxicological implication both to the creatures themselves, and to the humans who eat them and, in the case of fish meal, for animals.

Sediments have the property of absorbing organic contaminants from water within their bulk (accumulation), and indeed, it has been shown that the concentration, for example, of some types of insecticides in river sediment is some 10,000 times greater than that occurring in the surrounding water.

To date, insufficient attention has been given to the analysis of sediments, and one of the objects of this book is to draw the focus of analysts and others concerned to the methods that are available and their sensitivities and limitations.

Information on metals and organic substances combines all the exciting features of analytical chemistry. First, the analysis must be successful and in many cases must be completed quickly. Often, the nature of the substances to be analysed is unknown, might occur at exceedingly low concentrations, and might, indeed, be a complex mixture. To be successful in such experiments requires analytical skills of a high order and the availability of sophisticated instrumentation.

This book was written with the interests of the following groups in mind: managers and scientists in all aspects of the water industry, river management, fishery industries, sewage effluent treatment and disposal, land drainage, and water supply. It will also be of interest to agricultural chemists and agriculturalists concerned with the ways in which organic chemicals used in crop or soil treatment permeate through the ecosystem; biologists and scientists involved in fish, insects, and plant life; and also to the medical profession, including toxicologists, public health workers, and public analysts. Other groups to whom this book will be of interest include oceanographers, environmentalists and, last but not least, members of the public who are concerned with the protection of our environment.

Finally, it is hoped that this book will act as a spur to students of all subjects mentioned and assist them in the challenge that awaits them in controlling the pollution of the environment so as to ensure that by the dawn of the new millennium we are left with a worthwhile environment to protect.

1 Sampling Devices and Sample Storage of Nonsaline Water and Seawater

Strictly speaking, although methods used in the sampling and storage of seawater are not the subject matter of this book, they are included here since, on occasion, borderline samples such as river water occur, which is heavily diluted with seawater, for example, in an estuary.

1.1 SAMPLING DEVICES

Common to all analytical methods is the need for correct sampling. It is still the most critical stage with respect to risks of accuracy in aquatic trace metal chemistry, owing to the potential introduction of contamination. Systematic errors introduced here will make the analysis unreliable. Severe errors were commonly made during sampling by most laboratories, owing to ignorance or at least underestimation of the problems connected with sampling. This is the principal reason why nearly all trace metal data prior to 1975 for seawater and many freshwater systems have to be regarded as inaccurate or at the very least doubtful.

Surface water samples are usually collected manually in re-cleaned polyethylene bottles (from a rubber or plastic boat) from the sea, lakes, and rivers. Samples are collected in the bow, the front of boats, against the wind. In the ocean or in larger inland lakes, a sufficient distance (about 500 m) in an appropriate wind direction has to be maintained between the boat and the research vessel to avoid contamination. The collection of surface water samples from the vessel itself is impossible, considering the heavy metal contamination plume surrounding each ship. Surface water samples are usually taken at 0.3–1.0 m depth in order to be representative and to avoid interference by the air/water interfacial layer in which organics consequently bound by heavy metals accumulate. Usually, sample volumes between 0.5 L and 2 L are collected. Substantially larger volumes cannot be handled in a sufficiently contamination-free manner in subsequent sample pretreatment steps.

Reliable deep-water sampling is a special and demanding art. It usually has to be done from the research vessel. Special devices and techniques have been developed to provide reliable samples.

Samples for mercury analysis should preferably be taken in pre-cleaned flasks. If, as required for other ecotoxic heavy metals, polyethylene flasks are commonly used

for sampling, then an aliquot of the collected water sample or the mercury determination has to be transferred as soon as possible into glass bottles, because with time
mercury losses can be expected in polyethylene bottles.

Ashton and Chau [1] have reviewed and discussed the techniques for the collection of seawater samples: preservation, storage, and prevention of contamination. The
most appropriate measurement techniques, preconcentration and extraction, method
validation, and analytical control, have been covered. The apparent aluminium content of seawater stored in ordinary containers such as glass and polyethylene bottles
decreased gradually, for example, to half in 2.5 h. But if the samples are acidified
with 0.5 mL/L concentrated sulphuric acid, the aluminium content remains constant
for at least 1 month. Accordingly, samples should be acidified immediately after collection. However, aluminium could be recovered by acidifying stored samples and
leaving them for at least 5 h.

Shipboard analysis for the sampling of trace metals in seawater has been discussed by Schuessler and Kremling [2] and Dunn et al. [3]. Teasdale et al. [4] have
reviewed methods for the collection of sediment pore waters using *in situ* dialysis
samples. Bufflap and Allen [5] compared centrifugation, squeezing, vacuum filtration, and dialysis methods for sediment pore water sampling.

Kelly et al. [43] described a sampling apparatus constructed to collect 28 L of
seawater. This equipment was designed to minimise sample contamination derived
from the ships environment.

The job of the analyst begins with the taking of the sample. The choice of sampling gear can often determine the validity of the sample taken; if contamination is
introduced in the sampling process itself, no amount of care in the analysis can save
the results. Sampling devices and sample handling for hydrocarbon analysis have
been exhaustively reviewed by Green et al. [6,7].

1.2   ANALYSIS OF SURFACE SAMPLES

The highest concentrations of naturally occurring dissolved and particulate organic
matter in oceans are normally found in surface films. When organic pollutants are
present, they too tend to accumulate in this surface film, particularly if they are
either nonpolar or surface active. Much of the available information on these surface
films has been reviewed by Wangersky et al. [8,9].

A major problem in the sampling of surface films is the inclusion of water in
the film. In the ideal sample, only the film of the organic molecules, perhaps a few
molecular layers in thickness, floating on the water surface, would be removed; the
analytical results should then be expressed in terms of either volume taken or surface area sampled.

In practice, all samplers in common use at this time collect some portion of the
surface layer of water. Each sampler collects a different slice of the surface layer;
thus, results expressed in weights per unit volume are only compared when the samples have been taken with the same sampler. Of the surface-film samplers described
in the literature, the 'slurp' bottle used by Gordon and Keizer [10] takes perhaps the
deepest cut, sampling as far as 3 mm into the water. The sampling apparatus consists simply of an evacuated container from which floating tubes with holes extend.

The main advantage of this sampler is its simplicity and low cost. Its disadvantages are the thickness of the water layer sampled and the inherent inability to translate the results of the analysis into units of weight per surface area swept.

A thinner and more uniform slice of the surface can be collected with the Harvey [11] rotating drum sampler. This consists of a floating drum, which is rotated by an electric motor, the film adhering to the drum being removed by a windshield wiper blade. The thickness of the layer sampled depends on the coating on the drum, the speed with which it rotates, and the water temperature. The slice taken will usually run between 60 and 100 μm. This method has the advantage of sampling a known area of ocean surface and of collecting a large sample very quickly. However, like most of the other surface-film samples, it must be operated from a small boat and can, therefore, only be used in calm weather.

The collector most commonly used for surface films is the Garrett screen [12], a monel metal or stainless steel screen, which is dipped vertically below the surface and then raised in a horizontal position through the surface film. The material clinging to the screen is then drained into a collection bottle. This sampler collects a slice of surface somewhere between 150 and 450 μm thick. A relatively small sample, about 20 mL, is collected on each dip, so that the collection of a sample of reasonable size is very time consuming. This method is also limited to calm weather, since the sampling must be conducted from a small boat. While the Garrett screen has been adopted by many investigators because of its simplicity and low cost, as well as its relative freedom from contamination, it is far from satisfactory solution to the problem of surface-film sampling. The small size of the sample taken greatly restricts the kinds of information which can be extracted from the sample. Also, the uncertainty as to the thickness of the film samples makes comparison between samplers difficult.

Daumas et al. [13] compared the Harvey drum sampler to the Garrett screen and found greater organic enrichment in both dissolved and particulate matter in the drum samples. The size of the difference between the two samplers suggests that the Garrett screen included two to three times as much water as did the Harvey drum.

Several samplers have been constructed on the principle of the use of a specially treated surface to collect surface-active materials. Harvey and Burzell [14] used a glass plate, which was inserted and removed vertically with the material adhering to the plate and then collected with a wiper. This method of sampling still included some water; if a surface which preferentially absorbs hydrophobic or surface-active material is used, for example, Teflon®, either normal or specially treated, only the organic materials will be removed, and the water will drain away. Such samplers have been described by Garrett and Burger [15], Larsson et al. [16], and Miget et al. [17], among others. Anufrieva et al. [18] used polyurethane sheets, rather than Teflon, as the absorbent. While this sort of sampler seems, at least theoretically, to have many advantages, since the surface swept is easily defined and the underlying water is excluded, the sample taken is very small. In addition, it must be removed from the sampler by elution with an organic solvent, so the chances of contamination from the reagents used are sharply increased.

An interesting method which combines sampling and analysis in one step has been described by Baier [19]. A germanium probe is dipped into the water and carefully withdrawn, bringing with it a layer of surface-active material.

Organics in the surface layer are then analysed directly by internal reflectance infrared spectroscopy. Since there is no handling of the sample, contamination is reduced to a minimum. However, only infrared spectral analysis is possible with this system; since the material absorbed on the germanium prism is always a mixture of compounds, and since the spectrophotometer used for the production of the spectra is not a high-precision unit, the information coming from this technique is limited. While the identification of specific compounds is not usually possible, changes in spectra, which can be related to the time of day, season, or to singular events, can be observed.

Overall, there is still not really a satisfactory method for sampling surface film. Of the methods now in use, the method used by most researchers is favoured for practical reasons and not for any inherent superiority as a collator; unless a method is available for the determination of total organic carbon, it is difficult to estimate the efficiency of any surface-film collection.

1.3   SURFACE SAMPLING

Sampling subsurface waters, although simpler than sampling surface film, also presents some not completely obvious problems. For example, the material from which the sampler is constructed must not add any organic matter or metal to the sample. To be completely safe, then the sampler should be constructed either of glass or metal. All glass samples have been used successfully at shallow depths [20,21]. To avoid contamination from material in the surface film, these samplers are often designed to be closed while they are lowered through the surface, and then opened at the depth of sampling. The pressure differential limits the depth of sampling to the upper 100 m; below this depth, implosion of the sampler becomes a problem.

Implosion at a greater depth can be prevented either by strengthening the container or by supplying pressure compensation. The first solution was applied in the Blumer sampler [22]. The glass container is actually a liner inside an aluminium pressure housing; the evacuated sampler is lowered to the required depth, where a rupture disc breaks, allowing the sampler to fill. Even with the aluminium pressure casing, however, the sampler cannot be used below a few thousand metres without damage to the glass liner. Another approach to the construction of glass sampling containers involves equalisation of pressure during the lowering of the sampler. Such a sampler has been described by Bertoni and Melchiorri-Santolini [23]. Gas pressure is supplied by a standard diver's gas cylinder through an automatic delivery valve similar to the type used by scuba divers. When the sampler is opened to the water, the pressuring gas is allowed to flow out as the water flows in. The sampler in its original form was designed for use in Lago Maggiore, Italy, where the maximum depth is about 200 m, but in principle it can be built to operate at any depth.

Stainless steel samplers have been devised largely to prevent organic contamination. Some have been produced commercially. The Bodega–Bodman sampler and the stainless steel Niskin bottle, formerly manufactured by General Oceanics, Inc., are examples. These bottles are both heavy and expensive. The Bodega–Bodman bottle, designed to take very large samples, can only be attached to the bottom of the sampling wire; therefore, the number of samples taken on a single station is limited by the wire time available, and depth profiles require a great deal of station time.

The limitations of glass and stainless steel samplers have led many researchers to use the more readily available plastic samples, sometimes with the full knowledge of the risks and sometimes with the hope that the effects resulting from the choice of sampler will be small compared with the amounts of organic matter present. The effects of the containers can be of three classes:

1. Organic materials may be contributed to the sample, usually from the plasticisers used in the manufacture of the samples.
2. Organic materials, particularly hydrophobic compounds, may be absorbed from a solution on the walls of the sampler.
3. Organic materials may be absorbed from the surface film or surface waters, then desorbed into the water samplers at depth, thereby smearing the real vertical distributions.

The first case is the most likely to be a problem with the new plastic samplers. Although there is little in the literature to substantiate the belief, folklore has it that aging most plastic samplers in seawater markedly reduces the subsequent leaching of plasticisers. The second case is known to be a problem; in fact, the effect is used in the various Teflon surface-film samplers already mentioned. This problem alone would see to militate against the use of Teflon for any sampling of organic materials, unless a solvent wash of the sampler is included routinely. With such a solvent, we introduce all the problems of impurities in the reagent.

The third and largely unexpected case appeared as a problem especially in the analysis of petroleum hydrocarbons in seawater [24]. In this case, petroleum hydrocarbons, picked up presumably in the surface layers or surface film, were carried down by the sampling bottles and were measured as part of the pollutant load of the deep waters. While the possibility of absorption and subsequent release is obviously most acute with hydrophobic compounds and plastic samplers, it does raise a question as to whether any form of sampler, which is open on its passage through the water column, can be used for the collection of surface-active materials. The effects of such transfer of material may be unimportant in the analysis of total organic carbon, but could be a major factor in the analysis of single compounds or classes of compounds.

Again, as in the case of surface samples, information on the comparative merits of the various water samplers is largely anecdotal. Although such studies are not inspiring and require an inordinate amount of time, both on the hydrographical wire and in the laboratory, they are as necessary for a proper interpretation of data as are intercalibration studies of the analytical methods. The lack of comparison studies of the various samplers increases the probability of polemics in the literature.

Smith [25] has described a device for sampling immediately above the sediment water interface of the ocean. The device consists of a nozzle supported by a benthic sled, a hose, and a centrifuged deck pump and is operated from a floating platform. Water immediately above the sediment surface is drawn through the nozzle and pumped through the hose to the floating platform, where samples are taken. The benthic sled is manipulated by means of a hand winch and a hydrowire.

1.4   INTERCOMPARISON OF SAMPLING DEVICES AND ANALYTICAL TECHNIQUES USING WATER FROM A CEPEX (CONTROLLED ECOSYSTEM POLLUTION EXPERIMENT) ENCLOSURE

Wong et al. [26] conducted an intercomparison of sampling devices using seawater at 9 m in a plastic enclosure of 65 m in Saanich Inlet, BC, Canada. The sampling methods were as follows:

1. Plastic pumping with Teflon tubing
2. Niskin polyvinyl (PVC) sampler
3. GO-FLO® sampler
4. Close–open–close sampler
5. Teflon-piston sampler

Sampling was conducted for 4 days:

- Day 1 for mercury
- Day 2 for lead, cadmium, copper, cobalt, and nickel by Chelex extraction and differential pulse polarography, as well as manganese by Chelex and flameless atomic absorptiometry
- Day 3 for lead by isotope dilution
- Day 4 for cadmium, copper, iron, lead, and zinc by Freon extraction and flameless atomic absorptiometry

Samples were processed in clean rooms in the shore laboratory within 30 min of sampling. Results indicated (1) the feasibility of intercalibration using the enclosure approach; (2) the availability of chemical techniques of sufficient precision in the case of copper, nickel, lead, and cobalt for sampler intercomparison and storage tests; and (3) a problem in subsampling from the captured water in a sampler and the difficulty of commonly used samplers to sample water in an uncontaminated way at the desired depth.

The Teflon tubing used in the pumping system, the Niskin sampler, and the GO-FLO sampler were cleaned by immersion in 0.05% nitric acid for the tubing and by soaking the inside of the samplers in 0.05% nitric acid overnight, rinsing with distilled water and repeating the dilute acid/distilled water cycle. The close–open–close sampler was cleaned by 0.1 N nitric acid overnight and then rinsed with distilled water until the blank was acceptable. The Teflon-piston sampler was cleaned by sucking in 0.05% nitric acid and standing overnight (in the case of polythene bag liner used in the Teflon-piston sampler, hydrochloric acid was used instead of nitric acid).

The storage bottles were cleaned as follows. The Pyrex bottles (2 L) were used for mercury samples only. They were cleaned by filling with a solution of 0.1% $KMnO_4$, 0.1% $K_2S_2O_8$, and 2% nitric acid, heated to 80°C for 2 h, and after cooling and rinsing, stored filled with 2% nitric acid containing 0.01% $K_2C_2O_7$ until ready for use. Conventional 1 or 2 L polyethylene bottles were used for other metal samples. All bottles were stored inside two or three plastic bags to prevent contamination.

Mercury was determined after suitable digestion by the cold vapour atomic absorption method [27]. Lead was determined after digestion by a stable isotope dilution technique [28–30]. Copper, lead, cadmium, nickel, and cobalt were determined by differential pulse polarography following concentration by Chelex 100 ion-exchange resin [31,32] and also by the Freon TF extraction technique [33]. Manganese was determined by flameless atomic absorption spectrometry. Recoveries were between 83% and 100%. The average mercury content obtained by pumping Nykin samplers, GO-FLO sampler, and the close–open–close devices was 9.09 ± 0.01 ± 8.08 ± 0.8 and 0.01 ± 0.02 nmol/ kg, respectively.

For mercury, the best recoveries were obtained by peristaltic isotope dilution, and mass spectrometry showed the lead values to be 0.73 ± 0.02, 0.72 ± 0.03, 0.75 ± 0.02, 0.78 ± 0.05, and 0.81 ± 0.03 nmol/kg for sampling by peristaltic pump, Niskin sampler, GO-FLO sampler, close–open–close sampler, and Teflon-piston sampler, respectively. For the other two techniques, the Teflon-piston sampler results showed considerable higher values. In the case of mercury, the best recovery was obtained by the peristaltic pump [26].

The Freon extraction and cold vapour atomic absorption approach showed the same range of values as the isotope dilution approach, that is, 0.071 ± 0.36, 0.76 ± 0.13, 0.73 ± 0.13 nmol/kg for the pumping with the Niskin sampler and close–open–close sampler, respectively, with the exception of the GO-FLO sampler with the low value of 0.58 ± 0.15 nmol/kg. However, the range of values was wide, for example, for peristaltic pumping it was 1.12 nmol/kg for the first cast, dropping to 0.46 nmol/kg for the third cast. Chelex extraction and differential pulse polarography showed an even larger spread from 1.09 ± 0.26 nmol/kg for the close–open–close sampler.

Wong et al. [26] concluded the following:

1. It is feasible to capture a large volume of water in the range of 65,000 L by the CEPEX approach for the purpose of sampler intercomparison. It is possible by artificial stimulation of a plankton bloom and detritus removal to produce a reasonably homogeneous body of seawater for the study. The proximity of the *in situ* enclosure for the experiment and the onshore, clean laboratory facilities eliminates errors introduced by shipboard contamination under less than ideal conditions on cruises.
2. The following analytical techniques seem to be adequate for the concentrations under consideration: copper and nickel by Freon extraction and FAA cold vapour atomic absorption spectrometry, cobalt by Chelex extraction and differential pulse polarography, mercury by cold vapour atomic absorption absorptiometry, lead by isotope dilution plus clean room manipulation and mass spectrometry. These techniques may be used to detect changes in the aforementioned elements for storage tests: Cu at 8 nmol/kg, Ni at 5 nmol/kg, Co at 0.5 nmol/kg, Hg at 0.1 nmol/kg, and Pb at 0.7 nmol/kg.
3. The salinity of seawater captured by various sampling devices in the CEPEX enclosure indicates problems not revealed in the usual oceanographic sampling situation. Relative to peristaltic pumping, all samples exhibited some salinity anomalies. Inadequate flushing to rinse the sampler of any concentrated brine or entrapped seawater is thought to be a problem.

4. Logistics and cleaning procedures are important factors in successful sampler intercomparison. It is not desirable or possible to endorse or to condemn the performance of a certain type of sampler or analytical technique based on the results of one set of tests, especially if procedures are changed.
5. The problems of subsampling from the same seawater sample have to be studied in greater detail.
6. A long-term but sustained effort on sampler intercomparison would be advantageous in identifying problems.

Bruland et al. [34] discussed sampling procedures used in their preconcentration–electrothermal atomic absorption spectrometric method for the determination of cadmium, copper, nickel, and zinc in estuary seawater. In this work, surface samples were collected from a small raft rowed crosswind and more than 200 m away from the research vessel. Acid-cleaned polyethylene bottles (500 and 100 m) were submerged off the bow, rinsed and filled with seawater.

Deeper samples were collected by using two different sampling systems. A Teflon-coated 30-1 PVC ball-valve sampler (General Oceanics, GO-FLO) was modified with the replacement of the standard stopcock with a Teflon valve. The GO-FLO samplers are designed to enter the water in a closed, sealed position. At a depth of 10 m, a pressure release allows the ball valves to open and the sampler to fill with water. It then free flushes until tripped by a Teflon messenger. The sampler was clamped on Dacron-sheathed, plastic, Phillystran hydroline 10 m above a polypropylene-enclosed lead weight. The hydroline was led through a stainless steel snatch block-meter wheel to a portable winch with a stainless steel drum. The other sampling system used in this comparative study was the CIT deep-water, common-lead sampler designed and constructed by Schaule and Patterson [35]. In this investigation, sample handling and processing were done onboard the ship inside a modular Porta-lab equipped with a positive pressure filtered air supply and specifically designed for trace metal analysis. After collection, the GO-FLO samplers were secured in a rack on the outside of the Porta-Lab™ where the Teflon valve was rinsed with ultra-clean water and connected to a length of Teflon tubing. This tubing led through the wall of the Porta-Lab to a polypropylene ball valve used to control the flow of seawater. In addition, trace metal samples from the CIT sampler were taken by Schaule [35] while collecting samples for lead analysis.

The seawater samples were preserved after collection by the addition of 4 mL of 6 M quartz-redistilled (Q-)hydrochloric acid per litre.

Because of the potential for contamination, most of the seawater samples were not filtered. The primary purpose of this study was to compare sampling techniques; this comparison could be performed more easily without the extra processing step of filtration. The samples collected closest together (100, 110 m; 400, 410 m) can be compared to demonstrate the consistency between the GO-FLO and CIT samplers. The copper, cadmium, and zinc results obtained by solvent extraction for the different samplers appear identical. The nickel values are in fair agreement. The samplers can also be compared indirectly by means of the vertical profiles obtained. Samples collected by the various sampling techniques yielded consistent data, which fit smoothly into the observed profiles.

Schaule and Patterson [35] found these two samples to be contaminated for lead. On the basis of these results, the copper and nickel concentrations were increased by 20–60 ng/L and 100–1500 ng/L, respectively, but the cadmium and zinc results show no evidence of such contamination. The problem with this seal was subsequently corrected by Schaule, and it appears that these two are the only questionable CIT samples. In addition, the first GO-FLO sample collected on the cruise (50 m) appeared to be contaminated for zinc by approximately 150 ng/L; this was most likely a result of either inadequate flushing of the sampler before collection, or contamination during transfer from sampler to storage bottle. Whatever the cause, this was the only GO-FLO sample that showed a discrepancy from the other samples collected.

Ashton and Chau [1] reviewed the techniques for the collection of seawater samples, preservation, storage, and prevention of contamination of these samples. The most appropriate measurement techniques, preconcentration and extraction, methods validation, analytical control, and reporting formats are covered.

Schuessler and Kremling [2] have discussed the samples of dissolved and particulate trace elements in shipboard analysis.

1.5 INTERCOMPARISON OF SEAWATER SAMPLING FOR TRACE METALS

Several round-robin intercalibrations for trace metals in seawater [26–30] have demonstrated a marked improvement in both analytical precision and numerical agreement of results among different laboratories. However, it has often been claimed that spurious results for the determination of metals in seawater can arise unless certain devices and practical methods of sampler deployment are applied to the collection of seawater samples. It is, therefore, desirable that the biases arising through the use of different, commonly used sampling techniques to be assessed to decide upon the most appropriate technique(s) for both oceanic baseline and nearshore pollution studies.

Two international organisations, the International Council for the Exploration of the Sea (ICES) and the Intergovernmental Oceanographic Commission (IOC), have sponsored activities aimed at improving the determination of trace constituents in seawater through intercalibrations. Since 1975, ICES has conducted a series of trace metal intercalibrations to assess the compatibility of data from a number of laboratories. These exercises have included the analysis of both standard solutions and real seawater samples [36–41]. The considerable improvement in the precision and relevant agreement between laboratories has been reflected in the results of these intercalibrations. By 1979, it had been concluded that sufficient laboratories were capable of conducting high-precision analyses of seawater for several metals to allow an examination to be made of the difference between commonly used sampling techniques for seawater sample collection.

In the early 1980s, the ICO, with the support of the World Meteorological Organisation (WMO) and the United National Environment Program (UNEP), organised a workshop on the intercalibration of sampling procedures at the Bermuda Biological Station, during which the most commonly used sampling bottles and hydrowires were to be intercompared. This exercise formed part of the IOC/WMO/UNEP Pilot Project on monitoring background levels of selected pollutants in open-ocean waters. Windom [42] had already conducted a survey of the seawater sampling and analytical techniques used by marine laboratories, and the conclusions of the survey were largely used for the selection of sampling devices to be intercompared. The bottles selected for comparison in Bermuda were modified and unmodified GO-FLO samples, modified Niskin bottles, and unmodified Hydro-Bios® bottles. GO-FLO samples are the most widely used sampling device for trace metals in seawater. The other two devices continue to be used by several marine laboratories. Windom's [42] 1979 survey established that the most common method of sampler deployment was on hydrowires, as opposed to the use of rosette systems. The hydrowires selected for intercomparison were Kevlar®, stainless steel, and plastic-coated steel. Kevlar and plastic-coated steel were selected because they are widely used in continental shelf and nearshore environments and are believed to be relevantly 'clean'.

The method of intercomparison of the various devices was to deploy pairs of sampler types on different hydrowires to collect water samples from a homogeneous body of deep water at Ocean Station S (Panulirus Station) near Bermuda. The water at this depth has characteristics of $3.97°C \pm 0.05°C$ temperature and $35.01\% \pm 0.02\%$ salinity for January [33]. The restricted length of Kevlar hydrowire available necessitated the collection of samples in the lower thermocline at depths between 1150 and 1250 m.

Data analysis was reduced to a separate one-way analysis of variance on the data from individual laboratories in order to examine the difference between types of sampling bottle on a single (common) hydrowire and to determine the influences of the three types of hydrowires using a single type of sampling bottle (modified GO-FLO). Samples were replicated so that there were, in all cases, two or more replicates to determine the lowest level and analytical error.

Replicate unfiltered water samples were collected for each participant for the comparison of pairs of sampling bottles on different hydrowires. Modified GO-FLO bottles were employed on each of the three hydrowires, and this permitted a comparison of the three types of hydrowires. Only in the cases of iron and manganese were there indications of inhomogeneity at levels that might invalidate the intercomparison. This is assumed to be due to inhomogeneity in the distribution of suspended particulate material, which will influence metals that have major fractions in the particulate phase.

The concentration of trace metals reported in the literature such as plastic-coated steel, hydrowires, stainless steel hydrowires, Kevlar hydrowire, and samples such as modified GO-FLO samples or modified Niskin samplers are not large, and in no way can they account for the recent decline in the oceanic ion extraction of trace metals reported in the literature.

Nevertheless, for several metals, most notably copper, nickel, and zinc, significant differences are different between both bottles and hydrowires. For deep-ocean studies, the best combination of those tested is undoubtedly modified GO-FLO samplers and plastic-coated steel hydrowire. Except in the cases of mercury and manganese, Hydro-Bios samplers appear to yield higher values than modified GO-FLO samplers.

Kevlar and stainless steel hydrowires generally yield measurably greater concentrations of those metals than does plastic-coated steel. These differences, however, are small enough to suggest that these hydrowires are still suitable for trace metal studies of all but the most metal-depleted waters if proper precautions are taken [44].

A major conclusion of the Bermuda experiment is that the use of differing sampling devices and hydrowires only accounts for a small portion of the differences between trace metal results from different laboratories. It appears that the major contributions to such differences are analytical artefacts. It is important to stress that although the sampling tools available to the marine geochemist appear adequate for the measurement of metal distribution in the ocean, the execution of cooperative monitoring programs for metals should be preceded by mandatory intercomparisons of sample storage and analytical procedures.

REFERENCES

1. R. Ashton and R. Chau, *Analyst,* 1987, 112, 841.
2. U. Schuessler and K. Kremling, *Deep-Sea Research Part 1*, 1993, 40, 257.
3. J. Dunn, C.D. Hall, M.R. Heath, R.B. Mitchell, and B.J. Ritchie, *Deep-Sea Research Part 1*, 1993, 40, 867.
4. P. Teasdale, G.E. Batley, S.C. Apte, and I.T. Webster, *Trends in Analytical Chemistry*, 1995, 14, 250.
5. S.E. Bufflap and H.E. Allen, *Water Research*, 1995, 29, 2051.
6. D.R. Green, Committee of Marine Analytical Chemistry, Canadian Centre for Inland Waters in the Environment, Ottawa, Quebec, Canada, 1977, p. 30.
7. D.R. Green, Committee of Marine Analytical Chemistry, Canadian Centre for Inland Waters in the Environment, Ottawa, Quebec, Canada, 1979, p. 63.
8. P.J. Wangersky, *Annual Review of Ecology and Sys*, 1976, 7, 161.
9. P.J. Wangersky, *Deep Sea Research*, 1976, 23, 457.
10. D. Gordon and P.O. Keizer, Technical Report No. 481, Fisheries Research Board, Canada, 1974.
11. G.W. Harvey, *Limnology and Oceanography*, 1966, 11, 608.
12. W.D. Garrett, *Limnology and Oceanography*, 1965, 10, 602.
13. R.A. Daumas, P.L. Laborde, J. Marty, and A. Saliot, *Limnology and Oceanography*, 1976, 319, 319.
14. G.W. Harvey and L.A. Burzell, *Limnology and Oceanography*, 1972, 19, 156.
15. W.D. Garrett and R. Burger, Sampling and determining the concentration of film-forming organic constituents of the air-water interface, U.S. NRL Memo Report 2852, Naval Research Laboratory, Washington, DC.
16. K. Larsson, G. Odham, and A. Sodergrem, *Marine Chemistry*, 1974, 2, 49.
17. R. Miget, H. Kator, and C. Oppenheimer, *Analytical Chemistry*, 1974, 45, 1154.
18. N.M. Anufrieva, A.B. Gornitsky, M.P. Nesterova, and I.A. Neirovskaya, *Okeanologiya*, 1976, 16, 255.
19. R.E. Baier, *Journal of Geophysical Research*, 1972, 77, 5062.
20. B.H. Gump, H.A. Hertz, W.E. May, S.N. Chesler, S.M. Dyszel, and D.P. Enagnio, *Analytical Chemistry*, 1975, 47, 1223.
21. P.D. Keizer, D.C. Gordon Jr., and J. Dale, *Journal of the Fisheries Research Board*, Ottawa, Canada, 1977, 34, 37.
22. R.C. Clark Jr., M. Blumer, and S.O. Raymond, *Deep-Sea Research*, 1969, 14, 125.
23. R. Bertoni and M. Melchriorrye-Santoline, *Memorie Dell'Istituto Italiano Di Idrobiologia*, 1972, 29, 97.

24. D.C. Gordon, D.D. Keizer, Technical Report No. 481. Fisheries Research Board, Ottawa, Canada, 1974.
25. K.L. Smith, *Limnology and Oceanography*, 1971, 16, 675.
26. C.S. Wong, K. Kremling, J.P. Riley et al., Ocean Chemistry Division, Institute of Ocean Sciences, Sidney, British Columbia, Canada; Ocean Chemistry Division contract to SEAKEM Oceanography Ltd., Sidney, British Columbia, Canada; Marine Chemistry Department, Institute für Meereskunde und der Universitat, Kiel, Germany; Department of Oceanography, University of Liverpool, Liverpool, UK, 1985.
27. Participants in the IDOE Workshops, *Marine Chemistry*, 1976, 4, 389.
28. V.J. Stukas and C.D. Wong, *Science*, 1970, 211, 1424.
29. C.S. Wong, K. Kremling, J.P. Riley et al., Measurement of trace metals in seawater, and intercomparison of sampling devices and analytical techniques using CEPAX enclosure in seawater, Unpublished M/S Report, NATO Study, Funded by NATO Scientific Affairs Division, Brussels, Belgium, 1979.
30. M.I. Abdullah, O.A. Ay-Rasin, and J.P. Riley, *Analytica Chimica Acta*, 1976, 84, 363.
31. M.Q. Abdullah and J.G. Royale, *Analytica Chimica Acta*, 1972, 80, 58.
32. J.G. Danielsson, B. Magnusson, and S. Westerland, *Analytica Chimica Acta*, 1978, 98, 47.
33. J.G. Danielsson, B. Magnusson, and S. Westerland, *Analytica Chimica Acta*, 1978, 98, 49.
34. K.W. Bruland, G.A. Knauer, and J.H. Marin, *Liminology and Oceanography*, 1978, 23, 618.
35. B.K. Shaule and C.C. Patterson, Private communication.
36. J.M. Bewers, J. Dalziel, P.A. Yeats, and J.L. Barron, *Marine Chemistry*, 1981, 10, 175.
37. J. Olaffson, *Marine Chemistry*, 1978, 87, 87.
38. J. Olaffson, A preliminary report on ICES intercalibration of mercury in seawater for the Joint Monitoring Group of the Oslo and Paris Commissions, submitted to the Marine Chemistry Working Group of OCS, Paris, February 1980.
39. Y. Thibaud, Exercise d'Intercalibration CIEM, 1979, Cadmium en eau der mer, Report submitted to the Marine Chemistry Working Group of ICES, Paris, 1980.
40. P.G.W. Jones, A preliminary report on the ICES intercalibration of seawater samples for the analyses of trace metals, ICES CM 1977.E, 16, 1977.
41. P.G.W. Jones, An ICES intercalibration exercise for trace metal standard solutions, ICES CM1979.E, 15, 1976.
42. H.L. Windom, 1979, Report on the results of the ICES questionnaire on sampling and analysis of water for trace elements, Submitted to the First Meeting of the Marine Chemistry Working Group, Lisbon, Portugal.
43. A.G. Kelly, L. Cruz, and D.E. Wells, *Analytica Chimica Acta*, 1993, 276, 3.
44. F.R. Selater, A.E. Boyle, and J.M. Edmond, *Earth Planet Science Letters*, 1976, 31, 119.

2 Sample Preservation during Shipment and Storage

Analytical chemists have long been aware of the necessity for the purification of any organic solvents used in trace analysis. The advent of gas and liquid–liquid chromatographs has made it clear just how many impurities can hide behind a 'high-purity' label. Re-distillation of organic solvents just before use is a commonplace in most analytical laboratories.

The actual amount of trace metal contamination in organic reagents present, for example, in reagent grade sodium chloride may be low enough so that it is not listed on the label, but still high enough to produce an artificial water containing more of the substance it is required to determine than the real article. The presence of these compounds becomes serious when the analyst wishes to concoct an artificial water sample for standards and blanks. If the chemical in question can withstand oxidation, either at a high temperature or in the presence of active oxygen, organic material may be eliminated. However, many compounds used in routine analysis cannot be treated in this manner. If such chemicals must be used, the calculation of true methods blank can become a major analytical problem.

Another problem, equally unrecognised, is the organic and inorganic content of distilled water. For most analytical procedures, simple distillation is sufficient treatment; perhaps in special cases, extremes such as distillation from permanganate or distillation in quartz are considered necessary. The experience of many researchers has been that no form of chemical pretreatment will remove all impurities from distilled water and that some form of high-temperature oxidation of organic impurities in the water must be used [1–3]. Depending on the original source of the water, normal distillation will leave between 0.25 and 0.6 mg/L organic matter in the distillate. The amounts and the kinds of compounds may vary with the seasons and with the dominant phytoplankton species in the reservoir.

While at least partial solutions have been found for most of the problems of contamination, these solutions have been largely adopted piecemeal by the various laboratories. The reasons for the adoption of halfway measures are largely historical; the need for purification of reagent has, at last, been accepted.

If we were to choose the ideal method for the analysis of any component of water, it would naturally be an *in situ* method. Where such a method is possible, the problems of sampling and sample handling are eliminated and in many cases we can obtain continuous profiles rather than a limited number of discrete samples. A 'real-time' analysis not only permits us to choose our next sampling station on the basis of the results of the last station but also avoids the problem of the storage of samples until analysis.

The problems of storage and sample preservation are both important since the quantity we wish to measure is the *in situ* value and not the amount remaining after some period of biological and chemical activity.

The complete oxidation of organic contamination in a water sample takes longer than a year [4–6]; however, decomposition great enough to free most inorganic micronutrients takes place in the first 2 weeks of storage. Important changes in micronutrient levels resulting from bacterial utilisation of organic compounds can be seen after 1 day. Therefore, some method of preservation of the organic compounds must be sought if the samples are to be taken back to a laboratory.

The two most popular methods of sample preservation are quick-freezing and the addition of inorganic poisons. Samples that are frozen without the addition of preservatives are less likely to pick up contamination from sample handling.

The determination of traces of heavy metals in natural waters can be greatly affected by contamination (positive and negative) during filtration and storage of samples [3–6]. Until the work of Scarponi et al. [7], this problem, especially in filtration, had not been studied adequately and systematically with respect to the determination of cadmium, lead, and copper in water. Frequently, in order to make the determination easier, synthetic matrix samples [7–9] and/or high metal concentrations [7–11] have been used; otherwise, in order to demonstrate possible filter contamination, washed and unwashed filters have been analysed after washing [11–18].

The same filter can release or absorb trace metals depending on the metal concentration level and the main constituents of the sample [8,19], so that results claimed must be considered with caution in working with natural samples. To avoid contamination, the following procedures have often been used. Filters have been cleaned by soaking them in acids [15,18,20–24] or complexing agents [7–16,20–22,25,26] and/or conditioned either by soaking in seawater [19,27] or by passing through the filter a 0.2–2l sample before aliquots are taken for analysis [18,20,22,28,29]. Sometimes, however, the washing procedure is not found to be fully satisfactory [25]. For the sample, it has been reported that strong adsorption of cadmium and lead occurs on purified unconditioned membrane filters when triple-distilled water is passed through the filter, while there is no change in the concentration with a river water sample after filtration of 500 mL [19]. Some investigators prefer to avoid filtration when the particulate matter does not interfere after sampling [18,19,23,24,27–31]. In the latter case, filtered and unfiltered samples do not seem to differ significantly in measureable metal content [18,32,33].

Scarponi et al. [7] used anodic stripping voltammetry to investigate the contamination of water by cadmium, lead, and copper in nitration samples and the storage of samples collected near an industrial area. Filtration was carried out quickly enough to prevent bacterial growth. Note that the sample bottles must be immersed in a freezing bath to facilitate heat exchange.

While it is normally considered that both biological and chemical reactions will be essentially halted by freezing, this is not necessarily true. It has been shown that some reactions of considerable biochemical importance are, in fact, enhanced in the frozen state [41–43]. In any given case, it cannot simply be taken for granted that freezing will be a sufficient preservative; the efficiency of the method must be tested for the compounds in question. The method is also limited to those analyses that can be performed on a sample, perhaps 100–200 mL. Larger volumes of seawater take too long to freeze. If the next step in the sample preparation is freeze-drying, a considerable saving in time, as well as a decrease in possible contamination, can result from freezing the sample in the container to be used in freeze-drying.

We are almost forced into hoping that freezing will prove satisfactory as a method of sample preservation, since none of the usually inorganic poisons works in every case. Mercuric chloride has been found to be effective with those marine organisms responsible for N_2O production [44].

The problems that can be encountered in sampling, sample, preservation, and analysis have been discussed by King and Ciaco [45], Grice et al. [46], and Bridie et al. [47] (non-hydrocarbon interference); Farrington et al. [48] (hydrocarbon analysis); Kaplin and Poskrebysheva [49] (organic impurities); Hume [50] and Riley et al. [51] (oceanographic analysis); Acheson et al. [52] (polynuclear hydrocarbons); Giam and Chan [53] (phthalates); Giam et al. [54] (organic pollutants); and Grasshoff et al. [55,56] (phosphates).

Mart [57] has described typical sample bottle cleaning routine for use when taking samples for very low-level metal determinations.

Sample bottles and plastic bags, both made of high-pressure polyethylene, were rinsed by the following procedure. They were first cleaned with detergent in a laboratory washing machine, rinsed with deionised water, soaked in hot (about 60°C) acid bath, beginning with 20% hydrochloric acid, reagent grade, followed by two further acid baths of lower concentrations, the latter being from Merck (Suprapur quality or equivalent). The bottles are then filled with dilute hydrochloric acid (Merck, Suprapur); this operation is carried out on a clean bench, rinsed, and then filled with very pure water (pH 2). Bottles are wrapped into two polyethylene bags. For transportation, lots of 10 bottles are enclosed hermetically in a larger bag.

Acidification with mineral acids is also often used as a method for preservation. Preservation of samples is still a major problem; there is no general foolproof method applicable to all samples and all methods of analysis. The most generally accepted method of sample preservation is storage under refrigeration in the dark, with a preservative. This is another area that still needs extensive investigation.

Table 2.1 shows a selection of reagents used for preserving or fixing various inorganic determinants. These reagents are placed in the sample bottle before it is filled with the sample; consequently, the sample is 'protected' from the moment it is taken. Other references for the use of sample preservatives are reviewed in Table 2.2.

The optimal conditions for uncontaminated long-term storage of diluted heavy metal solutions, particularly seawater, are now a topic of great interest with contradictory results being frequently reported. For example, with regard to the type of material to be used for the container, some researchers have recommended the use of linear (high density) polyethylene instead of conventional (low density) [20,22,32,34,35], while others have reported that linear polyethylene is totally unsuitable or inferior to low-density polyethylene [21,25,36–38]. Moreover, as a general rule, findings for particular conditions are not necessarily applicable to all elements when concentrations, matrices, containers, or experimental conditions are different from those tested. As the macro- and micro constituents of natural waters

TABLE 2.1
Preservation Reagents for Inorganic Determinants

Parameter	Suggested Methods	Remarks
Acidity/alkalinity	Refrigeration	Gains or losses of CO_2 affect the result. Microbial action can affect this; should be one of the first analyses of the sample.
Metals (except mercury)	*Dissolved* Filter on-site acid to pH 1–2 *Suspended* Filter on-site *Total* Acidify to pH 1–2 with hydrochloric or nitric acid Use 250 mL bottle Use 10 mL 25% analar or aristar nitric acid stable for 1 month	This procedure should be used wherever possible at sewage works when compositing effluents. DoE recommendations vary between 2 and 10 ML depending on the element. Polyethylene bottle required. For potable water, acidification must be carried out on receipt in the laboratory; acidified bottles should not be left with private householders. If soluble metals are to be determined, filter sample before acid addition. Clean plastic Cassella by soaking in the acid.
Silica	Polyethylene or glass bottles, 125 mL	
Mercury	Nitric acid (5 mL) + 5% potassium dichromate (5 mL) per litre 125 mL sample bottle, add 1.5 mL 5% $K_2Cr_2O_7$ plus 4 mL conc. H_2SO_4 make up as needed	This treatment must be rendered immediately on sampling; otherwise, a large proportion can be lost within minutes. Glass bottle must be used (DoE recommendation). Bottles filled with 2 N nitric acid when not in use; rinsed out before adding reagents.
Preservatives		
Nitrogen (ammonia, nitrate, organic)	1. Sulphuric acid 2. Chloroform 3. Refrigeration	Chloroform has been used effectively. Sulphuric acid is effective as is frequently recommended. Acidification may be recommended by DoE.
Nitrite	Chloroform	Best analysed as soon as possible.
Sulphide	Zinc acetate/sodium carbonate, that is, 5 mL zinc acetate (110 g zinc acetate + 1 mL acetic acid /L) (A) + 5 mL sodium carbonate	Must be preserved immediately. Care needed for samples containing much suspended matter. To a bottle already containing 5 mL solution (A) add sample as quickly as possible, avoiding entrainment of air bubbles. 5 mL of solution (B) is then added to fill the bottle, which should be stopped and thoroughly mixed.

(Continued)

TABLE 2.1 (*Continued*)
Preservation Reagents for Inorganic Determinants

Parameter	Suggested Methods	Remarks
Cyanide	Glass DO bottles 250 mL; make up as needed	2 NaOH pellets.
Phosphate	1. Sulphuric acid conditioning 2. Use of prepared bottle/ refrigeration	Intensive investigation has been carried out on this subject. The concentration of the determinant may affect the method. The DoE recommendation will be for low levels, that is, less than 25 μg/L, iodised plastic bottle for high levels, acid conditioning glass bottle.
BOD	Refrigeration	Refrigeration is only partially effective. Analysis is best done immediately. Large changes can occur over a few hours. Glass bottles preferred.
COD	1. Refrigeration 2. Acidification with sulphuric acid to pH 2	Changes can occur quite rapidly with some samples. Glass bottles preferred. DoE recommendation. Hydrolysis can occur over longer periods.
Chlorine (residual)	Immediate analysis	No storage possible.
Cyanide	Addition of ascorbic acid, then render alkaline to pH 11–12 with sodium hydroxide	This determinant must be preserved immediately. Need for separate container. Any residual chlorine must be removed with ascorbic acid.
Dissolved oxygen	Winkler reagents	Addition of reagent immediately after sampling. Fixed sample should be kept in the dark and analysed as soon as possible. Where measurements are needed on liquids containing high levels of easily oxidisable organic matter, a portable meter should be used; otherwise, the copper sulphate/ sulphamic acid mix should be used. DoE recommendation.
Metals (except mercury)	*Dissolved* Filter on-site into acid to pH 1–2 *Suspended* Filter on-site *Total* Acidify to pH 1–2 with hydrochloric or nitric acid	This procedure should be used wherever possible at sewage works when compositing effluents. DoE recommendations vary between 2 and 20 mL, depending on the element. Polythene bottle required.
Phenolics	Copper sulphate 1 gL/L + phosphate acid to reduce pH to less than 4, refrigerated	

(*Continued*)

TABLE 2.1 (*Continued*)
Preservation Reagents for Inorganic Determinants

Parameter	Suggested Methods	Remarks
Surfactants, anionic	Refrigeration and either 1. Mercuric chloride 1% (2 mL/L) 2. Chloroform (2 mL/L)	These materials can be quite unstable over periods longer than a few hours.
Surfactants, non-ionic pH	Formaldehyde (5 mL/L) Refrigeration	Immediate analysis advisable, changes can occur quite rapidly, particularly if the container is opened. Well-buffered samples are less susceptible to changes.

can differ widely [39], extreme caution must be used in handling published results. Possibly, as recommended [20,22,40], it is best to ascertain the effectiveness of the storage system adopted in one's own laboratory.

The importance that sample containers can have on water sample composition is illustrated below by examples concerning the storage of metal solutions in glass and plastic bottles.

2.1   LOSS OF SILVER, ARSENIC, CADMIUM, SELENIUM, AND ZINC FROM SEAWATER BY SORPTION ON VARIOUS CONTAINER SURFACES [66]

The following container materials were studied: polyethylene, polytetrafluoroethylene, and borosilicate glass. The effect of varying the specific surface R (cm^{-1}) (ratio of the inner container surface in contact with the solution to the volume of the solution) on the adsorption of metals on the container surface was studied. New bottles were used exclusively. The differences in R values were achieved by adding the pieces of the material considered. To avoid the possibility of high active sites for sorption arising from fresh fractures, the edges of the added pieces of borosilicate glass were sealed by a flame. Prior to the use of all materials, the surfaces were cleaned by shaking with 8 M nitric acid for at least 3 days and by washing five times with distilled water.

Working solutions (1 L), which were 10^{-7} mol/L in one of the elements to be studied, were prepared by appropriate addition of the radioactive stock solutions to pH-adjusted artificial water sample. After the pH had been checked, 100 mL portions were transferred to the bottles to be tested. The filled bottles were shaken continuously and gently in an upright position, at room temperature, and in the dark. At certain time intervals, ranging from 1 min to 28 days, 0.1 mL aliquots were taken. The aliquots were counted in a 3 × 3 NaI(TI) well-type scintillation detector, coupled to a single-channel analyser with a window setting corresponding to the rays to be measured.

TABLE 2.2
Preservatives Used in Water Analysis

		References
Delivery of acid preservatives for trace metal determination in water		[58]
Suitability of polyethylene or glass containers for the storage of water samples for the determination of phosphate, nitrate, ammonium fluoride, silica, aluminium, arsenic, barium, cadmium, cobalt, copper, chromium(III), chromium(IV), iron(II), iron(III), lead, manganese(II), manganese(VII), molybdenum, nickel, silver, selenium(V), and zinc	Only low-density polyethylene is suitable for phosphate; glass containers are unsuitable for silica, aluminium, and zinc	[59]
Humic acid as a preservative for mercury(II) solutions in polyethylene containers	Addition of 50 mL/L humic acid reduces losses from 1 ng/L mercury solutions to less than 0.01 ng/L in 15 days	[60]
Comparison of mercury(II) chloride and sulphuric acid as preservative for nitrogen forms in water samples	$HgCl_2$ is more effective in preventing biological changes	[61]
Comparison of deep freezing, hydrochloric acid, or sodium hydroxide or chloroform or mercurous chloride addition for the preservation of sewage samples		[62]
Use of nitric acid for the preservation of lake water samples prior to the determination of zinc, lead, manganese, and iron	Duration of storage time did not affect results obtained	[63]
Method of preventing losses of heavy metals by adsorption on glass, polyethylene, or polypropylene container walls	Losses minimised by pretreating containers with dissolved aluminium or by shock-freezing at liquid nitrogen temperature	[64]
Evaluation of methods for stabilising nitrogen and phosphorus species in lake water samples, mercury ion, sulphuric acid, and chloroform compared	Only mercuric ion and chilling to 4°C preserved samples for 16 days	[65]

The counting times were chosen in such a way that at least 15,000 pulses were counted. The sorption losses were calculated from the activities of the aliquots and the activity of the aliquot taken at time zero. Taking into account the various sources of errors, mainly counting statistics, the maximum imprecision is about 3%. Therefore, calculated sorption losses of 3% and lower are omitted from the listings as being not significant.

Florence et al. [66] measured the percentage loss as a function of time for silver and cadmium from an artificial water sample stored in polyethylene, borosilicate glass, and PTFE at various pH and R values. For silver, for example, in a polyethylene container at pH 8.5, the percentage recovery was 7% for a 1 min contact time to 46% for a 28-day contact time. In the case of storage in a borosilicate glass container under the same conditions, recoveries were, respectively, 3% and 67%. Similar low recoveries were obtained for cadmium.

For arsenic (added as sodium arsenate) and selenium (added as sodium selenite), losses were insignificant in all the container materials considered, irrespective of matrix composition.

The sorption behaviour of the trace elements depends on a variety of factors which, taken together, make sorption losses rather difficult to predict. However, the data from this study and from the literature indicate the elements for which sorption losses may be expected as a function of a number of factors, such as trace element concentration, container material, pH, and salinity.

A reduction in contact time and the specific surface may be helpful in lowering sorption losses, and acidification with a strong acid will generally prevent the problems of loss by sorption. However, it must be emphasised that the use of acids may drastically change the initial composition of the aqueous sample, making unambiguous interpretation of the analytical results cumbersome or even impossible [66].

For cases of sample storage where losses cannot be excluded *a priori*, some sort of check is required. This should be done under conditions which are representative of the actual sampling, sample storage, and sample analysis. As this study indicates, the use of radiotracers is helpful in making such checks.

Various factors in sorption losses may be classified into four categories. The first is concerned with the analyte itself, especially the chemical form and concentration. The second category includes the characteristics of the solution, such as the presence of acids (pH), dissolved material (e.g., salinity, hardness), complexing agents, dissolved gases (especially oxygen, which may influence the oxidation state), suspended matter (competitor in the sorption process), and microorganisms (e.g., trace element take-up by algae). The third category comprises the properties of the container, such as its chemical composition, surface roughness, surface cleanliness, and, as this study demonstrates, the specific surface. Cleaning by prolonged soaking in 8 M nitric acid [67] is to be recommended. The history of the containers (e.g., age, method of cleaning, previous samples, exposure to heat) is important because it may be of direct influence on the type and number of active sites for sorption. Finally, the fourth category consists of external factors, such as temperature, contact time, access of light, and occurrence of agitation. All these factors must be considered in assessing the likelihood of sorption losses during a complete analysis.

2.2   LOSS OF CADMIUM, LEAD, AND COPPER FROM SEAWATER IN LOW-DENSITY POLYETHYLENE CONTAINERS

Scarponi and Capodaglio Poescon [68] used anodic stripping voltammetry to investigate the contamination of water samples by cadmium, lead, and copper during filtration and storage of samples collected near an industrial area. Filtrations were carried out under clean nitrogen to avoid sample contamination. Water leaches metals from uncleaned membrane filters, but after 1 L of water has passed through, the contamination becomes negligible. Samples stored in conventional polyethylene containers (properly cleaned and conditioned with pre-filtered water) at 4°C and natural pH remained uncontaminated for 3 months (5 months for cadmium); loss of lead and copper occurred after 5 months of storage. Reproducibility (95% confidence interval) was 8%–10%, 3%–8%, and 5%–6% at concentration levels of about 0.6, 2.5, and 6.0 µg/L for cadmium, lead, and copper, respectively.

The first aim of this work was to study the influence of an unwashed membrane filter on the cadmium, lead, and copper concentrations of filtered water samples. It was also desirable to ascertain whether, after passage of a reasonable quantity of water, the filter itself would be uncontaminated. If this were the case, it should be possible to eliminate the cleaning procedure and its associated contamination risk. The second purpose of the work was to test the possibility of long-term storage of samples at their natural pH (about 8) at 4°C, kept in low-density polyethylene containers, which had been cleaned with acid and conditioned with seawater.

Before use, new containers were cleaned by soaking in 2 M hydrochloric acid for 4 days and conditioned with pre-filtered seawater for a week, all at room temperature [19,22]. Teflon-covered stirring bars (required for the voltammetric measurements) were introduced into the containers at the beginning of the cleaning procedure. All the containers used in one procedure were rinsed and left filled with pre-filtered water sample until reuse. Containers used in another procedure and in the study of long-term storage could be regarded as having been conditioned for about 1 month and more than 2 months, respectively. Other plastic ware used in the sampling and filtration processes and the components of the voltammetric cell that came in contact with the sample solution underwent the same cleaning procedures as the containers.

Scarponi and Capodaglio Poescon [68] investigated dependence on the volume of samples filtered on the recovery of copper, cadmium, and lead in the filtrate metal levels, which became constant after 1–1.4 L.

There were no changes in the metal concentrations for 3 months for lead and copper, or for 5 months for cadmium. Also, after 5 months of storage, some loss of lead and copper (25% and 24%, respectively) was observed, possibly because of the formation and slow adsorption on container surfaces of hydroxo- and carbonato-complexes [22,69]. Hence, at 4°C in polyethylene containers, no significant changes of heavy metal concentrations occurred over a 3-month period [28,30,70].

Scarponi and Capodaglio Poescon [68] concluded that filtration of seawater through uncleaned membrane filters showed positive contamination by cadmium, lead, and copper.

Pellenberg and Church [71] have discussed the storage and processing of estuarine water samples for analysis by atomic absorption spectrometry.

Flegal and Stukas [72] have described a sample-processing technique necessary for avoiding lead contamination of seawater samples prior to lead isotope measurements by thermal ionisation mass spectrometry. Levels down to 0.2 ng/kg were determined.

To preserve mercury-containing samples [73,74], Coyne and Collins [74] recommended pre-acidification of the sample bottle with concentrated nitric acid to yield a final pH of 1 in the sample solution. However, when this procedure was used for the storage of water samples and distilled deionised water in low-density polyethylene storage containers, abnormally high absorption was observed. This absorption at the mercury wavelength (253.7 nM) is due to the presence of volatile organic plasticiser material and any polyethylene residue leached by the concentrated acidified nitric acid and by the acid solution at pH 1. These procedures employ the storage of samples in polyethylene bottles, and gas-phase mercury detection may be subject to artificial mercury absorption due to the presence of organic material.

Reported mercury values in the sea determined since 1971, span three orders of magnitude, due at least in part to errors induced by incorrect sampling [75–77]. Olafsson et al. [78] have attempted to establish reliable data on mercury concentrations in water samples obtained on cruises in North Atlantic waters.

The sampling, storage, and analytical methods used by Olafsson et al. [79] in this study have been evaluated. The Hydro-Bios water bottles used were modified by replacing internal rubber rings with silicone rubber equivalents. The water bottles were cleaned by filling with a solution of the detergent Deacon 90. Samples for analysis of mercury were drawn into 500 mL Pyrex vessels and acidified to pH 1 with nitric acid (Merck 457), containing less than 0.05 nmol/L mercury impurities. The Pyrex bottles were pre-cleaned with both nitric acid and a solution of nitric and hydrofluoric acids (10:1). Then a small volume of nitric acid is added, and the bottle is subsequently stored up to the time of sampling. Ashore, reactive mercury was determined by cold vapour atomic absorption after preconcentration by amalgamation on gold [79]. The total mercury concentration was similar to that obtained using 500 W low-pressure mercury lamp (Hanovia) and immersion irradiation equipment. The precision of the mercury determination assessed by analysing 19 replicates over a period of 107 days was found to be ±2.0 pmol/L for a concentration of 12.5 pmol/L.

2.3 LOSS OF ZINC, CADMIUM, STRONTIUM, ANTIMONY, IRON, SILVER, COPPER, COBALT, RUBIDIUM, SCANDIUM, AND URANIUM FROM WATER SAMPLES IN POLYETHYLENE AND GLASS CONTAINERS

Robertson [80] has measured the adsorption of zinc, caesium, strontium, antimony, indium, iron, silver, copper, cobalt, rubidium, scandium, and uranium onto glass and polyethylene containers. Radioactive tracers of these elements were added to the water. The samples were adjusted to the original pH of 8.0, and aliquots were poured into polyethylene bottles, Pyrex glass bottles, and polyethylene bottles containing enough concentrated hydrochloric acid to bring the pH to about 1.5 (usually about 1 mL). Adsorption on the containers was observed for storage periods of up to 75 days with the use of NaI(TI) well crystal. Negligible adsorption on all containers was registered for zinc, caesium, strontium, and antimony. Losses of indium, iron, silver, copper, rubidium, scandium, and uranium occurred from water at pH 8.0 in polyethylene (except rubidium) or Pyrex glass (except silver). With indium, iron, silver, and cobalt, acidification to pH 1.5 eliminated adsorption on polyethylene, but this was only partly effective with scandium and uranium.

Hawley and Ingle [81] studied the breakdown of a 1.0 μg/L methylmercury chloride solution caused by 1.0% (v/v) nitric acid alone, 0.01% (w/v) potassium dichromate alone, and a mixture of 1.0% (v/v) nitric acid and 0.1% (w/v) potassium dichromate, which are used as preservatives for total mercury. Measurements of inorganic and methylmercury content were made within hours of preparation and after 1.3 and 8 days of standing in 100 mL glass volumetric flasks at room temperature. The results were compared to those obtained with an unpreserved 1 μg/L methylmercury chloride solution.

Over a period of 8 days, mercuric ion concentration increased from 0 to 0.6 ppm, while methylmercury concentration decreased from 1 to about 0.35 ppm.

Twenty percent of the methylmercury was observed to be converted to inorganic mercury (the form easily reducible by stannous chloride) under these conditions in slightly over a day. The total amount of mercury (inorganic and organic) in solution remained fairly constant over 3 days with an approximate 25% ± 8% loss over a period of 8 days. Comparison of decomposition induced by 0.01% potassium dichromate alone and 1.0% nitric acid alone to that caused by the combination of the two reagents indicates that the major factor appears to be the presence of nitric acid. Nitric acid alone converts almost half of the methylmercury to mercuric ions in just about 3 days and losses in terms of total mercury from the solution amount for about 26% ± 5% over 8 days. Potassium dichromate is not nearly as destructive although about 15% of the CH_3Hg^+ is decomposed in 3 days while maintaining greater than 90% effectiveness in retaining total mercury for more than a week.

It should be noted that, in preparing any of the test solutions, the acid or dichromate was diluted to 50–75 mL with water before the addition of the methylmercury chloride solution so that the organomercury compound was never in direct contact with the concentrated preservation reagent. The unpreserved methylmercury retained its concentration remarkably well over a 3-day period, and as expected a minimum of methylmercury breakdown is observed over that time. However, losses of about 33% ± 12% of the total mercury concentration after 8 days were noted, again as expected since no preservatives were present. The presence of inorganic mercury at the end of the test period may be partially attributed to photon-induced decomposition since the methylmercury is somewhat sensitive to light.

This study indicates that, where speciation of mercury is the primary objective, the use of nitric acid should be avoided to minimise decomposition or if added, the analysis must be run as soon as possible, preferably within hours of the addition. Although with extended periods of preservation, acid and dichromate should be used and total mercury can be obtained with a fair amount of accuracy since losses would be minimised. However, the original speciation information for the sample is no longer determinable unless analysis is carried out immediately after preservation.

REFERENCES

1. P.J. Wangersky, *American Science*, 1965, 53, 358.
2. K. Hickman, I. White, and E. Stark, *Science*, 1973, 180, 15.
3. B.E. Conway, H. Angerstein-Kozlowska, and W.B.A. Sharp, *Analytical Chemistry*, 1973, 45, 1331.

4. A. Otsuki and T. Hanya, *Limnology and Oceanography*, 1972, 87, 248.
5. N.W. Jannasch and P.H. Pritchard, *Memorie Dell'Istituto Italiano Di Idrobiologia*, 1972, 24, Suppl., 289–306.
6. W.S. Wiebe and L.R. Pomeroy, *Memorie Dell'Istituto Italiano Di Idrobiologia*, 1972, 29, Suppl., 325–352.
7. G. Scarponi, G. Capodaglio, P. Cescon, B. Cosma, and R. Frache, *Analytica Chimica Acta*, 1982, 135, 263.
8. R.E. Truitt and J.H. Weber, *Analytical Chemistry*, 1979, 51, 2057.
9. J.H. Weber and R.E. Truitt, Research Report 21, Water Resource Research Centre, University of New Hampshire, Durham, NH, 1979.
10. K.T. Marvin, R.R. Proctor Jr., and R.A. Neal, *Limnology Oceanography*, 1970, 15, 320.
11. J. Gardiner, *Water Research*, 1974, 8, 157.
12. D.W. Spencer and F.T. Manheim, U.S. Geological Survey Professional Paper, 650-D, 1969, p. 288.
13. D.W. Spencer, P. Brewer, and P.L. Sachs, *Geochimica Cosmochimica Acta*, 1972, 36, 71.
14. R. Dams, K.A. Rahn, and J.W. Winchester, *Environmental Science and Technology*, 1972, 6, 441.
15. G.T. Wallace Jr., I.S. Fletcher, and R.A. Duce, *Environmental Science and Health*, 1977, A12, 493.
16. G. Duychserts and G. Gillain, In *Essays on Analytical Chemistry* (ed. W. Wanninen), Pergamon, Oxford, UK, 1977, p. 417.
17. R.G. Smith, *Talanta*, 1978, 25, 173.
18. L. Mart, *Fresenius Zeitschrift für Analytische Chemie*, 1979, 296, 352.
19. H.W. Nürnberg, R. Valenta, L. Mart, B. Raspor, and L. Sipos, *Fresenius Zeitschrift Für Analytische Chemie*, 1976, 282, 357.
20. J.P. Riley, D.E. Robertson, J.W.R. Dutton et al., In *Chemical Oceanography* (eds. J.R. Riley and G. Skirrow), 2nd edn., Vol. 3, Academic Press, London, UK, 1975, p. 193.
21. M. Zief and J.W. Mitchell, *Contamination Control in Trace Element Analysis*, Wiley, New York, 1976.
22. G.E. Batley and D. Gardner, *Water Research*, 1977, 11, 745.
23. K.W. Bruland, R.P. Franks, and G.A. Knauer, *Analytica Chimica Acta*, 1979, 105, 233.
24. D.C. Burrell, *Marine Science Communications*, 1979, 5, 283.
25. D.E. Robertson, In *Ultrapurity: Methods and Techniques* (eds. M. Zief and R. Speights), Marcel Dekker, New York, 1972, p. 207.
26. R. Salim and G. Cooksey, *Journal of Electroanalytical Chemistry*, 1980, 106, 251.
27. P. Figura and B. McDufie, *Analytical Chemistry*, 1980, 52, 1433.
28. R. Fukai and L. Huynh-Ngoc, *Marine Pollution Bulletin*, 1976, 7, 9.
29. D.C. Burrell and M.L. Lee, In *Water Quality Parameters*, ASTM, STP-573, ASTM, Philadelphia, PA, p. 58.
30. G.E. Batley and D. Gardner, *Estuarine, Coastal and Marine Science*, 1978, 7, 59.
31. A. DeForest, R.W. Pettis, and G. Fabris, *Australian Journal of Marine and Freshwater Research*, 1978, 29, 193.
32. J. Gardiner and M.J. Stiff, *Water Research*, 1975, 9, 517.
33. A. Zirino, S.H. Lieberman, and C. Clavell, *Environmental Science and Technology* 1978, 12, 73.
34. V.T. Bowen, P. Strohal, M. Saiki et al., In *Reference Methods for Marine Radioactivity Studies* (eds. Y. Nishiwaki and R. Fukai), International Atomic Energy Agency, Vienna, Austria, 1970, pp. 12–14.
35. D.C. Bowditch, C.R. Edmond, P.J. Dunstan, and J. McGlynn, Technical Paper No. 16, Australian Water Resources Council, 1976, p. 22.
36. G. Tolg, *Talanta*, 1972, 19, 1489.

37. G. Tolg, In *Comprehensive Analytical Chemistry* (ed. G. Svehia), Vol. 3, Elsevier, Amsterdam, the Netherlands, 1975, p. 1.
38. J.R. Moody and R.M. Lindstrom, *Analytical Chemistry*, 1977, 49, 2264.
39. W. Davison and M. Whitfield, *Journal of Electroanalytical Chemistry*, 1977, 75, 763.
40. G. Scarponi, E. Miccoli, and R. Frache, In *Proceedings of the Third Congress of the Association of Italian Oceanology and Limnology*, Sorrento, Italy, Pergamon Press, Oxford, UK, 1978, p. 433.
41. H.E. Alburn and N.H. Grant, *Journal of the American Chemical Society*, 1965, 87, 4174.
42. N.H. Grant and H.E. Alburn, *Biochemistry*, 1965, 4, 1913.
43. N.H. Grant and H.E. Alburn, *Archives of Biochemistry and Biophysics*, 1967, 118, 292.
44. I. Yoshinari, *Marine Chemistry*, 1976, 4, 189.
45. D.L. King and I.L. Ciaco (eds.), *Sampling of Natural Waters and Waste Effluent*, Marcel Dekker, New York, p. 45.
46. G.D. Grice, C.R. Harvey, and R.H. Backus, *Bulletin of Environmental Toxicology*, 1972, 7, 125.
47. A.L. Bridie, J. Box, and S. Herzberg, *Journal of the Institute of Petroleum*, 1973, 59, 263.
48. J.W. Farrington, A D Report No. 777695/GA, U.S. National Technical Information Service.
49. A.A. Kaplin and L.M. Poskrebysheva, *Izuestia Tomal Politckh Institut*, 1974, 233, 91.
50. D.N. Hume, In *Analytical Methods in Oceanography* (ed. R.P. Gibb Jr.), ACS, Washington, DC, 1971, p. 1.
51. J.P. Riley, In *Chemical Oceanography* (eds. J.P. Riley and G. Skirrow), Vol. 3, Academic Press, London, UK, 1975, p. 193.
52. M.A. Acheson, R.M. Harrison, R. Perry, and R.A. Wellings, *Water Research*, 1975, 10, 207.
53. C.S. Giam and H.S. Chan, *Special Publication*, National Bureau Standards (US) No. 422, 1976, p. 761.
54. C.S. Giam, H.S. Chan, and G.S. Neff, In *Proceedings of International Conference Environmental Sensing Assessment ACS*, Washington, DC, 1976.
55. K. Grasshoff, *Analytical Chemistry*, 1966, 220, 89.
56. K. Grasshoff, *Verlag Chemie*, 1976, 1, 50.
57. L. Mart, *Analytical Chemistry*, 1979, 296, 350.
58. R.L. Guest and H. Blutstein, *Analytical Chemistry*, 1981, 53, 727.
59. D.C. Bowditch, C.R. Edmond, P.J. Dunstan, and J.A. McGlynn, Australian Water Resources Council Technical Paper No. 16, Research Project No. 71/35B, Australian Government Publishing Service, Canberra, Australian Capital Territory, Australia, 1976, 37 pp.
60. R.W. Heiden and D.A. Aikens, *Analytical Chemistry*, 1983, 55, 2327.
61. L.H. Howe and W.C. Holley, *Environmental Science and Technology*, 1969, 3, 478.
62. F.J. Spaengler, *Zeitschrift Für Analytische Chemie*, 1978, 11, 128.
63. A. Henrikson and K. Blamer, *Vatten*, 1977, 77, 33.
64. H. Schuermann and H. Hartkamp, *Fresenius Zeitschrift Für Analytische Chemie*, 1983, 315, 430.
65. M.J. Fishmann, L.J. Schroder, and M.W. Shockey, *International Journal of Environmental Studies*, 1986, 26, 231.
66. T.M. Florence and G.E. Bailey, *CRC Capital Reviews on Analytical Chemistry*, August 1980, 219.
67. R.W. Karin, J.A. Buone, and J.L. Fashing, *Analytical Chemistry*, 1975, 47, 2296.
68. G.S. Scarponi and G. Capodaglio Poescon, *Analytica Chimica Acta*, 1982, 135, 268.
69. K.S. Subranian, E.L. Chakraborti, J.E. Sheiras, and I.S. Maines, *Analytical Chemistry*, 1978, 50, 444.

70. J.H. Carpenter, W.L. Bradford, and V. Grant, *Estuarine Research* (ed. L.E. Cronin), Academic Press, New York, 1975.
71. R.E. Pellenberg and T.M. Church, *Analytica Chimica Acta*, 1978, 97, 81.
72. A. Flegal and V. Stukas, *Journal of Marine Chemistry*, 1987, 22, 163.
73. D. Degobbis, *Limnology Oceanography*, 1973, 18, 146.
74. R. Coyne and J.C. Collins, *Analytical Chemistry*, 1972, 44, 1093.
75. R.A. Carr and P.E. Wilkniss, *Environmental Science and Technology*, 1973, 7, 62.
76. C. Feldman, *Analytical Chemistry*, 1974, 46, 99.
77. W.F. Fitzgerald, Distribution of mercury in natural waters, In *The Biochemistry of Mercury in the Environment* (ed. J.O. Nriagu), Elsevier, Amsterdam, the Netherlands 1979.
78. J. Olaffson, 1981, Trace metals in seawater, In *Proceedings of a NATO Advanced Research Institute on the Trace Metals in Seawater* (eds. C.S. Wong et al.), 30 March–3 April 1981, Sicily, Italy, Plenum Press, New York.
79. J. Olaffson, *Analytica Chimica Acta*, 1974, 68, 207.
80. D.E. Robertson, *Analytica Chimica Acta*, 1968, 42, 533.
81. J.E. Hawley and J.D.F. Ingle, *Analytical Chemistry*, 1975, 47, 719.

3 Determination of Cations in Seawater

As shown in Table 3.1, a number of cations have been found in seawater samples taken over the entire globe. Table 3.2 illustrates the trace element concentration found in north seawater, obtained by Newton activation analysis, which is a very useful technique for element surveys in water.

3.1 ALUMINIUM

3.1.1 SPECTROPHOTOMETRIC METHODS

Aluminium has been determined by spectrophotometric methods using aluminon [41,42], oxine [43,44], Eriochrome Cyanine R [45], and Chrome Azurol [46]; by fluorometric methods using Pontachrome Blue Black R [47,48], Lumogallion [49–51], and salicylaldehyde semicarbazone [52–54]; and also by gas chromatographic methods [55,56], emission spectroscopy [57], and neutron activation analysis [58,59].

Korenaga et al. [60] have described an extraction procedure for the spectrophotometric determination of trace amounts of aluminium in seawater with pyrocatechol violet. The extraction of ion associate between the aluminium/pyrocatechol violet complex and the quaternary ammonium salt, zephiramine (tetradecyldimethylbenzyl ammonium chloride), is carried out with 100 mL seawater and 10 mL chloroform. The excess of reagent extracted is removed by back washing with 0.25 M sodium bromide solution at pH 9.5. The calibration graph at 590 nm obeyed Beer's law over the range of 0.13–1.34 µg aluminium.

Several ions such as manganese, iron(II), iron(III), cobalt, nickel, copper, zinc, cadmium, lead, and uranyl react with pyrocatechol violet and, to some extent, are extracted together with aluminium. The interference from these ions and other metal ions generally present in seawater could be eliminated by extraction with diethyldithiocarbamate as a masking agent. Typical aluminium contents obtained by this method of seawater report ranges from 6 to 30 µg/L.

3.1.2 SPECTROFLUOROMETRIC METHODS

Howard et al. [61] determined aluminium dissolved in seawater by the micelle-enhanced fluorescence of its lumogallion complex. Triton X-100 surfactant was used for the determination of aluminium at very low concentrations.

Salgado Ordonez et al. [62] used di-2-pyridylketone 2-furoyl-hydrazone as a reagent for the fluorometric determination of down to 0.2 µg aluminium in seawater. A buffer solution at pH 6.3 and 1 mL of the reagent solution were added to samples containing between 0.25 and 2.50 µg aluminium. The fluorescence was measured at 465 nm.

TABLE 3.1
Range of Metal Concentrations in Open Ocean Seawater

Element	Location	Concentration (μg/L)	Consensus Value (μg/L)	References
Aluminium	Open seawater surface	0.1		[1]
	Open seawater, 3 km depth	0.6		[1]
Bismuth	Pacific, surface	<0.00005	–	[2]
	Pacific, 2500 m depth	<0.000003	–	[3]
Cadmium	Open ocean, salinity 35%	0.03	–	[4]
	Arctic Sea	0.010–0.045	–	[5]
	Arctic Sea, surface	0.0127	–	[6]
	Arctic Sea, 2000 m depth	0.023	–	[6]
	Arctic Sea	0.018	–	[7]
	Pacific	0.02–0.04	–	[8]
	Kattergat/Skaggerat	0.022	–	[9]
	Norwegian Sea	0.02–0.025 (surface)	–	[10]
		0.02–0.025 (3000 m)	–	
	Sargasso Sea	0.035–0.042 (216 m)	–	[11]
		0.109–0.0126 (4926 m)	–	[11]
	Baltic Sea	0.03–0.06	–	[12]
	Open sea	0.03	–	[4]
	Open sea	0.079	–	[13]
	Open sea	0.12–0.30	–	[14]
	Open sea	0.03–0.17	–	[15]
Chromium	Pacific	Cr^{III} 0.005–0.52	–	[16,17]
		Cr^{VI} 0.03–0.96	–	[16,17]
		Organic Cr 0.07–0.32	–	[16,17]
		Total Cr 0.06–1.26	0.03	[16,17]
	Mediterranean	Cr^{III} 0.02–0.05	–	[18]
		Cr^{VI} 0.05–0.38	–	[18]
	Open ocean	Total 0.07–0.97	0.03	[19]
		Cr^{III} 0.08–0.22	–	[20]
		Cr^{VI} 0.13–0.68	–	[20]
		Total 0.18–0.19	0.03	[4]
Cobalt	North Sea	0.07–0.16	0.005	[21]
	Open ocean (salinity 35%)	0.003		[4]
	Open sea	0.04	–	[22]
	Open sea	0.003	–	[4]
	Open sea	0.15–0.16	–	[14]
Copper	Pacific	0.3–2.8	0.05	[8]
	Open ocean (salinity 35%)	0.121		[4]
	Good quality seawater	0.36–8.6		[23]
	Sargasso Sea	0.072–0.081 (216 m)		[11]
	Sargasso Sea	0.26–0.33 (4926 m)		[11]

(Continued)

TABLE 3.1 (*Continued*)
Range of Metal Concentrations in Open Ocean Seawater

Element	Location	Concentration (µg/L)	Consensus Value (µg/L)	References
	Baltic Sea	0.59–0.99		[9]
	Baltic Sea	0.6–1.0		[12]
	Baltic Sea	0.0063–0.0252 (organic)		[24]
	0.6–0.751 (total)	24		
	North Sea	0.208		[24]
	Norwegian Sea	0.08–0.10 (surface)		[10]
	Norwegian Sea	0.08–0.10 (300 m)		[10]
	Danish Sound	0.48		[9]
	Arctic Sea	0.097		[24]
	Open sea	0.341		[13]
	Open sea	0.48–1.51		[14]
Iron	Pacific	140–320		[8]
	Open ocean (salinity 35%)	0.2		[4]
	Open seawater	2.1		[22]
	Open seawater	3.25		[13]
	Pacific	<0.01–0.7		[25]
Lead	Pacific	0.6–0.8	–	[8]
	Arctic Sea	0.01–28	–	[5]
	Arctic Sea	0.019–0.021	–	[7]
	Open ocean (salinity 35%)	0.095		[4]
	Sargasso Sea	0.000041 (surface)	–	[26]
		0.0083–0.012 (4800 m)		[26]
	West North Atlantic	0.00017–0.0003		[27]
	Norwegian Sea	<0.0002 (3000 m)	11	
		0.025–0.065 (surface)		
	Open sea	0.095	4	
	Arctic Sea	0.015		
	Open sea	0.0083		
	Open sea	0.03–9.0		
	Open sea	<0.04–0.28		
Manganese	Open ocean (salinity 35%)	0.018	0.02	[4]
Mercury	Atlantic open sea	0.021–0.078	<0.2	[27]
	Open ocean	0.002–0.011		[28]
	Off Iceland	0.04		[29]
Molybdenum	Pacific	11.2–12.0		[8]
	Non-central Pacific	3.2		[30]
	Seawater, Japan	11.5		[31]
	Open sea	5.3		[22]
Nickel	Pacific, 4000 m depth	0.45–0.84	0.17	[32]
	Pacific	0.15–0.93		[8]

(*Continued*)

TABLE 3.1 (*Continued*)
Range of Metal Concentrations in Open Ocean Seawater

Element	Location	Concentration (µg/L)	Consensus Value (µg/L)	References
	Pacific, surface	0.16–0.29		[32]
	Open ocean (salinity 35%)	0.341–0.608		[33]
	Open ocean	0.38–0.46		[34]
	Open ocean	0.27		[4]
	Norwegian Sea	0.175–0.20 (surface)		[10]
		0.175–0.20 (3000 m)		[10]
	Sargasso Sea	0.26–0.27 (216 m)		[11]
		0.45–0.47 (4926 m)		[11]
	Baltic Sea	0.6–0.9		[12]
	Arctic Sea	0.099		[24]
	Open sea	0.545		
Nickel	Open sea	0.76–1.58		[14]
Rare earths	North Atlantic, below mixed layer	La 13.0×10^{-12} mol/kg		[35]
		Ce 16.8	–	
		Nd 12.8	–	
		Sm 2.67	–	
		Eu 0.644	–	
		Gd 3.4	–	
		Dy 4.78	–	
		Er 4.07	–	
		Yb 3.55		
Rhenium	Atlantic	6–8	–	[36]
Selenium	Seawater	0.021–0.029	–	[37]
	Open ocean	0.00095	–	[22]
Silver	Open sea	0.08		[14]
Thorium	Open sea	<0.0002		[22]
Tin	Open sea	0.02 (SnIV)		[38]
		0.05 (SnII)		[38]
Uranium	Seawater	1.9	–	[39]
		2.6	–	[34]
Vanadium	Pacific	1.73–2.00	2.5	[40]
		1.29–1.87		[8]
	Adriatic Sea	1.64–1.73		[40]
	Open sea	0.45		[2]
Zinc	Pacific	1.9–3.0	0.49	[8]
	Arctic Sea	0.125–0.16		[7]
		0.05–0.34		[5]
	Open sea (salinity 35%)	0.28		[4]
		4.9		[22]

(*Continued*)

TABLE 3.1 (*Continued*)
Range of Metal Concentrations in Open Ocean Seawater

Element	Location	Concentration (µg/L)	Consensus Value (µg/L)	References
	Norwegian Sea	0.08–0.30 (surface)		[10]
		0.10–0.18 (300 m)		[10]
	Open sea	0.074		[13]
		0.3–10.9		[15]
		2.6–10.1		[14]

TABLE 3.2
Determination of Trace Elements in Seawater (Mean ± SD)

Element	Water without Suspended Material; 5 Determinations (g/L)
Ag	$(8.8 + 0.4) \cdot 10^{-9}$
As	$(3.5 + 0.3) \cdot 10^{-7}$
Au	$(3.5 + 0.3) \cdot 10^{-10}$
Ba	$(5.7 + 0.1) \cdot 10^{-7}$
Br	$(5.5 + 0.1) \cdot 10^{-7}$
Ca	$(3.6 + 0.2) \cdot 10^{-5}$
Cd	$<10^{-6}$
Ce	$(3.4 + 0.3) \cdot 10^{-4}$
Co	$(4.5 + 0.3) \cdot 10^{-4}$
Cr	$(1.4 + 0.1) \cdot 10^{-7}$
Eu	$(8.2 + 1.4) \cdot 10^{-10}$
Fe	$(1.5 + 0.2) \cdot 10^{-5}$
Hg	$(2.2 + 0.2) \cdot 10^{-6}$
K	$(3.6 + 1.9) \cdot 10^{-5}$
La	$(3.2 + 0.1) \cdot 10^{-9}$
Mo	$(4.4 + 1.4) \cdot 10^{-8}$
Na	$(3.6 + 0.3) \cdot 10^{-4}$
Sb	$(5.7 + 0.4) \cdot 10^{-9}$
Sc	$(4.5 + 0.3) \cdot 10^{-8}$
Se	$(4.5 + 0.3) \cdot 10^{-8}$
U	$(3.3 + 1.0) \cdot 10^{-8}$
Zn	$(2.3 + 0.1) \cdot 10^{-6}$

Note: Samples taken at 54° 3' north, latitude 6° 30' east, longitude in the North Sea.

Suarez et al. [63] described a fully automated fluorimetric determination of aluminium in seawater using syringe dispersive liquid–liquid metroextraction (DLLME) and lumogallion as the fluorescing agent.

The complete analytical procedure was automised and carried out within 4 min. Dispersive liquid–liquid microextraction was done using *n*-hexanol as an extracting solvent and ethanol as a dispersing solvent in a 1:8 v/v percent mixture. The lumogallion complex was extracted by an organic solvent and separated from the aqueous phase within the syringe of an automated pump. The limits of detection (3σ) and quantification (10σ) were 8.0 ± 0.5 nmol/L and 26.7 ± 1.6 nmol/L, respectively. The relative standard deviation for eight replicate determinations of 200 nmol/L Al was 0.5%. Ambient concentrations of samples were quantifiable with the concentrations ranging from 43 to 142 nmol/L. Standard additions gave analyte recoveries from 97% to 113%, proving the general applicability and adequateness of the analyser system to real sample analysis.

3.1.3 STRIPPING VOLTAMMETRY

Van der Berg et al. [64] determined aluminium in seawater by anodic stripping voltammetry. The method involved complexation with 1,2-dihydroxyanthraquinone-3-sulphonic acid, collection of the complex on the hanging mercury drop electrode, and determination by cathodic stripping voltammetry.

Detection was 1 nmol/L aluminium for an adsorption time of 45 s. No serious interferences were found, but ultraviolet irradiation was recommended for samples with a high organic content.

Van den Berg et al. [64] determined aluminium in freshwater and seawater by cathodic stripping voltammetry after absorptive collection of the aluminium complex of 1,2-dihydroxyanthraquinone-3-sulphonic acid. The limit of detection was 1 nM aluminium for an adsorption time of 45 s.

3.1.4 OTHER PROCEDURES

Methods based on atomic absorption spectrometry [65] and gas chromatography [65] have been described for the determination of aluminium in seawater. The determination of aluminium by inductively coupled plasma mass spectrometry and neutron activation analysis is discussed respectively, in Section 3.70, "Multimetal Analysis" (Table 3.4F and 3.4N).

3.2 AMMONIUM

3.2.1 SPECTROPHOTOMETRIC METHODS

The determination of ammonium in seawater has long been recognised as an important measurement in environmental and ecological studies. The procedures used for such determination fall into three categories:

1. Based on the formation of bis-(3-methyl-1-phenyl-5-) pyrazolone, described by Prochazkova [84] and modified by Johnston [85] and Strickland and Austin [86]

2. Based on a reaction with sulphanilamide coupled to N-(1-naphthyl)ethylene diamine [87,88]
3. Based on the Berthelot reaction between ammonia, phenol, and hypochlorite at alkaline pH to form indophenol blue

Although the first method is more sensitive, it is generally multistage and often uses organic solvent extraction so that its automation is rendered less successful. Numerous procedures have been described for the production of indophenol blue [66–83].

The conversion of amino acids to ammonia is a process known to occur extensively at higher temperatures [82]. All conditions, including photon flux and ionic activities, are precisely controlled to give stability and reproducibility in a kinetic system. Serious colour suppression had been noted when indophenol blue type methods such as those of Koroleff [69] and Solorzano [70] had been applied to certain types of samples. Hampson [81] investigated the causes of such colour suppression. In this method, colour development is complete in 40 min and remains constant for many hours. About 3% precision is obtainable at ammonia concentrations above 0.1 mg/L N.

The methyl and ethyl primary, secondary, and tertiary amines were examined for possible interference in the ammonia–indophenol blue reaction by allowing the reaction to take place in pure seawater containing a known addition of ammonia convenient for measurement (0.40 mg/L ammonia N), in parallel paired experiments with and without a known addition of each amine in turn. The amines were present at 100-fold excess over the ammonia, all concentrations being expressed on an N atom basis. All the amines suppressed the indophenol blue reaction with ammonia very strongly and in the general order: primary amines < secondary amines < tertiary amines.

To overcome the suppression effect of amines in the determination of ammonia, Hampson [81] investigated the effect of nitrite ions added either as nitrite or as a nitrous acid. Very considerable suppression by nitrite does occur, although it is not as strong as with any of the amines. Again, it is not great so long as the nitrite N concentration is less than the ammonia N concentration but rapidly increases as the nitrite concentration exceeds the ammonia concentration. In fact, the nitrite-modified method was found to be satisfactory in open seawater samples and polluted estuary waters.

Berg and Abdullah [80] described a spectrophotometric autoanalyser method based on phenol, sodium hypochlorite, and sodium nitroprusside for the determination of ammonia in sea and estuarine water (i.e. the indophenol blue method).

The manifold design allows for the determination of ammonia concentration in the range of 0.2–20 µg/L as NH_4 over a salinity range of 35%–10% with negligible interference from amino acids.

The interference from amino acids was investigated and found to be negligible as reported by Solorzano [70] and Harwood and Huyser [76], who did not employ heating for indophenol blue colour development. Solutions containing 50 µg N/L of urea, histidine, lysine, glycine, and alanine were analysed. The NH_4–N detected, ranged between 0.4% (for urea) and 2.2% (for alanine) of the added nitrogen.

Hampson [81] used ultraviolet photon activation energy of appropriate frequency and a ferrocyanide catalyst to activate selectively the reaction between ammonia, phenol, and sodium hypochlorite. The reaction is carried out at an optimum pH of 10.5 since urea may break down at a pH over 11 and at a low temperature of 30°C ± 1°C to avoid alkaline hydrolysis.

Le Corre and Treguer [89] developed an automated procedure based on the oxidation of the ammonium ion to nitrite by hypochlorite in the presence of sodium bromide followed by spectrophotometric determination of the nitrite. The validity of automatic analysis of ammonium nitrogen in seawater was tested. Le Corre and Treguer [89] discussed the effect of salinity on the determination of ammonia and described a suitable correction procedure.

Brzezinska [90] described a spectrophotometric method for the determination of nanomolar concentrations of ammonium in seawater. A sequence of reagent solutions in deionised distilled water was added to seawater samples (180 mL) as follows: phenol (2.4 mL, 10%), sodium aquopentacyanoferrate (1 mL of a freshly prepared solution containing 0.03 g sodium aquopentacyanoferrate in 104 mL of double distilled water), and sodium hypochlorite (6 mL, 5.5%). The sodium aquopentacyanoferrate acted as a coupling reagent in the formation of indophenol. Reaction mixtures were kept in the dark for 2 h at 40°C and allowed to cool for 1.5 h before adding phosphoric acid (1.65 mL of 1.0 M) and *n*-hexanol (6.6 mL). The organic phase containing indophenol was pipetted into a clean tube and methylene chloride (10 mL) added followed by pH 12 buffer (10 mL). Indophenol blue was re-extracted into the aqueous phase, and its concentration was determined colorimetrically at 640 nm. Interference effects by metals, nitrite, urea, and amino acids present in seawater were discussed. Calibration curves were linear to 2 μM ammonium.

3.2.2 Spectrofluorometric Metals

Amornthammorory and Zhang [91] described a shipboard fluorometric analyser for high-resolution underway measurement of ammonium in seawater. This method is based on the reaction of ammonium with *o*-phthaldialdehyde (OPA) and sulphite. This method shows no refractive index and salinity effect from seawater samples. The potential interferences in seawater have been reduced. The instrument response is linear over a wide range of ammonium concentration. A limit of detection of 1.1 nM was estimated. The application of this method to low-level ammonium measurement requires a correction of interference species, such as amino acids. The system can be used for both freshwater and seawater samples.

3.2.3 Flow Injection Analysis

Willason and Johnson [92] described a modified low injection analysis procedure for ammonia in seawater. Ammonium ions in the sample were converted to ammonia, which diffused across a hydrophobic membrane and reacted with an acid–base indicator. A change in light transmittance of the acceptor steam produced by the ammonia was measured by a light-emitting diode photometer. The automated method had a detection limit of 0.05 µmol/L and a sampling rate of 60 or more measurement per hour.

3.2.4   HIGH-PERFORMANCE LIQUID CHROMATOGRAPHY

Gardner et al. [95,96] applied this technique for determining ammonium ion in seawater. The liquid chromatographic method involved fluorometric detection, after post-column labelling with *o*-phthalaldehyde/2-mercaptoethanol reagent. This method was developed to directly quantify $^{15}NH_4[^{14}NH_4 + ^{15}NH_4]$ ion ratios in aqueous samples that had been enriched with $^{15}NH_4$ for isotope dilution experiments. Cation-exchange chromatography, with a sodium borate buffer mobile phase, was selected as the separation mode because the two isotopes have slightly different constants in the equilibrium reaction between ammonium. When the two forms of ammonium were passed separately through a high-performance cation-exchange column under precisely controlled chromatographic conditions, the retention time (RT) of $^{15}NH_4$ was 1.1012 times the RT of $^{14}NH_4$. The two isotopic forms of ammonium ion were not resolved into separate peaks when they were injected together, but the retention time of the combined peak, as defined by an integrator, increased with increasing percentages of $^{15}NH_4$ in the mixture. The relationship of retention time shift versus percentage of NH_4 relative to total ammonium followed a sigmoid-shaped curve with the maximum retention time shifts per change in isotopic composition occurring between 25% and 75% $^{15}NH_4$. Using a calibration curve based on this relationship and a solution of separately injected $^{14}NH_4$ in mobile-phase buffer as an 'internal standard', Gardner et al. [95,96] were able to directly determine the concentrations and ratios of the two isotopes in enriched seawater.

3.2.5   OTHER PROCEDURES

Other procedures that have been applied for determining ammonium in seawater include polarography [94] and ion-selective electrodes [93]. Degobbis [78] studied the storage of seawater samples for ammonia determination. The effects of freezing, rate of freezing, filtration, addition of preservatives, and type of container on the concentration of ammonium ion samples stored for up to a few weeks were investigated. Both rapid and slow freezing were equally effective in stabilising ammonium ion concentration, and the addition of phenol as a preservative was effective in stabilising non-frozen samples for up to 2 weeks.

3.3   ANTIMONY

3.3.1   SPECTROPHOTOMETRIC METHODS

Afanasev et al. [97] described an extraction and spectrophotometric procedure for determining antimony(V) in seawater. The coefficient of variation is −2% to −5% at antimony concentrations of 1.5–5.2 µg/L. The results obtained are in agreement with the data of neutron activation analysis.

3.3.2 HYDRIDE GENERATION ATOMIC ABSORPTION SPECTROMETRY

Bertine and Lee [98] described hydride generation techniques for determining total antimony, Sb(V), Sb(III), SB–S species, and organoantimony species in frozen seawater samples.

Total Antimony: Total antimony content was analysed by a hydride-generation technique utilising a quartz burner with a hydrogen flame. The water sample was made 2 N in hydrochloric acid. Two millilitres of 20% (w/v) potassium iodide was added, and the sample was degassed using helium as a carrier gas. A silanized glass wool trap to collect SbH_3 was then placed on liquid nitrogen, and 2 mL of 5% sodium borohydride was slowly injected over a period of 100 s. The sample was stripped for 300 s; the trap was then removed from the liquid nitrogen, and the hydride was carried to an electrically heated quartz burner with a hydrogen flame. The antimony concentration was measured using atomic absorption spectrophotometry. Detection limits of about 0.01 ng were obtained.

Both the hydrochloric acid and sodium borohydride contributed to the blank. Sb(V) in the 12 N hydrochloric acid was removed by uptake on a Dowex 1–X8 anion-exchange resin. Sodium borohydride was purified, after dissolution, by the addition of 0.5 mL sodium hydroxide (50%) to 200 mL of 5% sodium borohydride, and subsequent filtration through a hydrochloric acid pre-cleaned 0.45 μm millipore membrane.

Hydrogen sulphide gas interfered in the determination of antimony Sb(III). Two millilitres of citrate buffer (purified by Fe–APDC precipitation) was added to maintain a pH of 5–6. The sample was degassed for 100 s prior to the injection of 2 mL of 5% sodium borohydride. The sample was stripped for 300 s. This procedure gave a complete extraction of antimony(III), and there was no extraction of antimony(V) even in the presence of 100-fold higher concentrations of antimony(V).

At the pH of 5–6, hydrogen sulphide evolution by degassing proceeded at a much slower rate. A background correction using a hydrogen lamp or lead acetate placed in line was able to remove any interference from the amount remaining after 400 s. However, it was found that the extraction yield of antimony(III) standards prepared in sodium sulphide was not only incomplete but the yield was also inversely proportional to the amount of sulphide added to the standards. It has been hypothesised that in sulphide-rich waters, an antimony sulphide complex may exist. A complete yield of antimony(III) in sodium sulphide could be attained by making a 1–2 mL sample 2 N in hydrochloric acid, degassing for 5 min, bringing the volume to 100 mL, adding sufficient Tris buffer to bring the pH to 6, then proceeding with the hydride generation method as earlier. No antimony(V), even in 1000-fold excess, was detected by this method.

Sturgeon et al. [99] described a hydride generation atomic absorption spectrometric method for the determination of antimony in seawater. The method uses the formation of stibine using sodium borohydride. Stibine gas was trapped on the surface of a pyrolytic graphite–coated tube at 250°C, and antimony was determined by atomic absorption spectrometry. An absolute detection limit of 0.2 ng was obtained, and a concentration detection limit of 0.04 μg/L was obtained for 5 mL sample volumes.

3.3.3 Other Procedures

Other procedures which can be used for the determination of antimony in seawater, discussed in Section 3.70, include graphite furnace atomic absorption spectrometry (Table 3.4A), hydride generation atomic absorption spectrometry (Table 3.4D) by inductively coupled plasma mass spectrometry (Table 3.4F), anodic stripping voltammetry (Table 3.4I), X-ray fluorescence spectroscopy (Table 3.4N), and photoluminescence spectroscopy (Table 3.4Q).

3.4 ARSENIC

3.4.1 Spectrophotometric Methods

Afansev et al. [100] described an extraction photometric method for the determination of arsenic at the µg/L range in seawater. This method uses diantipyrylmethane as the chromogen reagent. The coefficient of variation is 2.5% for arsenic concentrations in the 1.5–5 µg/L range. A good agreement was obtained with the results obtained by neutron activation analysis. A UK standard official method [101] has been published for the spectrophotometric determination of arsenic in seawater. The determination is affected by the conversion of arsine using sodium borohydride, which is added slowly to the acidified samples by a peristaltic pump. The liberated arsine is trapped in an iodine/potassium iodide solution, and the resultant arsenate was determined spectrophotometrically as the arsenomolybdenum blue complex at 866 nm. The method is applicable down to 0.19 µg arsenic.

Haywood and Riley [111] described procedures for the determination of arsenic in seawater. While this method does not include organic arsenic species, these can be rendered reactive either by photolysis with ultraviolet radiation or by oxidation with potassium permanganate or a mixture of nitric acid and sulphuric acid. Arsenic(V) can be determined separately from total inorganic arsenic after extracting As(III) as a pyrrolidine dithiocarbamate complex into chloroform.

In the method for inorganic arsenic, the sample is treated with sodium borohydride added at a controlled rate. The arsine evolved is absorbed in a solution of iodine, and the resultant arsenate ion is determined photometrically by a molybdenum blue method. For seawater, the range, standard deviation, and detection limit are 1–4 µg/L, 1.4%, and 0.14 µg/L, respectively. Silver and copper cause serious interference at concentrations of a few tens of milligrams per litre; however, these elements can be removed either by preliminary extraction with a solution of dithizone in chloroform or by ion exchange.

Haywood and Riley [111] showed that arsenic in the form of tetraphenylarsonium chloride, (o-arsonophenylazo)2-naphthol 3,6 disulphonic acid, and arsonophenylazo-p-methaminobenzene is quantitatively decomposed in seawater by ultraviolet radiation.

3.4.2 Hydride Generation Atomic Absorption Spectrometry

Howard and Comber [102] converted arsenic in seawater to its hydride prior to determination by atomic absorption spectrometry.

Burton [103] studied the distribution of arsenic in the Atlantic Ocean. Samples from 1000 m and above were filtered through acid-washed 0.45 μm Sartorius membrane filters. Analyses of samples from depth below 1000 m were made on unfiltered water.

The sample was treated with ascorbic acid, hydrochloric acid, and potassium iodide to release arsenic. The solution was stood for 30 min to allow reduction of arsenic(V) to arsenic(III), which was necessary to ensure quantitative recovery of inorganic arsenic as arsine. With nitrogen passing through the flask, 8% w/v sodium borohydride solution was added. The arsine that evolved was trapped in a solution containing 0.7% w/v potassium iodide and excess iodine over a period of 3 min.

The concentrates were subsequently analysed for arsenic by atomic absorption spectrophotometry fitted with a carbon furnace, linked to a zinc reductor column for the generation of arsine. A continuous stream of argon was allowed to flow with the column connected into the inert gas line between the carbon furnace control unit and the inlet to the furnace.

A wide range of elements were tested for interfering effects; significant interferences were found only at concentrations much higher than those encountered in seawater. No significant difference was found in the results when a sample of seawater was analysed in the way described and also by the same procedure but using the method of standard additions. The determination of arsenic by this technique is also discussed in Section 3.70, "Multimetal Analysis" (Table 3.4D).

3.4.3 ANODIC STRIPPING VOLTAMMETRY

Jaya et al. [105] carried out an anodic stripping voltammetric determination of arsenic(III) at a copper-coated glassy electrode. The deposition of copper on the electrode made it sensitive to the presence of arsenic(III) and suitable for use by anodic stripping voltammetry analysis. The height of the stripping peak was linearly dependent on the concentration of arsenic(III) in the solution in the range of 7.5–750 μg/L arsenic(III). Lead, zinc, cadmium, manganese, and thallium did not cause significant interference, but bismuth did. The method gave 92%–106% arsenic recovery when tested on synthetic seawater samples and on natural arsenic-free seawater spiked with arsenic at levels of 10 and 20 ng/L.

Hua et al. [106] carried out an automated determination of total arsenic in seawater by flow constant current stripping analysis with gold fibre electrodes in which the sample was acidified and pentavalent arsenic was reduced to the trivalent state with iodide. The arsenic was then deposited potentiostatically for 4 min on a 25 μm gold fibre electrode and subsequently stripped with constant current in 5 M hydrochloric acid. Cleaning and regeneration of the gold electrode were fully automated.

Huiliang et al. [107] described a flow potentiometric and constant current stripping analysis for arsenic(IV) without prior chemical reduction to arsenic(III). Details of a procedure for the determination of pentavalent arsenic by means of flow potentiometry and constant current stripping analysis are given. It involved reduction of arsenic to the elemental state on a gold-plated platinum fibre electrode at very low reduction potential and subsequent re-oxidation either by means of a constant current or chemically using gold as an oxidant. Methods for applying this technique for determining total arsenic in acidified seawater are presented. The determination of arsenic is also discussed under multimetal analysis in Section 3.70 (Table 3.4G).

3.4.4 X-RAY FLUORESCENCE SPECTROSCOPY

Becker et al. [108] described a method for the determination of dissolved arsenic in seawater at µg/L levels by precipitation energy-dispersive X-ray fluorescence spectroscopy. Arsenic was precipitated as magnesium ammonium arsenate with magnesium ammonium phosphate as a carrier. The precipitate was collected on a glass fibre filter. An energy-dispersive X-ray spectrometer was the rhodium primary target operated at 60 kV and 2 mA, and a silver secondary target was used to measure arsenic. Arsenic recovery was greatest when 3 mL of phosphate carrier was used. The limit of detection was 0.7 µg/L. The method was suitable for all types of natural waters, including seawater.

3.4.5 NEUTRON ACTIVATION ANALYSIS

Ryabinin et al. [109] used neutron activation analysis to determine arsenic in seawater. Yusov et al. [110] separated arsenic(III) and arsenic(IV) in seawater using a chloroform solution of ammonium pyrrolidine diethyldithiocarbamate. The separated functions were then analysed by neutron activation analysis.

3.4.6 OTHER PROCEDURES

Inductively coupled plasma mass spectrometry [104] has also been employed to determine arsenic in seawater. The determination of arsenic in seawater is also discussed under multimetal analysis in Section 3.70, using photoluminescence spectroscopy (Table 3.4Q), graphite furnace atomic absorption spectrometry (Table 3.4B), and neutron activation analysis (Table 3.4O).

3.5 BARIUM

Barium is of oceanographic interest since it is a non-conservative stable trace element. In spite of the relatively short 11,000-year oceanic residence time for barium, ocean biology largely determines its distribution in the ocean interior. Dissolved concentrations in the major oceans – mapped as part of the GEOSECS program – range between 40 and 200 nmol/kg (5.5–28 µg/L), and profiles show the lowest concentrations near the surface and enrichment at depth in a fashion similar (but not identical) to the distribution of the nutrient element silicon. Its determination to a precision of better than 1% by isotope dilution mass spectrometry has earned barium the distinction of being the 'best measured' nutrient-like trace metal in seawater [112].

3.5.1 GRAPHITE FURNACE ATOMIC ABSORPTION SPECTROMETRY

Epstein and Zander [113] used graphite furnace atomic absorption spectrometry for the direct determination of barium in sea and estuarine water. Roe and Froelich [114] achieved a detection of 30 pg barium for 50 µL injections of seawater using direct injection graphite furnace atomic absorption spectrometry.

Dehairs et al. [115] described a method for the routine determination of barium in seawater by graphite furnace atomic absorption spectrophotometry. Barium is separated from major cations by collection on a cation-exchange resin. It is removed from the resin with nitric acid with recoveries exceeding 99%.

Bishop [116] determined barium in seawater by the direct injection Zeeman modulated graphite furnace atomic absorption spectrometry. The V_2O_5/Si modifier added to undiluted seawater samples promotes injection, sample drying, graphite tube life, and the elimination of most seawater components in a slow char at 115°C–1200°C. Atomisation is at 2600°C. Detection is at 553.6 nm, and calibration is by peak area. Sensitivity is 0.8 absorbance s/ng ($M_0 = 5.6$ pg/0.0044 Abs) at an internal argon flow of 60 mL/min. The detection limit is 2.5 pg barium in a 25 mL sample or 0.5 pg using a 135 mL sample. The precision is 1.2% and accuracy is 2%–3% for natural seawater (5.6–18 µg/L). The method works well in organic-rich seawater matrices and sediment pore waters.

3.5.2 OTHER PROCEDURES

The determination of barium has also been discussed under multimetal analysis in Section 3.70. These methods include X-ray fluorescence spectroscopy (Table 3.4N) and neutron activation analysis (Table 3.4O).

3.6 BERYLLIUM

3.6.1 GRAPHITE FURNACE ATOMIC ABSORPTION SPECTROMETRY

Okutani et al. [17] achieved a rapid and simple preconcentration of beryllium by selective adsorption using activated carbon as an adsorbent and acetylacetone as a complexing agent. The method has been used for the determination of a trace amount of beryllium by graphite furnace atomic absorption spectrometry. The beryllium–acetylacetonate complex is absorbed easily onto activated carbon at pH 8–10. The activated carbon which adsorbed the beryllium–acetylacetonate complex was separated and dispersed in pure water. The resulting suspension was introduced directly into the graphite furnace atomiser. The determination limit was 0.6 ng/L (S/N = 3), and the relative standard deviation at 0.25 µg/L was 3.0%–4.0% (n = 6). Not only was there no interference from the major ions such as sodium, potassium, magnesium, calcium, chloride, and sulphate in seawater, but there was also no interference from other minor ions. The method was applied for determining nanograms per millilitre levels of beryllium in seawater and rainwater.

3.6.2 OTHER PROCEDURES

Other methods reported for the determination of beryllium include UV–visible spectrophotometry [118–120], gas chromatography [121], flame atomic absorption spectrometry (AAS) [122–126], and graphite furnace (GF) AAS [127–135]. The ligand acetylacetone (acac) reacts with beryllium to form a beryllium–acac complex and has been extensively used as an extracting reagent of beryllium. Indeed, the solvent extraction of beryllium as the acetylacetonate complex in the presence of EDTA has been used as a pretreatment method prior to atomic absorption spectrometry [123–125]. Less than 1 µg of beryllium can be separated from milligram levels of iron, aluminium, chromium, zinc, copper, manganese, silver, selenium, and uranium by this method.

The application of inductively coupled plasma mass spectrometry is discussed in Section 3.10 under multication analysis (Table 3.4F).

3.7 BISMUTH

3.7.1 ATOMIC ABSORPTION SPECTROMETRY

Shijo et al. [133] converted bismuth in seawater into its dithiocarbamate complex and then extracted the complex into xylene prior to determination in amounts down to 0.3 ppt by electrothermal atomic absorption spectrometry. The application in this technique is also discussed in Section 3.70 (Table 3.4A).

3.7.2 HYDRIDE GENERATION ATOMIC ABSORPTION SPECTROMETRY

Soo Lee [134] determined picogram amounts of bismuth in seawater by flameless atomic absorption spectrometry with hydride generation. Bismuth is reduced in solution by sodium borohydride to bismuthine, stripped with helium gas, and collected *in situ* in a modified carbon rod atomiser. The collected bismuth is subsequently atomised by increasing the atomiser temperature and detected by an atomic absorption spectrophotometer. The absolute detection limit is 3 pg of bismuth. The precision of the method is 2.2% for 150 pg and 6.7% for 25 pg of bismuth.

Bismuth at levels of 0.05–0.25 ng/L (dissolved) and 0.13–0.29 ng/L (total) was found in Pacific Ocean samples. Bay water contained significantly higher concentrations: 0.4–0.6 ng/L (dissolved) and 0.4–2.0 (total) bismuth.

The application of this technique is also discussed in Section 3.70, "Multimetal Analysis" (Table 3.4D).

3.7.3 ANODIC STRIPPING VOLTAMMETRY

Numerous researchers have investigated this technique for determining bismuth in seawater [135–144]. In the method of Gilbert and Hume [135], the sample containing a silica cell was purged and stirred by the passage of purified nitrogen. A platinum counter electrode was used. The reference electrode consisted of a silver wire, previously anodised in seawater, held in a borosilicate glass tube containing a small, untreated portion of the sample that was separated from the sample being analysed by a plug of unfused Vycor. To diminish the effect of the steeply rising background current on the stripping peaks, a compensating circuit was devised. Bismuth was deposited at −0.4 V from seawater made 1 M in hydrochloric acid and gave a stripping peak of −0.2 V, the height of which was proportional to concentration without interference from antimony or metals normally present. Antimony was deposited at −0.5 V from seawater made 4 M in hydrochloric acid and gave a stripping peak at −0.3 V, the area of which was proportional to the sum of antimony and bismuth. By using the standard addition technique, satisfactory results were obtained for the concentration ranges 2.09–2.00 µg/kg for bismuth and 0.2–0.5 µg/kg for antimony.

Computerised potentiometric stripping analysis [137–139] is, in many respects, a simpler analytical technique than linear sweep or differential pulse anodic stripping voltammetry. Only data for the determination of bismuth(III) in seawater are reported by Florence [136].

It became apparent to Florence [136] that the seawater analysed by the computerised technique was approximately one order of magnitude lower than the results obtained by electroanalytical [135,136,141] and ion-exchange [140] techniques.

Because the direct determination of such low concentrations of bismuth in seawater by means of potentiometric stripping analysis was time consuming, a simple preconcentration technique was used. This technique was based on the coprecipitation of bismuth(III) with magnesium hydroxide, thus taking advantage of the naturally high magnesium concentration in seawater [142,143].

The efficiency of the bismuth(III) coprecipitation procedure was investigated by adding 1,2,3,4,5, and 10 pmol bismuth(III) to 200 mL subsamples of a Kattegatt surface water prior to magnesium hydroxide precipitation. The recovery was in the range of 80%–110% for all samples.

Square-wave anodic stripping voltammetry was employed by Kormorsky-Lovric [144] for the determination of bismuth in seawater. A bare glass–C rotating disk electrode was preconditioned at −0.8 V versus Ag/AgCl, prior to the concentration of bismuth. The method was applied to seawater in the 12 ng/L range.

3.8   BORON

3.8.1   Spectrophotometric Methods

Various chromogenic reagents have been used for the spectrophotometric determination of boron in seawater. These include curcumin [145,146], nile blue [147], and 3,5 di-tert butylcatechol and ethyl violet [148]. Uppstroem [145] added anhydrous acetic (1 mL) and propionic anhydride to the aqueous sample containing up to 5 mg of boron per litre as boric acid (H_3BO_3) in a polyethylene beaker. After mixing and the dropwise addition of oxalyl chloride to catalyse the removal of water, the mixture was cooled to room temperature. Subsequently, concentrated sulphuric acid–anhydrous acetic acid (1:1) and curcumin reagent are added, and the mixed solution is set aside for at least 30 min. Finally, 20 mL (90 mL of 96% ethanol, 180 g ammonium acetate buffer) was added to destroy the excess protonated curcumin, and 135 mL anhydrous acetic acid diluted to 1 L with water was added, the mixture was cooled to room temperature and the extinction was measured at 545 nm.

3.8.2 Phosphorimetric Methods

Marcantoncetos et al. [149] have described a phosphorimetric method for the determination of traces of boron in seawater. This method is based on the observation that in the 'glass' formed by ethyl ether containing 8% of sulphuric acid at 77 K boric acid gives luminescent complexes with dibenzoylmethane. A 0.5 mL sample is diluted with 10 mL 96% sulphuric acid; to 0.05–0.3 mL of this solution, 0.1 mL 0.04 M-dibenzoylmethane in 96% sulphuric acid is added. The solution was diluted to 0.4 mL with 96% sulphuric acid, heated at 70°C for 1 h, cooled, and ethyl ether is added in small portions to give a total volume of 5 mL. The emission is measured at 77 K at 508 nm, with excitation at 402 nm. At the level of 22 ng boron per millilitre, 100-fold excesses of 33 ionic species give errors of less than 10%. However, both tungsten and molybdenum interfere.

3.8.3 Coulometric Methods

Tsaikov [152] described a coulometric method for the determination of boron in coastal seawaters. This method is based on the potentiometric titration of boron with electro-generated hydroxyl ions, after the removal of the cation component by ion exchange. The method has good reproducibility and is more accurate than other methods; it is fairly rapid (25–30 min per determination).

3.8.4 Other Procedures

Horta and Curtins [150,151] discussed an atomic absorption spectrophotometric method for the determination of boron in seawater.

3.9 CADMIUM

In the determination of cadmium in seawater for both operational reasons and ease of interpretation of results, it is necessary to separate particulate material from the sample immediately after collection. The 'dissolved' trace metal remaining will usually exit in a variety of states of complexation and possibly also of oxidation. These may respond differently in the method, except where direct analysis is possible with a technique using high-energy excitation, such that there is no discrimination between different states of the metal. The only technique of this type with sufficiently low detection limits is carbon furnace atomic absorption spectrometry, which is subject to interference effects from the large and varying content of dissolved salts.

3.9.1 Atomic Absorption Spectrometry

Batley and Farrah [153] and Gardner and Yates [154] used ozone to decompose organic matter in samples and thus break down metal complexes prior to atomic absorption spectrometry. With this treatment, metal complexes of humic acid and EDTA were broken down in less than 2 min. These observations led Gardner and

Yates [154] to propose the following method for the determination of cadmium in seawater.

The filtered sample is acidified to about pH 2. Prior to extraction, the sample is ozonised for 30 min. Nitrogen is passed through the sample for 5 min to remove excess ozone. Then the pH is raised to about 5 by the addition of ammonia solution, and 5 mL Chelex–100 resin in the ammonia form is added. After stirring for at least 1 h, the resin is collected in a pyrex chromatography column and washed with the calculated quantity of an appropriate buffer to elute calcium and magnesium. After further washing with 50 mL deionised water, the resin is eluted with 2 M nitric acid to a volume of 25 mL. The elute is analysed by graphite furnace atomic absorption spectrometry.

Han et al. [155] reported on an electrothermal atomic absorption spectrometric technique for the determination of cadmium in natural water, using sodium phosphate for matrix modification. The method was applied for determining cadmium in seawater, and comparable results were obtained by anodic stripping voltammetry.

The application of this technique is also discussed in Section 3.70, "Multimetal Analysis" (Table 3.4A).

3.9.2 GRAPHITE FURNACE ATOMIC ABSORPTION SPECTROMETRY

Various researchers have discussed the application of graphite furnace atomic absorption spectrometry in the determination of cadmium in seawater [153,154, 156–174].

Danielsson et al. [160] have described a graphite furnace method for the determination of cadmium in seawater in solvent extracts of seawater. Levels of cadmium found in Arctic seawater ranged from 0.13 nmol/L at the surface to 0.205 nmol/L at 2000 m depth.

As cadmium is one of the most sensitive graphite furnace atomic absorption, it is not surprising that this is the method of choice for the determination of cadmium in seawater. Earlier researchers separated cadmium from the seawater salt matrix prior to analysis. Chelation and extraction [161–168], ion exchange [162,165,169], and electrodeposition [170,171] have all been studied.

The direct determination of cadmium in seawater is particularly difficult because the alkali and alkaline earth salts cannot be fully charred away at temperatures that will also not volatilise cadmium. Most researchers in the past [165,172–175] who have attempted a direct method have volatilised cadmium at temperatures which would leave sea salts in the furnace. This required careful setting of temperatures and was disturbed by situations that caused temperature settings to change with the life of the furnace tubes.

Lundgren et al. [172] showed that the cadmium signal could be separated from a 2% sodium chloride signal by atomising at 820°C below the temperature where sodium chloride was vaporised. This technique has been called *selective volatilisation*. They detected 0.03 µg/L cadmium in the 2% sodium chloride solution. They used an infrared optical temperature monitor to set the atomisation temperature accurately.

Campbell and Ottaway [156] also used selective volatilisation of the cadmium analyte to determine cadmium in seawater. They could detect 0.04 µg/L cadmium (2 pg in 50 µL) in seawater. They dried at 100°C and atomised at 1500°C with no char step. Cadmium was lost above 350°C.

Guevremont et al. [158] used a direct, selective volatilisation determination of cadmium in seawater. They used 20 µL seawater samples, 1 g/L of EDTA, an atomisation ramp from 250°C to 2500°C in 5 s, and the method of additions. Their detection limit was 0.01 µg/L (0.2 pg in 20 µL), and the characteristic amount was 0.7 pg/0.0044 Abs. The EDTA promoted the early atomisation of cadmium below 600°C. Their test seawater sample (0.053 µg/L) was confirmed by other methods. These authors were unable to separate reliably the cadmium and background signals by using the method of Campbell and Ottaway [156]; the EDTA made this possible.

Guevremont et al. [158] also studied the use of different matrix modifiers in the graphite furnace gas method for the determination of cadmium in seawater. These included citric acid, lactic acid, aspartic acid, histidine, and EDTA. The addition of less than 1 mg of any of these compounds to 1 mL seawater significantly decreased matrix interference. Citric acid achieved the highest sensitivity and reduction of interference with a detection limit of 0.01 µg cadmium per litre.

In a similar work, Sturgeon et al. [165] compared direct furnace methods with the extraction method for cadmium in coastal seawater samples. They found 0.2 µg/L cadmium and measured cadmium down to 0.01 µg/L. They used 10 µg/L ascorbic acid as a matrix modifier. Various organic matrix modifiers were studied by Guevremont [157] for this analysis. He found critic acid to be somewhat preferable to EDTA, aspartic acid, lactic acid, and histidine. The method of standard additions was required. The standard deviation was better than 0.01 µg/L in a seawater sample containing 0.07 µg/L. Generally, he charred at 300°C and atomised at 1500°C. The method required compromise between char and atomisation temperatures, sensitivity, heating rates, and so on, but the analytical results seemed precise and accurate. Nitrate added as sodium nitrate delayed the cadmium peak and suppressed the cadmium signal.

Sperling [173] reported extensively on the determination of cadmium in seawater. He added ammonium persulphate, which permitted charring of seawater at 430°C without the loss of cadmium. For work below 2 µg/L cadmium in seawater, he recommended extraction of cadmium to separate it from the matrix [166,174,175]. He found no change in the measured levels over many months when the seawater was stored in high-density polyethylene or polypropylene.

Guevremont et al. [157,158] discussed the use of organic modifiers for direct graphite furnace atomic absorption determination of cadmium in seawater. The application of this technique is also discussed in Section 3.70 (Table 3.4B).

3.9.3 ZEEMAN ATOMIC ABSORPTION SPECTROMETRY

Pruszkowska et al. [175] described a simple and direct method for the determination of cadmium in coastal water utilising a platform graphite furnace and Zeeman background correction. These researchers obtained a detection limit of 0.013 µg/L in 12 µL samples or about 0.16 pg cadmium in the coastal seawater sample.

The characteristic integrated amount was 0.35 pg cadmium per 0.0044 Abs. A matrix modifier containing diammonium hydrogen phosphate and nitric acid was used. Standards contained sodium chloride and the same matrix modifier as the samples. No interference was observed from the matrix.

Seawater samples usually contain 2%–3% of several alkali and alkaline earth salts with sodium chloride as the main constituent. A 2 µL sample of seawater charred at 700°C has a background signal so high (more than 2 A) that even the Zeeman correction system cannot handle it. The large amounts of sodium chloride present in the seawater are reportedly volatilised below 950°C [176], but even with ammonium phosphate, the matrix modifier recommended for cadmium, it is not possible to char at so high a temperature. Here, 200 µg of diammonium hydrogen phosphate reduced the background signal beam of 2 µL seawater to 0.5 A, but 500 µg ammonium nitrate reduced the background more effectively to 0.16 A. No reduction in the cadmium signal occurred in the presence of ammonium nitrate if the char temperature was below 600°C and phosphate was used as a matrix modifier. If ammonium nitrate was used without phosphate, cadmium was lost at temperatures below 500°C. The addition of phosphate stabilised the cadmium, while ammonium nitrate promoted the release or conversion of the bulk of the background-producing material. The addition of the phosphate produced a background signal that appeared much later than the cadmium peak.

It was shown that 1.25 mg ammonium nitrate is enough to keep the background signal below 1.5 A, and there are no large differences in background absorbances for amounts from 1.25 to 7.5 mg ammonium nitrate.

It was also shown that 2% nitric acid reduced the background to a level that can be handled by the Zeeman correction system. From 4% to 8% nitric acid, the changes in background signal shapes were not very large.

Brewer [177] used electrically vaporised thin gold film atomic emission spectrometry to determine cadmium at the 10 µg/L level in highly acidic saline solutions following preconcentration with a microload of strong base anion-exchange resin.

Knowles [178] used extraction with ammonium pyrollidine dithiocarbamate dissolved in methyl isobutyl ketone to extract cadmium from seawater prior to analysis by Zeeman atomic absorption spectrometry. The method was capable of determining 0.04 µg/L of cadmium in seawater when concentration factors of 100 were used.

Lum and Callaghan [179] did not use matrix modification in the electrothermal atomic absorption spectrophotometric determination of cadmium in seawater. The undiluted seawater was analysed directly with the aid of Zeeman to effect background correction. The limit of detection was 2 ng/L.

Electrothermal atomic absorption spectrophotometry with Zeeman background correction was used by Zhang et al. [180] for the determination of cadmium in seawater.

The application of this technique is also discussed in Section 3.70 (Table 3.4C).

3.9.4   X-RAY FLUORESCENCE SPECTROSCOPY

Margul et al. [181] described a method based on high-energy polarised beam energy-dispersive X-ray fluorescence combined with activated thin layers for the determination of traces of cadmium in seawater. Quantitation of cadmium contained in the thin layers was performed by high-energy polarised-beam-dispersive X-ray fluorescence.

The analysis of different impregnated layers contacted with solutions ranging from 5 to 8000 µg/L cadmium showed a linear response between the cadmium concentration in the aqueous solutions and the metal present in the thin layer, with a detection limit of 0.7 µg/L. The accuracy of the method was confirmed by analysing spiked seawater samples and a synthetic water sample containing, besides cadmium, high amounts of other metal pollutants such as nickel, copper, and lead. The attained results were comparable to those obtained by anodic stripping voltammetry or inductively coupled plasma spectrometry [182].

The application of X-ray fluorescence spectroscopy in the determination of cadmium is also discussed in Section 3.70 (Table 3.4D).

3.9.5 POLAROGRAPHY

Kounaves and Zirino [183] studied cadmium–EDTA complex formation in seawater using computer-assisted stripping polarography. They showed that the method was capable of determining the chemical speciation of calcium in seawater at concentrations down to 10^{-8} M.

Turner et al. [184] studied the automated electrochemical stripping of cadmium in seawater.

Stolzberg [185] reviewed the potential inaccuracies of anodic stripping voltammetry and differential pulse polarography in determining trace metal speciation and thereby the bioavailability and transport properties of trace metals in natural waters. In particular, it is stressed that non-uniform distribution of metal–ligand species within the polarographic cell represents another limitation inherent in electrochemical measurement of speciation. Examples relate to the differential pulse polarographic behaviour of cadmium complexes of NTA and EDTA in seawater.

In a method described by Yoshimura and Uzawa [186], cadmium in seawater is coprecipitated with zirconium hydroxide prior to determination by square-wave polarography. The precipitate is dissolved by hydrochloric acid, and cadmium concentration is determined from the peak height of the polarogram at −0.64 V. The calibration curve was linear for concentrations ≤5.0 µg of cadmium.

3.9.6 MISCELLANEOUS

Capodaglio et al. [624] studied the evolution of cadmium (and lead and copper) complexation by organic ligands along the water column during the 1990/1991 summer in the Gerlache Inlet (Antarctica). The complexation was estimated by determining the total dissolved concentration of metals, labile concentration, ligand concentrations, and relative conditional stability constants. The mean value of the total dissolved cadmium concentration was 0.83 nM until mid-December; the concentration was gradually depleted initially in the subsurface layer, then down to the bottom. The mean concentration along the water column in February was 0.15 nM. The labile fraction represented 90% of the total until December, and it was reduced to about 20% in the upper 50 m by February. Cadmium was complexed by one class of ligands detectable after mid-December. Initially, it was present only in the surface layers and later extended to the bottom. The ligand concentration reached a maximum (2.2 nM) during the phytoplankton bloom in December.

Capodaglio et al. [624] compared the results obtained from surface samples in Terra Nova Bay in 1987/1988 summer with those obtained in the 1989/1990 summer.

Frache et al. [625] studied the evolution of dissolved and particulate cadmium (and Cu and Pb) profiles in the coastal waters of Gerlache Inlet (Terra Nova Bay, Western Ross Sea) during the 1997/1998 Austral Summer. In order to relate the distributions of trace metals with physical and biological processes, a series of temperature and salinity measures were made, and water samples were collected to determine nutrients and chlorophyll. Samples of pack and marine microlayers (50–150 µm) were also collected and analysed.

It was found that metal concentrations in the surface layer were mainly affected by the dynamics of the pack ice melting, allowing for phytoplankton activity. The first process influences both the input of metals from meltwaters and the covering of the seawater surface, allowing atmospheric dust input only when all ice has been melt or removed. The high amount of cadmium in the particulate included in the pack ice does not seem to affect the concentration in the surface particulate; on the contrary, the corresponding increase in dissolved cadmium indicates that it is released in a dissolved form when the pack ice melts. Surface distribution is further complicated by the effect of phytoplankton activity, which removed copper and cadmium from the water, incorporating them into the organic particulate. Finally, in the absence of pack ice, there is evidence of inputs of lead and copper due to atmospheric dust brought into the column water through a marine microlayer.

3.9.7 OTHER PROCEDURES

Other procedures used to determine cadmium in seawater are discussed in Section 3.70, "Multimetal Analysis", and include neutron activation analysis (Table 3.4O), isotope dilution methods (Table 3.4K), anodic stripping voltammetry (Table 3.4G), potentiometric stripping analysis (Table 3.4O), cathodic stripping voltammetry, politico metric stripping analysis (Table 3.4O), and inductively coupled plasma emission spectrometry (Table 3.4E).

3.10 CAESIUM

Nuclear activities such as electricity production by nuclear power plants or accidents such as the Chernobyl disaster release radionuclides, including caesium, into the environment. The caesium concentration in these matrices is very low, so that in addition to a sensitive analytical method, it is necessary to make use of an enrichment technique to bring the caesium concentration within the scope of the analytical method.

3.10.1 ATOMIC ABSORPTION SPECTROMETRY

Atomic absorption spectrometry is suitable as a method of analysis of the concentrate and is applicable to radioactive and nonradioactive forms of the element.

Atomic absorption spectrometry has been used to determine caesium in seawater [187]. The method uses preliminary chromatographic separation on a strong cation-exchange resin, ammonium hexacyanocobalt ferrate, followed by electrothermal atomic absorption spectrometry. The procedure is convenient, versatile, and reliable, although decomposition products from the exchanger, namely iron and cobalt, can cause interference.

Caesium is fully retained by a chromatographic column of ammonium hexacyanocobalt ferrate and can then be recovered by the dissolution of ammonium hexacyanocobalt ferrate in hot 12 M sulphuric acid.

As iron and cobalt both interfere with the determination of caesium, using the 852.1 nm caesium line, these elements were removed in a preliminary separation and then caesium was determined.

Ganzerli et al. [188] also used copper hexacyanoferrate(II) on a silica support to absorb caesium from both seawater and freshwater. A specific analytical method is not described, though atomic absorption spectrophotometry might be used.

Shen and Li [189] extracted caesium (and rubidium) from brine samples with 4-ter-butyl-2-(α methyl-benzyl) phenol prior to the atomic absorption determination of the metal.

3.10.2 Other Procedures

Other procedures used to determine caesium content are discussed in Section 3.70 and include neutron activation analysis (Table 3.4O).

3.11 CALCIUM

3.11.1 Titration Methods

Jagner [190] used a method described by computerised photometric titration in the high-precision determination of calcium in seawater.

Calcium is titrated with EGTA (1,2-bis-(2-aminoethoxyethane)-N,N,N',N'-tetraacetic acid) in the presence of the zinc complex of zincon as indirect indicator for calcium. Theoretical titration curves are calculated by means of the computer program HALTAFALL in order to assess accuracy and precision. The method gives a relative precision of 0.00028 when applied to estuarine water of 0.05%–0.35% salinity.

The complexometric titration is at present considered the best method for the determination of calcium, but investigators have differed in the end-point detection technique used and in their evaluation of interference by other alkaline earth elements. Studies using different end-point techniques [190–202], some of which also considered magnesium-to-calcium ratios in seawater, do not agree on the effect of magnesium on the titration of calcium with EGTA (1,2-bis-(2-aminoethoxyethane)-N,N,N',N'-tetraacetic acid). Table 3.3 lists the findings of some of these studies; the cited references report that magnesium has no effect, causes a positive interference, and in one case, has a negative interference.

In most cases where strontium interference was evaluated, a positive interference was found, but the degree of correction (of the calcium titre) varied from about

TABLE 3.3

Recently Reported Studies on the Determination of Calcium Using EGTA Titration

References	Method	Conclusion
[191]	Hg–electrode	No Mg interference at seawater ratios
[192]	Zn–Zincon	Positive Mg interference from Mg:Ca = 1:5 and higher
[193]	Zn–Zincon	No Mg interference at seawater ratios
[194]	Theoretical, Zn–Zincon	Titration error if Mg > Ca
[195]	Various chemical visual indicators	No Mg interference at seawater ratios when end-point sharp
[196]	Zn–Zincon	Mg interference of +0.729% on Ca titre; Sr interference of +0.388 and on Ca titre
[197]	GHA	Mg interference of –0.23% on Ca titre; Sr interference of +0.77% on Ca titre
[198]	Stability constants 'conditional constants'	Sr interference; increased titration error at seawater ratios – dependent on end-point sensitivity
[199]	Ca–red	Sr interference of +0.37% on Ca titre
[190]	Computer-simulated curves of Zn–Zincon	Mg interference at seawater ratios No Mg interference; Sr interference of +0.9%
[200]	Amalgamated–Ag electrode	On Ca titre
[201]	Ca-ion electrode	No Mg interference
[202]	Ca-ion electrode	No Mg interference; no Sr interference

−0.38% in several studies to −0.77% and −0.88% in other investigations, which claim that all or nearly all strontium is coprecipitated.

In light of these observations, Olson and Chen [203] decided to use a correction factor in their visual end-point calcium titration method involving titration with EGTA. They found that interferences by magnesium and strontium were insignificant at the molar ratios normally found in seawater but were more serious in samples containing higher ratios of magnesium or strontium to calcium. An average value of 0.02103 was obtained for the ratio of calcium to chlorinity in samples of standard seawater.

The presence of normal concentrations of sodium, magnesium, and strontium has no net effect on the determination of calcium above the approximate level of accuracy of about 0.1% so that no correction factor seems necessary. A sufficient amount of titrant must be added to complex at least 98% of dissolved calcium before buffer is added; this apparently reduces the loss of calcium by coprecipitation with magnesium hydroxide.

Interference effects begin to appear at higher magnesium or strontium molar ratios. Tsunogai et al. [197] found the interference of magnesium to be negative, and for strontium, it was related to the extraction into the organic layer of the calcium GHA complex. They found a positive interference for strontium at twice the seawater molar ratios. Therefore, the interferences of the individual alkaline earth elements on the calcium titration found by Olson and Chen [203] are consistent in direction, though clearly not in magnitude, with those that were reported by Tsunogai et al. [197]. The presence of sodium (chloride) in the solutions also seems to diminish these interference effects in both cases. No explanation was found for the reduced interference effect when sodium is present.

Van't Riet and Wynn [204] carried out potentiometric titrations of calcium (and magnesium) in seawater. Calcium is first determined by direct titration with tetrasodium 1,2-bis-(2-aminoethoxy)-ethane-N,N,N',N'-tetra-acetate at pH 8.5 in the presence of trisodium citrate to mask magnesium. The end-point is detected as a sharp upward break in the curve. Magnesium is then determined by continuing the titration with EDTA (tetrasodium salt) until a second break occurs. The sensitivity for each metal is 0.5 μg in a maximum volume of 100 mL; the accuracy is high.

3.11.2 ATOMIC ABSORPTION SPECTROMETRY

The calcium content of seawater-suspended particulate matter has been determined by atomic absorption spectrometry by Ezat [205] and others [210,211]. The particulate material is collected on a 0.45 μm millipore membrane filter and subsequently dissolved in hydrochloric acid and nitric acid.

3.11.3 FLAME PHOTOMETRY

Blake et al. [206] described a flame photometric method for the determination of calcium in solutions of high sodium content. A neutral solution (100 mL) containing up to 50 ppm of calcium and up to 4% of sodium is passed through a column of Chelex–100 resin (Na form). A specified amount of hydrochloric acid (pH 2.4) is passed through, and the percolate containing the sodium is discarded. Elution is then effected with 2 N hydrochloric acid, and the column is washed with water. The combined eluate and washings are diluted to 100 mL, and calcium is determined by flame photometry at 622 nm. There is no interference from magnesium, zinc, nickel, barium, mercury, manganese, copper, or iron present separately in concentrations of 25 mg/L each. Aluminium depresses the amount of calcium found.

Chow and Thompson [207] also used flame photometry to determine calcium in seawater.

3.11.4 CALCIUM-SELECTIVE ELECTRODES

Whitfield et al. [208] used a calcium-selective electrode to monitor EGTA and DCTA titrations of aqueous mixtures of calcium, magnesium, and sodium. The pattern of titration curves observed with changing solution composition agreed qualitatively with that predicted theoretically, but the overall potential drop was usually lower than that predicted; end-points were determined by graphical and numerical methods. The technique is suitable for the determination of calcium and magnesium in seawater with an estimated accuracy of 0.5%. The electrode also responds to zinc, iron, lead, copper, nickel, and barium. In seawater that is free from coastal influences, the concentration of these elements is too low to cause interference.

3.11.5 INDUCTIVELY COUPLED PLASMA ATOMIC EMISSION SPECTROMETRY

Brenner et al. [209] applied inductively coupled plasma atomic emission spectrometry for determining calcium (and sulphate) in brines. The detection limit was 70 µg/L, and a linear dynamic range of 1 g/L was obtained.

3.11.6 OTHER PROCEDURES

Other procedures used to determine calcium in seawater include neutron activation analysis (Table 3.4O), graphite furnace atomic absorption spectrometry (Table 3.4B), and X-ray fluorescence spectroscopy (Table 3.4N).

3.12 CERIUM

3.12.1 SPECTROFLUOROMETRIC METHODS

Shigematsu et al. [212] determined cerium fluorometrically at the 1 µg/L level in seawater. Quadrivalent cerium is coprecipitated with ferric hydroxide, the precipitate is dissolved in hydrochloric acid, and interfering ions are removed by extraction with isobutyl methyl ketone.

The aqueous phase is evaporated almost to dryness with 70% perchloric acid, then diluted with water and passed through a column of bis-(2-ethylhexyl) phosphate on poly(vinyl chloride), from which Ce^{IV} is eluted with 0.3 M perchloric acid. The eluate is evaporated, then made 7 M in perchloric acid, and treated with Ti^{III} and the resulting Ce^{III} is determined spectrofluorometrically at 350 nm (excitation at 255 nm).

3.12.2 OTHER PROCEDURES

Other procedures are discussed in Section 3.70, including performance liquid chromatography (Table 3.4P) and neutron activation analysis (Table 3.4O).

3.13 CHROMIUM

Reported concentrations of chromium in open ocean waters range from 0.07 to 0.96 µg/L with a preponderance of values near the lower limit. Methods used for the determination of chromium at this concentration have generally used some form of matrix separation and analyte concentration prior to determination [213–216], electroreduction [217,218], and ion-exchange techniques [219,220].

Whereas it is desirous to utilise analytical schemes that permit elucidation of the various chromium species, particularly since chromium(VI) is acknowledged to be a toxic form of this element, it is useful to have the capability of rapid total chromium measurement.

3.13.1 SPECTROPHOTOMETRIC METHODS

Diphenylcarbazone and diphenylcarbazide have been widely used for the spectro-photometric determination of chromium [221]. Chromium(III) reacts with diphenyl-carbazone, whereas chromium(V) reacts (probably via a redox reaction combined with complexation) with diphenylcarbazide [222]. Although speciation would seem a likely prospect with such reactions, commercial diphenylcarbazone is a complex mixture of several components, including diphenylcarbazide, diphenylcarbazone, phenylsemicarbazide, and diphenylcarbadiazone, with no stoichiometric relation-ship between diphenylcarbazone and diphenylcarbazide [233]. As a consequence, the use of diphenylcarbazone to chelate chromium(III) selectively also results in the sequestration of some chromium(VI). Total chromium can be determined with diphenylcarbazone following reduction of all chromium to chromium(III).

The use of immobilised chelating agents for sequestering trace metals from aque-ous and saline media presents several significant advantages over chelation solvent extraction approaches to this problem [224,225]. With little sample manipulation, large preconcentration factors can generally be released in relatively short times with low analytical blanks.

As a consequence of these considerations, Willie et al. [226] developed a new approach to the determination of total chromium, which involves preliminary concentration of dissolved chromium from seawater by means of an immobilised diphenylcarbazone chelating agent, prior to determination by atomic absorption spectrometry. Chromium was first reduced to chromium(III) by the addition of aque-ous sulphur dioxide. Aliquots of seawater were then adjusted to pH 9.0 ± 0.2 by using high-purity ammonium hydroxide and gravity fed through a column of silica. The sequestered chromium was then eluted from the column with 0.2 M nitric acid. More than 93% of chromium was recovered in the first 5 mL of elute by this method.

Levels of chromium found in coastal waters (29.5% salinity) were between 0.05 and 0.100 µg/L and in open ocean waters (35% salinity) between 0.15 and 0.19 µg/L.

3.13.2 CHEMILUMINESCENCE METHODS

The chemiluminescence technique has been used to determine trivalent chromium in seawater. Chang et al. [227] showed that luminol techniques for the determination of chromium(III) were hampered by a salt interference, mainly from magnesium ions. Elimination of this interference is achieved by seawater dilution and utilising bromide ion chemiluminescence signal enhancement. The chemiluminescence results were comparable with those obtained by a graphite furnace flameless atomic absorption analysis for the total chromium present in samples. The detection limit is 3.3×10^{-9} (0.2 µg/L) for seawater with a salinity of 35% with 0.5 M bromide enhancement.

Dubovenko et al. [228] used chemiluminescence to determine chromium in brines and waste waters. The method is based on the enhancement of the chemilu-minescence by chromium in the reaction of 4-(diethylamino) phthalhydrazide with hydrogen peroxide. The detection limit is 0.025 µg of chromium per litre, and the chemiluminescence is directly proportional to chromium concentrations in the range of 5×10^{-10} to 10^{-6} M.

3.13.3   ATOMIC ABSORPTION SPECTROMETRY

3.13.3.1   Tri- and Hexa-Valent Chromium

Various researchers have discussed the separate determination of chromium(III) and chromium(VI) in seawater [214,229,230,234,242].

Cranston and Murray [214,230] obtained samples in polyethylene bottles that had been pre-cleaned at 20°C with 1% distilled hydrochloric acid. Total chromium $(Cr^{IV}) + Cr^{III} + Cr_p$ (particulate chromium) was precipitated with iron(II) hydroxide and reduced chromium was coprecipitated with iron(III) hydroxide; $(Cr^{III} + Cr_p)$ was coprecipitated with iron(II) hydroxide and reduced chromium was coprecipitated with iron(III) hydroxide.

The iron hydroxide precipitates were filtered through 0.4 μm nucleopore filters. Particulate chromium was also obtained by filtering unaltered samples through 0.4 μm filters. The iron hydroxide coprecipitates were dissolved in 6 M distilled hydrochloric acid and analysed by flameless atomic absorption. The limit of detection of this method is about 0.1–0.2 nmol/L. The precision was about 5%.

3.13.3.2   Organic Forms of Chromium

In the determination of the two oxidation states of chromium, the calculation of one oxidation state by difference presupposes that the two oxidation states in question were statistically the only chromium-containing contributors to the total concentration. Because of this, contributions from other possible species such as organic complexes were generally not considered. It has been suggested [231], however, that this presumption may not be warranted and that contributions from organically bound chromium should be considered. This arises from the reported presence of dissolved organic species in natural waters, which form stable soluble complexes with chromium and which may not readily be amenable to determination by procedures commonly in use. The results of research into the valency of chromium present in seawater have not always been consistent. For instance, Grimaud and Michard [232] reported that chromium(III) predominates in the equatorial region of the Pacific Ocean, whereas Cranston and Murray [214] found that practically all chromium is in the hexavalent state in the northeast Pacific. Organic chromium(III) complexes may be formed under the conditions prevailing in seawater as well as inorganic chromium(III) and (VI) forms. Inconsistencies in earlier research may, therefore, be at least partly due to the fact that the possibility of organic chromium species was ignored [231,233].

Nakayama et al. [234] described a method for the determination of chromium(III), chromium(VI), and organically bound chromium in seawater. They found that seawater in the Sea of Japan contained about 9×10^{-9} M dissolved chromium. This was shown to be divided into about 15% inorganic chromium(III), about 25% inorganic chromium(VI), and about 60% organically bound chromium.

These researchers studied the coprecipitation behaviours of chromium species with hydrated iron(III) and bismuth oxides.

3.13.3.3   Collection of Chromium(III) and Chromium(V)
            with Hydrated Iron(III) or Bismuth Oxide

Only chromium(III) coprecipitates quantitatively with hydrated iron(III) oxide at the pH of seawater, around 8. To collect chromium(VI) directly without pretreatment, for example reduction of chromium(III), hydrated bismuth oxide, which forms an insoluble compound with chromium(VI), was used. Chromium(III) is collected with hydrated bismuth oxide (50 mg, 400 mL/L seawater). To collect chromium(VI) in seawater, a pH of about 4 was used. Both chromium(III) and chromium(VI) are thus collected quantitatively at the pH of seawater around 8.

3.13.3.4   Collection of Chromium(III) Organic Complexes with Hydrated Iron(III) or Bismuth Oxide

The percentage collection of chromium(III) with hydrated iron(III) oxide may decrease considerably in the neutral pH range when organic materials capable of combining with chromium(III), such as citric acid and certain amino acids, are added to seawater [235]. Moreover, synthesised organic chromium(III) complexes are scarcely collected with hydrated iron(III) oxide over a wide pH range.

As it was not known what kind of organic matter acts as the major ligand for chromium in seawater, Nakayama et al. [234] used EDTA and 8-quinolinol-5-sulphonic acid to examine the collection and decomposition of organic chromium species, because these ligands form quite stable water-soluble complexes with chromium(III), although they are not actually present in seawater. Both these chromium(III) chelates are stable in seawater at pH 8.1 and are hardly collected with either of the hydrated oxides. The organic chromium species were then decomposed to inorganic chromium(III) and chromium(VI) species by boiling with 1 g ammonium persulphate per 400 mL/L seawater acidified to 0.1 M with hydrochloric acid. Iron and bismuth, which would interfere in atomic absorption spectrometry, were 99.9% removed by extraction from 2 M hydrochloric acid solution with a p-xylene solution of 5% tri-iso-octylamine. Chromium(III) remained almost quantitatively in the aqueous phase in the concentration range of 10^{-9} to 10^{-6} M, whether or not iron or bismuth was present. However, as about 95% of chromium(VI) was extracted by the same method, samples which may contain chromium(VI) should be treated with ascorbic acid before extraction so as to reduce chromium(VI) to chromium(III).

When the residue obtained by the evaporation of the aqueous phase after the extraction was dissolved in 0.1 M nitric acid and the resulting solution was used for electrothermal atomic absorption spectroscopy, a negative interference, which was seemingly due to residual organic matter, was observed. This interference was successfully removed by digesting the residue with 1 mL of concentrated hydrochloric acid and 3 mL of concentrated nitric acid. This process had the advantage that the interference of chloride in the atomic absorption spectroscopy was eliminated during the heating with nitric acid.

Various forms of chromium determined in various oceans between 1950 and 1980 [236–241] were in the following ranges: Cr^{III}, 0.02–0.52 µg/L; Cr^{V}, 0.13–0.96 µg/L; and organic chromium, 0.06–0.23 µg/L. In most of these methods, coprecipitation with hydrated iron(III) oxide was used to separate chromium(III) from chromium(VI), and the chromium(VI) concentration was subsequently determined by suitable reduction of chromium(VI) to chromium(III) before further coprecipitation. In other cases, hydrated iron(II) oxide served as both a reductant and carrier.

Mullins [242] described a procedure for determining the concentrations of dis-solved chromium species in seawater. Chromium(III) and chromium(VI) separated by coprecipitation with hydrated iron(III) oxide, and total dissolved chromium is determined separately by conversion to chromium(VI), extraction with ammonium pyrrolidine diethyl dithiocarbamate into methyl isobutyl ketone, and determina-tion by atomic absorption spectroscopy. The detection limit is 40 g/L chromium. In the waters investigated, total concentrations were relatively high (1–5 μg/L), with chromium(VI) being the predominant species in all the areas sampled. Typically, seawater samples taken in the Sydney area were 0.27–0.88 μg/L Cr^{III}, 0.49–3.17 μg/L Cr^{VI}, and 0.56–0.82 μg/L of organically bound chromium.

3.13.3.5   Zeeman Atomic Absorption Spectrometry

Moffett [243] determined chromium in seawater by Zeeman-corrected graphite tube atomisation atomic absorption spectrometry. Chromium is first complexed with a pentane-2,4-dione solution of ammonium-1-pyrrolidine carbodithioc acid. Then this complex is extracted from water with a ketonic solvent, such as methyl isobutyl ketone, 4-methylpentan-2-one, or diisobutyl ketone.

Down to 1 ng/mL of chromium can be determined by this procedure. Recoveries obtained on standard samples are adequate.

The application of this technique is discussed in Section 3.70.

3.13.3.6   Gas Chromatography

Mugo and Orians [244] discussed shipboard methods for the determination of chromium(III) and total chromium in seawater by derivatisation with trifluoroacety-lacetone followed by gas chromatography using an election capture detector.

3.13.3.7   High-Performance Liquid Chromatography

High-performance liquid chromatography, coupled with an inductively coupled plasma mass spectrometric detector, has been used to determine microgram per litre concentrations of chromium(III) and chromium(VI) in seawater [245].

3.13.3.8   Isotope Dilution Gas Chromatography–Mass Spectrometry

Isotope dilution gas chromatography–mass spectrometry has also been used for the determination of microgram per litre levels of total chromium in seawater [246–248,250–252]. The samples were reduced to produce chromium(III) and then extracted and concentrated as tri(1,1,1-trifluoro-2,4-pentanediono) chromium(III) [$Cr(tfa)_3$] into hexane. The $Cr(tfa)_2$ + mass fragments were monitored into a selected ion monitoring (SIM) mode.

Isotope dilution techniques are attractive because they do not require quantitative recovery of the analyte. One must, however, be able to monitor specific isotopes, which is possible by mass spectrometry.

In this method, chromium is extracted and preconcentrated from seawater with trifluoroacetylacetone [H(tfa)], which complexes with trivalent but not hexavalent chromium. Chromium reacts with trifluoroacetylacetone in a 1:3 ratio to form an octahedral complex, $Cr(tfa)_3$. The isotope abundance of its most abundant mass frag-ment, $Cr(tfa)_2$+, was monitored by a quadruple mass spectrometer.

An excellent agreement was obtained for chromium in seawater by isotope dilution spark resource mass spectrometry and graphite furnace atomic absorption spectrometry.

3.13.3.9   Anodic Stripping Voltammetry

Boussemart and Van den Berg [251] absorbed chromium(III) in seawater onto silica and then re-oxidised it to chromium(VI) prior to the determination in amounts down to 1 pmol/L by a voltammetric procedure. The application of this technique is also discussed in Section 3.70 (Table 3.4G).

3.13.3.10   Speciation

Ahern et al. [252] discussed the speciation of chromium in seawater. The method employed coprecipitation of trivalent and hexavalent chromium, separately, from samples of surface seawater and determination of the chromium in the precipitates and particulate matter by thin film X-ray fluorescence spectrometry. An ultraviolet irradiation procedure was used to release bound metal. The ratios of labile trivalent chromium to total chromium were in the range of 0.4–0.5, and the total labile trivalent and hexavalent chromium were in the range of 0.3–0.6 µg/L. Bound chromium ranged from 0 to 3 µg/L and represented 0%–90% of total dissolved chromium.

3.13.3.11   Other Procedures

Other procedures for the determination of chromium in seawater discussed in Section 3.70 include graphite furnace atomic absorption spectrometry (Table 3.4B), neutron activation analysis (Table 3.4O), X-ray fluorescence spectroscopy (Table 3.4N), inductively coupled plasma emission spectrometry (Table 3.4E), and inductively coupled plasma mass spectrometry (Table 3.4F).

3.14   COBALT

3.14.1   Spectrometric Methods

3.14.1.1   Further Analysis

Kouimtzis et al. [254] described a spectrophotometric method for determining down to 1 µg/L cobalt in seawater in which cobalt is extracted with 2,2′ dipyridyl-2-pyridylhydrazone (DPPH) [255–259], and the cobalt complex is back-extracted into 20% perchloric acid and this solution is evaluated spectrophotometrically at 500 nm. Because of the low concentrations of cobalt present in seawater, large sample volumes of 1–2 L are required for this analysis.

Kentner and Zeitlin [253] added sodium citrate, hydrogen peroxide, and 1-nitro-2-naphthol to 1 L of seawater and extracted the coloured cobalt complex formed into chloroform prior to the measurement of the extinction at 410 nm.

In another spectrophotometric procedure, Motomizu [260] added to the sample (2:1) 40% (w/v) sodium citrate dehydrate solution and a 0.2% solution of 2-ethylamino-5-nitrosophenol in 0.01 M hydrochloric acid. After 30 min, 10% aqueous EDTA and 1,2-dichloroethane were added. The organic phase was washed successively with hydrochloric acid (1:2) as well as potassium hydroxide and hydrochloric acid (1:2). The extinction was measured at 462 nm.

3.14.2   ATOMIC FLUORESCENCE SPECTROMETRY

Yuzefovsky et al. [261] used C_{18} resin to preconcentrate cobalt from seawater prior to determination at the ppt level by laser-excited atomic fluorescence spectrometry with a graphite electrothermal atomiser.

3.14.3   FLOW INJECTION ANALYSIS

Sakamoto-Arnold [262] determined picomolar levels of cobalt in seawater by flow injection analysis with chemiluminescence detection. This method is based on the luminescence oxidation of gallic acid in alkaline hydrogen peroxide. A preconcentration/separate step in the flow injection analysis manifold with an in-line column of immobilised 8-hydroxyquinoline was included to separate cobalt from alkaline earth ions. The detection limit is approximately 8 pM. The average standard deviation was ±5%.

Malaholff et al. [263] used a shipboard flow injection spectrophotometric technique to determine ppt concentrations of cobalt in seawater.

Carole et al. [264] also determined the picomolar levels of cobalt in seawater by flow injection analysis with chemiluminescence detection.

3.14.4   CATHODIC STRIPPING VOLTAMMETRY

Vega and Van den Berg [265] described a procedure for the direct determination of picomolar levels of cobalt in seawater. Cathodic stripping voltammetry is preceded by the adsorptive accumulation of the cobalt–nioxime (cyclohexane-2.2-dione dioxime) complex from seawater containing 6 µM nioxime and 80 mM ammonia at pH 9.1, onto a hanging mercury drop electrode, followed by reduction of the adsorbed species. The reduction current is catalytically enhanced by the presence of 0.5 M nitrite. Optimised conditions for cobalt include a 30 s adsorption period at −07.V and a voltammetric scan using differential pulse modulation. According to the proposed reaction mechanism, dissolved cobalt(II) is oxidised to cobalt(III) upon the addition of nioxime and high concentrations of ammonia and nitrite. A mixed cobalt(III)–ammonia–nitrite complex is adsorbed on the electrode surface, and cobalt(III) is reduced to cobalt(II) (complexed by nioxime) during the voltammetric scan, followed by its chemical re-oxidation by the nitrite, initiating a catalytically enhanced current. A detection limit of 3 pM cobalt (at an adsorption period of 60 s) enables the detection of this metal in uncontaminated seawater using a very short adsorption time. UV digestion of seawater is essential, as part of the cobalt may occur strongly complexed by organic matter and rendered nonlabile.

The application of this technique is also discussed in Section 3.70 (Table 3.4L).

3.14.5   POLAROGRAPHY

Harvey and Dutton [266] determined nanogram amounts of cobalt in seawater after concentration on manganese dioxide formed by the photochemical oxidation of divalent manganese in a photochemical reactor. The sample (1:1) was irradiated in a discharge lamp radiating at 254 and 185 nm. The manganese dioxide deposit that adhered to the silica of the reactor was dissolved in 0.15 M hydrochloric acid containing a trace of sulphur dioxide. The solution was evaporated to dryness, and the residue was dissolved in 0.625 M hydrochloric acid. Then 5 M aqueous ammonia and 0.1% dimethylgloxime in ethanol were added, and cobalt was determined by pulse polarography. The polarograph was operated in the derivative mode, starting at −1.0 V, and a 50 mV pulse height and a 1 s mercury drop life were used.

3.14.6 OTHER PROCEDURES

Other procedures used to determine cobalt in seawater, discussed in Section 3.70, "Multimetal Analysis", include atomic absorption spectrometry (Table 3.4A), graphite furnace atomic absorption spectrometry (Table 3.4B), X-ray fluorescence spectrometry (Table 3.4E), inductively coupled plasma atomic emission spectroscopy (Table 3.4E), chromopotentiometric analysis (Table 3.4I), inductively coupled plasma mass spectrometry (Table 3.4F), and neutron activation analysis (Table 3.4O).

3.15 COPPER

Copper(II) is present in natural waters in a variety of chemical forms. Pagenkopf et al. [267] and Sylva [268] indicated that the following species are found in freshwater systems: Cu^{2+}; $CuCO_3$; $Cu(CO_3)_2^{2-}$; $CuHCO_3^+$; $CuOH^+$; $Cu_2(OH)_2^{2+}$; and $CuCl^+$. It was also found that Cu^{2+} can be removed completely from aquatic systems by precipitation as $Cu(OH)_2$, $CuCO_3$, and $Cu(OH)_n(CO_3)_{1-n/2}$.

Sunda and Hanson [269] used ligand competition techniques for the analysis of free copper(II) in seawater. This work demonstrated that only 0.02%–2% of dissolved copper(II) is accounted for by inorganic species (i.e. Cu^{2+}, $CuCO_3$, $Cu(OH)^+$, Cu Cl+, etc.); the remainder is associated with organic complexes. Clearly, the speciation of copper(II) in seawater is markedly different from that in freshwater.

Importantly, Sunda and coworkers [269–271] demonstrated that free copper(II) – not total copper(II) – was responsible for copper(II) toxicity. Consequently, the impact of copper(II) on the marine environment can be ascertained only by the measurement of free copper(II) levels.

Copper may exist in particulate colloidal and dissolved forms in seawater. In the absence of organic ligands or particulate and colloidal species, carbonate and hydroxide complexes account for more than 98% of the inorganic copper in seawater [272–276,279]. The copper(II) concentration can be calculated if pH, ionic strength, and the necessary stability constants are known [274,286]. In most natural systems, the presence of organic materials and sorptive surfaces significantly alters speciation and decreased the utility of equilibrium calculations. Analytical difficulties in the measurement of copper(II) and copper associated with naturally occurring ligands have encouraged numerous researchers to introduce the 'complexation capacity' concept [267,277,278]. Functionally, the copper-complexing capacity of a water sample is the ability of the sample to remove added copper from the free ion pool [280]. Analytically, complexation capacity measurements depend on the quantitation of the complexing ability of an operationally defined group of ligands. It is assumed that unknown ligands may be classed into meaningful groups on the basis of the physical properties of their metallo-complexes (e.g. lability to anodic stripping voltammetry, chelating resins, or ultraviolet radiation). Schemes to determine the concentration of copper associated with different classes have been proposed as useful ways to address complexing capacity questions in natural systems [281,282] as have various titrimetric techniques [283,284]. Different analytical procedures measure the copper-chelating capacity of slightly different classes of ligands, and there is some overlap in the complexes included in classes defined by different techniques. For example, while there is a small fraction of organic material in seawater which forms ASV-labile complexes not dissociated by Chelex resin [285], most ASV-labile complexes are also labile to chelating resins [286].

Studies on the determination of copper in seawater have been predominantly concerned with speciation. Before discussing this, the limited amount of work on the determination of total copper, that is, work not concerned with speciation, is discussed as follows.

3.15.1 TITRATION PROCEDURE

Ruzic [287] considered the theoretical aspects of the direct titration of copper in seawater and the information this technique provides regarding copper speciation. The method is based on a graph of the ratio between the free and bound metal concentration versus the free metal concentration. The application of this method, which is based on a 1:1 complex formation model, is discussed with respect to trace metal speciation in natural waters. Procedures for the interpretation of experimental results are proposed for those cases where two types of complexes with different conditional stability constants are formed, or where the metal is adsorbed on colloidal particles.

Waite and Morel [288] described an amperometric titration procedure for the characterisation of organic copper–complexing ligands and applied it to a variety of synthetic and naturally occurring organic compounds. The procedure is based on the ability in the solutions of high chloride content to obtain a sensitive and reproducibility amperometric measurement of reducible copper(II) at positive voltages up to about 100 mV relative to an Ag/AgCl reference electrode. Copper(II) is reduced to copper(I), which is stabilised by chloride despite the presence of oxygen. The application of the titration technique to a high chloride content electrolyte containing various concentrations of nitrilotriacetic acid confirms that copper–ligand reduction and dissociation are not major problems provided that a sufficiently positive working electrode potential is chosen and that the concentration of the organic ligand is low. The application of the procedure to a variety of naturally occurring organic agents, including a fulvic acid, freshwater algal exudates, and a sample of Sargasso seawater, produces results that are consistent with those found by alternative methods.

3.15.2 SPECTROPHOTOMETRIC METHODS

Abraham et al. [289] determined total copper in seawater spectrophotometrically using quinaldehyde 2-quinolyl-hydrazone as a chromogenic reagent. This method is capable of determining copper at the ppb level.

3.15.3 Fluorescence-Based Fibre-Optic Biosensor

Zeng et al. [290] reported a determination of free copper ion at picomolar levels in seawater using a fluorescence-based fibre-optic biosensor. The sensor transducer is a protein molecule, site specifically labelled with a fluorophore attached to the distal end of an optical fibre. This binds free Cu(II) with high affinity and selectivity. The sensor demonstrates a detection limit of 0.1 pM free Cu(II) in a seawater model. The accuracy and precision of the sensor were comparable to cathodic ligand exchange/ adsorptive cathodic stripping voltammetry.

3.15.4 Atomic Absorption Spectrometry

Various researchers have studied the application of this technique in the determination of copper in seawater [291–295].

Cabon and Le Bihan [294] studied copper signals obtained by electrothermal atomic absorption spectrometric analysis of seawater matrices. The interference effects of sodium chloride, sodium nitrate, sodium sulphate, magnesium chloride, magnesium nitrate, and calcium chloride on the electrothermal atomic absorption spectrometric signal of copper in seawater were studied. Thermal treatment at temperatures between 600°C and 800°C caused the hydrolysis of magnesium chloride to magnesium oxide and minimised its interference. Ashing at higher temperatures of 1300°C, 1200°C, and 1100°C was carried out in the presence of sulphate, nitrate, and chloride salts, respectively, without the loss of copper. A study of the influence of two-component, chloride–nitrate or chloride–sulphate matrices illustrated the stabilising effect of the formation of metal oxides and metal sulphides on the copper signal. This stabilisation enhanced the decrease in interference connected with chloride removal in an acidic medium.

In further work, Cabon [295] proposed a model to describe the variations in copper signals caused by the principal inorganic ions in seawater (sodium, magnesium, calcium, chloride, and sulphate). Data obtained by ashing-simulated seawaters under different temperature conditions were used. Ashing at 800°C caused hydrolysis of magnesium chloride to magnesium oxide and the formation of sodium sulphide. Both these products enhanced the stability of copper in the furnace. A complementary decrease in interference occurred in the presence of magnesium when a small amount of nitrate (0.2 M) was added. The model was confirmed by results obtained using nitric or sulphuric acid as a modifier.

The application of this technique is also discussed in Section 3.70 (Table 3.4A).

3.15.5 Ion-Selective Electrodes

Prior to the introduction of ion-selective electrode techniques, *in situ* monitoring of free copper(II) in seawater was not possible due to the practical limitations of existing techniques (e.g. ligand competition and bacterial reactions). *Ex situ* analysis of free copper(II) is prone to experimental error as the removal of seawater from the ocean can lead to speciation of copper(II). Potentially, a copper(II) ion electrode is capable of rapid *in situ* monitoring of environmental free copper(II).

Ion-selective electrodes have been used for the potentiometric determination of the total cupric ion content of seawater [296]. Down to 2 µg/L cupric copper could be determined by this procedure.

Belli and Zirino [297] studied the behaviour and calibration of copper(II) ion-selective electrodes in waters. The Nerstian behaviour response was consistent over 10 orders of magnitude of chloride concentration and from pH 2 to 8.

De Marco [298] examined the response of copper(II) ion-selective electrodes in seawater. This researcher compared the behaviour of three types of copper(II) ion-selective electrodes (i.e. copper sulphide, copper selenide, and copper/silver sulphide) in seawater. X-ray photoelectric spectroscopy and X-ray diffraction showed that the unacceptably high detection limit of the copper sulphide electrode ($\sim 10^{-4}$ M Cu^{2+}) is due to membrane oxidation in cupric sulphide and $Cu_3(SO_4)(OH)_4$. Bare $Cu_{1.8}Se$ and CuS/Ag_2S electrodes displayed Nernstian response (i.e. ~100% Nernstian slope) in the range of 10^{-16} to 10^{-18} M free copper(II) with copper(II)-ethylenediamine buffers also containing 0.6 M sodium chloride. It is proposed that amelioration of the chloride interference at low levels of free copper(II), that is, $<10^{-18}$ M, is due to the kinetic limitations of the membrane reaction that is responsible for the chloride interference. Corrosion of the $Cu_{1.8}Se$ electrode contaminated seawater with a high level of copper(II) (~100 nM), while the CuS/Ag_2S electrode released a much lower amount of copper(II) (~2.4 nM). Electrode carryover and contamination of seawater by adsorbed free copper(II) are minimised by equilibration of the electrode in a sacrificial copper(II) buffer (i.e. $pCu_{free} = 15$) before analysis. The behaviour of copper(II) electrodes in seawater was interpreted in relation to free copper(II) levels, and results indicate a proportionality between free copper(I) and the electrode potential.

A layer mechanism has been proposed for the chloride interference in the CuS electrode [299,300], viz:

$$CuS(s) + Cu^2(aq) \rightarrow S(s) + 2Cu^+(aq)$$

$$Cu^+(aq) + nCl^-(aq) \rightarrow [CuCl_n]^{1-n}(aq)$$

3.15.6 ANODIC STRIPPING VOLTAMMETRY

Much of the work concerned with the speciation of copper and other trace metals in natural waters has been done using anodic stripping voltammetry [314–321]. This work has primarily progressed in two directions: studies of the shift in trace metal peak potentials with changing concentrations of ligands and the studies of change in metal peak height or peak area under differing experimental conditions.

Shuman and Michael [301] applied a rotating disc electrode to the measurement of copper complex dissociation rate constants in marine coastal waters. The technique was used to measure the extent of copper chelation in these samples and to establish an operational definition for labile and nonlabile metal complexes based on a kinetic criterion. Samples collected off the mid-Atlantic coast showed various degrees of chelation towards copper. A first-order dissociation rate constant for copper chelates was estimated to be of the order of 2 s^{-1}. It is suggested that this technique should be useful for metal toxicity studies because of its ability to measure both equilibrium concentration and kinetic availability of soluble metal.

Scorano et al. [302] determined copper at the 5×10^{-10} M level in seawater by anodic stripping voltammetry using ethylenediamine. These researchers investigated the properties of ethylenediamine (en) as a means of performing the analysis of ligand-exchangeable and labile (i.e. directly reducible at pH 8) copper in seawater at trace levels. Stripping polarographic or pseudopolarographic determinations show that the copper ethylenediamine complex behaves reversibly in seawater, exchanging two electrons at the mercury drop electrode. The role of chloride ions in competitive reactions with ethylenediamine for copper during the stripping step was also studied. In seawater made 2×10^{-3} M in ethylenediamine, copper(II) can be detected in the hanging mercury drop electrode by differential pulse anodic stripping voltammetry at the 5×10^{-10} M level with a deposition time of 10 min. A procedure for measuring pH 8 labile copper in seawater is obtained by coupling differential pulse anodic stripping voltammetry with a medium alteration method. The addition of ethylenediamine at the end of the electrolysis increases the peak height by more than twice by doubling the current yield per mole of copper and by removing interferences associated with the oxidation of copper in chloride media. This procedure facilitates the voltammetric study of copper in seawater under natural conditions.

Quentel et al. [303] complexed copper with 1,2-dihydroxyanthraquinon-3 sulphuric acid prior to determination by absorptive stripping voltammetry in amounts down to 0.3 nM in seawater.

Wang et al. [304] used a remote electrode, operated in the potentiometric stripping mode, for the continuous onboard measurement of copper distribution patterns in San Diego Bay.

Garcia-Monco Carrá et al. [305] described a 'hybrid' mercury film electrode for the voltammetric analysis of copper (and lead) in acidified seawater. Mercury plating conditions for preparing a consistently reproducible mercury film electrode on a glassy carbon substrate in acid media are evaluated. It is found that a 'hybrid electrode', that is, pre-plated with mercury and then re-plated with mercury *in situ* with the sample, gives very reproducible results in the analysis of copper in seawater. Consistently reproducible electrode performance allows for the calculation of a cell constant and prediction of the slopes of standard addition plots, useful parameters in the study of copper speciation in seawater.

Capodaglio et al. [306] studied the speciation of copper in the surface seawater of Terra Nova Bay by differential pulse anodic stripping voltammetry using samples collected during 1987/1988, 1988/1989, and 1989/1990 Italian expeditions. The total copper concentration ranges were between 0.5 and 4.8 nM and showed a uniform spatial distribution without evident differences between the three campaigns. The mean value for the labile fraction was 4% of the total for the first two expeditions, while it was below the detection limit for the last one. The results showed the presence of two classes of ligands, one stronger (mean concentration 1.5 nM), which showed a considerable variability due to seasonal and spatial factors, and one weaker (mean concentration 31 nM, average conditional stability constant 2.8×10^8 M^{-1}), which showed a homogeneous distribution in the studied area.

The application of this technique is also discussed in Section 3.70 (Table 3.4H).

3.15.7 Neutron Activation Analysis

Neutron activation analysis has been used [307] to determine total copper in seawater. The application of this technique is also discussed in Section 3.70 (Table 3.4O).

3.15.8 Miscellaneous

Other procedures used for the determination of copper in seawater include electrochemical stripping analysis [308], electron spin resonance technique [310], and adsorption on a Sep-Pac cartridge [311].

Buerge-Weirich and Suztzberger [312] applied solid-phase extraction for determining Cu^{11} in marine waters.

Li et al. [313] measured free copper ion activity in seawater using a passive-equilibrium sonic-assisted free ion recorder.

Marvin et al. [309] studied the effect of sample filtration on the determination of copper in seawater, demonstrating that glass filters seriously affect the reliability of subsequent analysis.

Frache et al. [625] studied the evolution of dissolved and particulate copper (also lead) profiles in the coastal waters of Gerlache Inlet (Terra Nova Bay, Western Ross Sea) during the 1997/1998 Austral Summer. In order to relate to the distribution of trace metals with physical and biological processes, a series of temperature and salinity measurements were made, and water samples were collected to determine nutrients and chlorophyll. Samples of pack ice and marine microlayers (50–150 μm) were also collected and analysed.

Concerning the surface layer, it was found that metal concentrations were mainly affected by the dynamics of the pack ice melting and phytoplankton activity. The first process influences both the input of metals from meltwaters and the covering of the seawater surface, allowing atmospheric dust input only when all the ice has been melted or removed. The direct release of particulate copper from the ice was shown by surface maxima and by the high concentrations of suspended particulate matter and particulate metals found in the ice core section interfaces with the seawater.

Capodaglio et al. [624] studied the evolution of copper (also lead and cadmium) by organic ligands along the water column during the 1990/1991 summer in the Gerlache Inlet in Antarctica. The total dissolved copper concentration ranged between 1.6 and 3.8 μM. The labile fraction was strongly dependent on depth and time, ranging from values lower than 1% surface water to a maximum value of 40% in deep water. Copper was complexed by two classes of ligands: the first was only present in surface layers and the second was homogeneously distributed along the water column. The results are compared to those obtained for surface samples collected in the whole of Terra Nova Bay between the 1987/1988 summer and the 1989/1990 summer.

3.15.9 Copper Speciation

Wood et al. [322] described an ion-exchange technique for measuring the copper-complexing capacity of seawater samples taken in the Sargasso Sea and continental shelf samples. This technique measures the copper-complexing capacity of relatively strong dissolved and colloidal organic complexing agents in natural seawater. The technique was used to compare the copper-complexing capacity of strong organic dissolved and colloidal complexing agents in these samples. They also analysed the relationship between the copper-complexing capacity of a specific group of complexing agents and the concentration of two large heterogenous pools of potential complexing agents: dissolved organic carbon and total particulate material. The copper-complexing capacity of those samples ranged from 0.014 to 1.681 µmol copper per litre in the inner shelf, from 0.043 to 0.095 mol copper per litre in the mid- and outer shelf waters, and from 0.010 to 0.036 µmol copper per litre at the Sargasso Sea stations.

Cathodic stripping voltammetry has been used to determine copper species in seawater [327,328]. Van der Berg [329] determined copper in seawater by cathodic stripping voltammetry of complexes with catechol.

A reduction current occurred when a solution of catechol and copper was subjected to cathodic stripping voltammetry at a hanging mercury drop electrode. The composition of the adsorbed film and the optimal conditions for its formation were investigated. The phenomenon was used to determine copper in seawater using AC polarography to measure complex adsorption. Currents were detected at the very low copper concentrations that occurred in uncontaminated seawater. Competition for copper ions by natural organic complexing ligands was evident at low concentrations of catechol. The method was more sensitive and had a shorter collection period than the rotating disc electrode DPASV technique, with comparable accuracy.

Zorkin et al. [323] developed a procedure to estimate the amount of biologically active copper in seawater based on the assumption that the divalent copper ion was the most toxic species and its concentration could be related to the amount of metal sorbed on a sulphonic acid cation-exchange resin. The method was applied to artificial seawater containing copper and the organic ligands EDTA, NTA, histidine, and glutamic acid. Other experiments showed that there was a correlation between the inorganic fraction determined by the ion-exchange procedure and the toxic fraction of copper quantified by a bioassay using the marine diatom.

Zirino and Kounaves [324] applied anodic stripping analysis to the study of the reduction of copper in seawater from San Diego Bay. The potential peak height plots obtained for copper at pH 8 featured broadly curving slopes, and no distinct limiting plateau was reached, even at the highest applied potential. The shapes of these 'waves' indicated an irreversible reduction. On the other hand, the reduction of copper at pH 3 is quasi-reversible with $E3/4 = E1/4 = 42$ mV.

Potentiometric stripping analysis has been applied by Shefrin and Williams [325] in the measurement of copper in seawater at environmental pH. The advantage of this technique is that it can be used to specifically measure the biologically active labile copper species in seawater samples at desired pH values. The method was applied to seawater samples that had passed a 0.45 µm millipore filter. Samples were studied both at high and at low pH values.

Shuman and Michael [326] introduced a technique that has sufficient sensitivity for kinetic measurement at very dilute solutions. It combines anodic scanning voltammetry with the rotating disk electrode and provides a method for measuring kinetic dissociation rates *in situ* and a method for distinguishing labile and nonlabile complexes kinetically, consistent with the way they are defined.

Odier and Plichon [327] used AC polarography to determine the chemical form and concentration of copper in seawater. The shift of the $E_{1/2}$ of reduction of copper(II) determined by AC polarography serves to identify the inorganic complexes of copper and to determine their formation constants. They showed that copper is present in seawater mainly as Cu^{2+}, $CuCl^+$, and $(Cu(HCO_3)_2(OH)^-$. Copper down to 3 µg/L was determined in seawater by AC polarography after acidifying to pH 5 by the passage of carbon dioxide.

Cathodic stripping voltammetry has been used to determine copper species in seawater [328]. Van der Berg [329] determined copper in seawater by cathodic stripping voltammetry of complexes with catechol.

Reduction of current occurred when a solution of catechol and copper was subjected to cathodic stripping voltammetry at a hanging mercury drop electrode. The composition of the adsorbed film and the optimal conditions for its formation were investigated. The phenomenon was used to determine copper in seawater using AC polarography to measure complex adsorption. Currents were detected at the very low copper concentrations that occurred in uncontaminated seawater. Competition for copper ions by natural organic complexing ligands was evident at low concentrations of catechol. The method was more sensitive and had a shorter collection period than the rotating disc electrode DPASV technique with comparable accuracy.

3.15.10 FURTHER TECHNIQUES

Other techniques for determining copper in seawater, discussed in Section 3.70, include cathodic stripping voltammetry (Table 3.4L), potentiometric stripping analysis (Table 3.4M), plasma emission spectrometry (Table 3.4J), isotope dilution methods (Table 3.4K), X-ray fluorescence spectroscopy (Table 3.4N), high-performance liquid chromatography (Table 3.4P), graphite furnace atomic absorption spectrometry (Table 3.4B), differential pulse anodic stripping voltammetry (Table 3.4H), Zeeman atomic absorption spectrometry (Table 3.4C), inductively coupled plasma atomic emission spectrometry (Table 3.4E), inductively coupled plasma mass spectrometry (Table 3.4F), cathodic stripping voltammetry (Table 3.4L), and chromopotentiometry (Table 3.4I).

3.16 DYSPROSIUM

3.16.1 ISOTOPE DILUTION ANALYSIS

The application of this technique is discussed in Section 3.70 (Table 3.4K).

3.17   ERBIUM

3.17.1   Isotope Dilution Analysis

The application of this technique is discussed in Section 3.70 (Table 3.4K).

3.18   EUROPIUM

3.18.1   Isotope Dilution Analysis

The application of this technique is discussed in Section 3.70 (Table 3.4K).

3.18.2   Neutron Activation Analysis

The application of this technique is discussed in Section 3.70 (Table 3.4O).

3.19   GADOLINIUM

3.19.1   Isotope Dilution Analysis

The application of this technique is discussed in Section 3.70 (Table 3.4K).

3.20   GALLIUM

3.20.1   X-ray Fluorescence Spectroscopy

The application of this technique is discussed in Section 3.70 (Table 3.4P).

3.21   GERMANIUM

3.21.1   Hydride Generation Furnace Atomic Absorption Spectrometry

Andreae and Froelich [330] described a procedure for the determination of germanium in seawater. In this method, the peak absorbance was somewhat dependent on atomisation temperature, rising sharply between 2400°C and 2500°C and remaining almost constant above this temperature. As the lifetime of the tube decreased with the burn temperature, it was decided to choose 2600°C as the analysis temperature. Under these conditions, tubes lasted for at least 100 determinations.

The sensitivity of this system is 430 pg/0.004 Abs. The standard deviation of the baseline noise is about 0.0007 Abs, resulting in a noise-limited detection limit of 140 pg of germanium at 95% confidence level. There is no detectable blank when the analysis is performed in deionised water, so that the noise-limited detection limit is the actual lower limit of determination at which quantitative analysis can be carried out. This corresponds to concentration detection limits of 1.4 ng/L for the 100 mL and 0.56 ng/L for the 250 mL reaction vessels, which are well below typical concentrations of germanium in natural waters.

The application of this technique is also discussed in Section 3.70 (Table 3.4N).

3.22   GOLD

3.22.1   INDUCTIVELY COUPLED PLASMA MASS SPECTROMETRY

Falkner and Edmond [331] determined gold at femtomolar (10^{-15} mL^{-1}) quantities in seawater by flow injection inductively coupled plasma quadruple mass spectrometry. The technique involves preconcentration by anion exchange of gold as a cyanide complex, [Au(CN)$_2^-$], using ^{195}Au radiotracer ($t_{1/2}$ = 183 days) to monitor recoveries. Samples are then introduced by flow injection into an inductively coupled plasma quadruple mass spectrometer for analysis. The method has a detection limit of ≈10 fM for 4 L of seawater preconcentrated to 1 mL and a relative precision of 15% at the 100 fM level.

3.22.2   NEUTRON ACTIVATION ANALYSIS

The application of this technique is discussed in Section 3.70 (Table 3.4O).

3.22.3   MISCELLANEOUS

Pilipenko and Pavlova [332] determined traces of gold in seawater using a 'spot' photometric method. This method is based on the catalysis (by AuCl$_2$SO$_4^-$) of the oxidation of iron(II) by silver(I) with the production of metallic silver. The sample is filtered through paper, and the paper is dried and decomposed with sulphuric acid, nitric acid, hydrofluoric acid, and water solution (1:1:1:1). The residue is dissolved in a few drops of aqua regia, and this solution is evaporated and the residue is dissolved in one drop of 0.5 M sulphuric acid. This solution is applied to a filter paper, and to the resulting spot are applied drops of phosphate–citric buffer solution (pH 2.4) of 0.72 M ferrous sulphate in 0.05 M sulphuric acid and drops of 0.1 M silver nitrate. The reflectance of the spot (due to metallic silver) is measured with a suitable instrument 15 s after the silver nitrate solution has been applied. The reflectance is proportional to the amount of gold on the paper from 3 to 60 pg. As little as 0.005 µg/L of gold can be determined in seawater.

Ogata and Kawasaki [333] studied the adsorption of Au$_{111}$ from aqueous solutions on calcined gibbsite.

3.23   HOLMIUM

3.23.1   ISOTOPE DILUTION ANALYSIS

The application of this technique is discussed in Section 3.70 (Table 3.4K).

3.24   INDIUM

3.24.1   NEUTRON ACTIVATION ANALYSIS

Matthews and Riley [334] determined indium in seawater at concentrations down to 1 ng/L. Preconcentrations of metals on a cation exchange were followed by the separation of alkali metals and alkaline earth metals by the retention of indium as a chloro-complex on an anion exchanger. Samples of indium containing eluate were then concentrated and irradiated in a thermal flux of 5×10^{12} neutrons/cm²/s for several weeks, and the resulting 1.98 MeV β-radiation of the long-lived indium isotope 114 m for several weeks nuclide was counted. Minor elements were removed by a series of post-irradiation solvent extraction stages.

3.24.2 OTHER PROCEDURES

Other procedures for the determination of indium in seawater discussed in Section 3.70 include inductively coupled plasma atomic emission spectrometry (Table 3.4E) and inductively coupled plasma mass spectrometry (Table 3.4F).

3.25 IRIDIUM

3.25.1 GRAPHITE FURNACE ATOMIC ABSORPTION SPECTROMETRY

The application of this technique is discussed in Section 3.70 (Table 3.4B).

3.26 IRON

Worldwide marine chemists and marine biologists have focused on the behaviour of iron in seawater since Martin et al. [335–339] pointed out that the phytoplankton growth in oceanic water was limited by the deficiency of iron derived from the atmosphere rather than the lack of nutrients in some oceanic regions, such as the equatorial Pacific, Gulf of Alaska, and Antarctic Ocean. This attractive hypothesis created a heated argument [340–343] and spurred the geochemical study of iron. For example, more recently Zhuang et al. [344] reported that half of the iron in Aeolian mineral dust existed in the form of iron(II), resulting in the enhancement of solubility of iron in surface water. In order to verify whether or not the iron deficiency contributes to the limitation of primary production and also to clarify the chemical species of iron, an accurate and rapid analytical method for determining iron in seawater is essential. A conventional analytical method, such as solvent extraction graphite furnace atomic absorption spectrometric detection, requires a contamination-free technique. Moreover, it is time consuming and troublesome, as litres of the sample solution must be treated, because the dissolved concentration of iron in oceanic waters is extremely low (1 nmol/L = 56 ng/L). Martin [339] found that the dissolved concentration of iron was less than 0.02 nmol/L in the shallow water (60 m) of the equatorial Pacific. The classical chemiluminescence method using a luminol–hydrogen peroxide system [345,346] is thought to be a promising method for the shipboard analysis of iron because it is highly sensitive to iron and requires only a small size detection device. However, iron(III) must be separated from the other heavy metal ions, such as manganese(II), chromium(III), cobalt(II), and copper(II) prior to detection, since the method is not specific to iron(III). To overcome these

problems, Obata et al. [347] developed an automated analytical method for determining iron in seawater using a closed flow system with a combination of a chelating
resin concentration and chemiluminescence detection.

3.26.1 SPECTROPHOTOMETRIC METHODS

Shirdar and Odzeki [348] determined iron in seawater by densitometry after enrichment as a bathophenanthroline disulphonate complex on a thin layer of anion-exchange
resin. Seawater samples (50 mL) containing iron(II) and iron(III) were diluted to
150 mL with water followed by sequential addition of 20% hydrochloric acid, 10%
hydroxylammonium chloride, 5 M ammonium solution (to pH 3.0 for iron(III) reduction), bathophenanthroline disulphonate solution, and 10% sodium acetate solution
to give a mixture with a final pH of 4.5. A macroreticular anion-exchange resin,
Amberlyst A27, in the chloride form was added, the resultant coloured thin layer was
scanned by a sensitometer and the absorbance was measured at 550 nm.

Blair and Tregeur [349] used a C_{18} column impregnated with ferrozine, a selective
ligand for iron(II), coupled to a spectrophotometer for online shipboard determination of iron at detection limits of 0.1 M.

3.26.2 CHEMILUMINESCENCE ANALYSIS

Eirod et al. [350] determined sub-nanomolar levels of iron(II) and total dissolved
iron in seawater by flow injection analysis with chemiluminescent detection in
amounts down to 0.45 nmol/L.

Obata et al. [347] devised an automated method for the determination of iron
in seawater by chelating resin concentration and chemiluminescence detection. In
this method, iron(III) in an acidified sample solution is selectively collected on
8-quinolinol immobilised chelating resin and then eluted with dilute hydrochloric
acid. The resulting eluent is mixed with luminol solution, aqueous ammonia, and
hydrogen peroxide solution, successively, and then the mixture is introduced into the
chemiluminescence cell. The iron concentration is obtained from the chemiluminescence intensity. The detection limit of iron(III) is 0.05 nmol/L when using an 18 mL
seawater sample. The method is applied to ordinary oceanic waters and hydrothermal waters collected in the North and South Pacific Oceans.

O'Sullivan et al. [351] determined down to 0.1 nmol/kg ferrous iron in seawater
by oxidation with oxygen followed by determination by a catalysed luminol chemiluminescence method.

3.26.3 ATOMIC ABSORPTION SPECTROMETRY

Atomic absorption spectrometry, coupled with solvent extraction of iron complexes,
has been used to determine down to 0.5 µg/L iron in seawater [352,353]. Hiiro et al.
[352] extracted iron as its 8-hydroxyquinoline complex. The sample was buffered to pH
3–6 and extracted with a 0.1% methyl isobutyl ketone solution of 8-hydroxyquinoline.
The extraction was aspirated into an air–acetylene flame and evaluated at 248.3 nm.

The application of this technique is also discussed in Section 3.70 (Table 3.4A).

3.26.4 GRAPHITE FURNACE ATOMIC ABSORPTION SPECTROMETRY

Moore [353] used a solvent extraction procedure to determine iron in frozen seawater. To a 200 mL aliquot of sample was added a solution of sodium diethyldithiocarbamate (1% w/v) and ammonium pyrrolidine dithiocarbamate (1% w/v). This was adjusted to pH 4 and extracted with 1.1.2 trifluoromethane. After evaporation to dryness, the residue was digested with hydrogen peroxide to decompose organic matter. The solution was exposed to a mercury vapour lamp prior to determination of iron.

The coefficient of variation was 21% for seven subsamples containing 1.6 nmol Fe/L and 30% for eight subsamples at 0.6 nmol Fe/L. The detection limit was estimated to be 0.2 nmol Fe/L.

The application of this technique is also discussed in Section 3.70 (Table 3.4B).

3.26.5 INDUCTIVELY COUPLED PLASMA MASS SPECTROMETRY

Lacan et al. [354] used double spike inductively coupled plasma mass spectrometry to determine the isotopic composition of dissolved iron in iron-depleted seawater and demonstrated the feasibility of the measurement of the isotopic composition of dissolved iron in seawater for an iron concentration range of 0.05–1 nmol/L. Iron is preconcentrated using a nitriloacetic acid super-flow resin and purified using an AG 1-x4 anion-exchange resin. The isotopic ratios are measured with a multi-collector inductively coupled plasma mass spectrometer (MC-ICPMS), coupled with a desolvator, using a $^{57}Fe-^{58}Fe$ double spike mass bias correction. An optima precision is obtained for a double spike composed of approximately 50% ^{57}Fe and 50% ^{58}Fe and a sample to double spike quantity ratio of approximately 1. Total procedural yield is $91 \pm 25\%$ (2SD, $n = 55$) for sample sizes from 20 to 2 L. The procedural blank ranges from 1.4 to 1.1 ng, for sample sizes ranging from 20 to 2 L, respectively, which, converted into Fe concentrations, corresponds to blank contributions of 0.001 and 0.010 nmol/L, respectively. Measurement precision determined from replicate measurements of seawater samples and standard solutions is 0.08% ($\sigma^{56}Fe$, 2SD).

3.26.6 FLUORESCENCE SIDEROPHORE SENSOR

With direct evidence that iron is the chemical limitation of phytoplankton growth, particularly in the Southern Ocean, it is increasingly important to develop a method that provides direct measurement of the bioavailable iron fraction in oceanic waters. Cathy et al. [355] reported the development of the fluorescence quenching–based siderophore biosensor capable of the *in situ* measurement of this ultra-trace Fe(III) fraction at ambient pH (~8). Parabactin was extracted from the cultures of *Paracoccus denitrificans*. The purified siderophore was encapsulated within a spin-coated sol–gel thin film, which was subsequently incorporated in a flow cell system. The parabactin biosensor has been fully characterised for the detection of Fe(III) in seawater samples. The biosensor can be regenerated by lowering the pH of the flowing solution, thereby releasing the chelated Fe(III), enabling multiple use. The limit of detection of the biosensor was determined to be 40 pM. For an Fe(III) concentration of 1 nM, a reproducibility with an RSD of 6% (n = 10) was obtained. The accuracy of the blossoming system has been determined through the analysis of a certified seawater reference sample. Samples from the Atlantic Ocean were analysed using the parabactin biosensor providing a concentration versus depth profile of the bioavailable Fe(III) fraction in the range of 50 pM to 1 nM.

3.26.7 MISCELLANEOUS

Radioisotope dilution using the chelating agent bathophenanthroline has been used to determine down to 5 μg/L iron in seawater [356].

Laglera and Van den Berg [357] examined the wavelength of the photochemical reduction of iron in Arctic seawater.

3.26.8 FURTHER TECHNIQUES

Other techniques used to determine iron in seawater, discussed in Section 3.70, include atomic emission spectrometry (Table 3.4A), inductively coupled plasma atomic emission spectrometry (Table 3.4E), anodic stripping voltammetry (Table 3.4G), cathodic stripping voltammetry (Table 3.4L), isotope dilution methods (Table 3.4K), X-ray fluorescence spectrometry (Table 3.4N), neutron activation analysis (Table 3.4O), and high-performance liquid chromatography (Table 3.4P).

3.27 LANTHANUM

3.27.1 ISOTOPE DILUTION METHOD

The application of this technique is discussed in Section 3.70 (Table 3.4K).

3.27.2 NEUTRON ACTIVATION ANALYSIS

The application of this technique is discussed in Section 3.70 (Table 3.4O).

Johnson et al. [358] have discussed the use of nanoporous sorbents for the collection of lanthanides and actinides in seawater.

3.28 LEAD

3.28.1 ATOMIC FLUORESCENCE SPECTROSCOPY

Bolshov et al. [359] used this technique to determine low lead concentrations. A detection limit of 0.05 pg/mL was achieved in studies with aqueous solutions as the reference using a graphite atomiser.

Cheam et al. [360] determined lead in seawater in amounts down to 1 ppt by laser-excited atomic fluorescence spectrometry.

3.28.2 ATOMIC ABSORPTION SPECTROMETRY

Ohta and Suzuki [361] investigated the electrothermal atomisation of lead for accurate determination of lead in water samples. Thiourea served to lower the atomisation temperature of lead and to eliminate the interference from chloride matrix. The addition of thiourea also allowed the accurate determination of lead irrespective of its chemical form. The absolute sensitivity (1% absorption) was 1.1×10^{-12} g of lead. The method permits the direct rapid determination of lead in water samples, including seawater.

No severe interference was noted in the method by arsenic, bismuth, calcium, copper, iron, magnesium, antimony, selenium, tin, and tellurium. The application of this technique is discussed in Section 3.70 (Table 3.4A).

3.28.3 GRAPHITE FURNACE ATOMIC ABSORPTION SPECTROMETRY

Various researchers [362–366] applied graphite furnace atomic absorption spectrometry in the determination of lead in seawater.

A large amount of sodium chloride in seawater samples causes nonspecific absorption [362–371], which can be only partially compensated by background correction. In addition, the seawater matrix may give rise to chemical as well as physical interferences related to the complex physicochemical phenomena [372–374] associated with the vaporisation of metals and the matrix itself.

Several matrix modifiers, which alter the drying or charring properties of the sample matrix, have been tested [362,375–378] to reduce nonspecific absorption. However, the matrix modification methods do not permit determinations of the indigenous lead in seawater because of the relatively high detection limit and poor precision. Yet, gross chemical manipulations of the samples should be avoided to prevent contaminations, which can be dramatic when the analyte is present at microgram per litre or sub-microgram per litre level.

In order to overcome the problem of high nonspecific absorption, alternative procedures have been tested, which involve prior separation of the trace metals from the salt matrix. Examples of extraction of trace metals from seawater as chelates with subsequent determination by electrothermal atomic absorption spectrometric procedures have been described [377,378], but these and similar methods are seldom effective and satisfactory when the matrix is very complex and the analyte concentration is very low.

In contrast, the coupling of electrochemical and spectroscopic techniques, that is, electrodeposition of a metal followed with detection by atomic absorption spectrometry, has received limited attention. Wire filaments, graphite rods, pyrolytic graphite tubes, and hanging drop mercury electrodes have been tested [379–389] for electrochemical preconcentration of the analyte to be determined by atomic absorption spectroscopy. However, these *ex situ* preconcentration methods are often characterised by unavoidable irreproducibility, contaminations arising from handling of the support, and detection limits unsuitable for lead detection at sub-ppb levels.

These drawbacks could be avoided by performing *in situ* deposition. Torsi et al. [366] and others [390] set up an apparatus which permitted both *in situ* electrochemical preconcentration of the analyte from a flowing solution and almost complete suppression of matrix effects because the matrix could be removed by suitable washing.

Halliday et al. [363] described a simple rapid graphite furnace method for the determination of lead in amounts down to 1 μg/L in polluted seawater. The filtered seawater is diluted with an equal volume of deionised water, ammonium nitrate is added as a matrix modifier, and aliquots of the solution are injected into a tantalum-coated graphite tube in a HGA–2200 furnace atomiser. The method eliminates the interference normally attributed to the ions commonly present in seawater. The results obtained on samples from the Firth of Forth were in good agreement with values determined by anodic stripping voltammetry. Hirao et al. [365] concentrated lead in seawater using a chloroform solution of dithizone and determined it in amounts down to 40 μg/L by graphite furnace atomic absorption spectrometry. Lead in 1 kg acidified seawater was equilibrated with lead-212 of a known radioactivity, extracted with dithizone in chloroform, back-extracted with 0.1 M hydrochloric acid, and subjected to graphite furnace atomic absorption spectrometry by a two-channel spectrometer. Recovery yield of lead was found to be 60%–90% from the radioactivity of lead-212 in the back-extract. Lead concentrations were thus determined with about 10% precision.

The accuracy of such analyses in the picogram to nanogram per gram of lead range depends primarily on the ability of the analyst to obtain a true estimate of contamination blanks introduced during the collection, transport, and handling of samples. In the laboratory, the latter step must be kept to an absolute minimum.

Sturgeon et al. [391] applied *in situ* metal trapping in the determination of lead in seawater. In this method, inorganic lead in seawater samples is converted to tetraethyl lead using sodium tetraethylboron ($NaB(C_2H_5)_4$), which is then trapped in a graphite furnace of 400°C. Quantitation is achieved by using a simple calibration graph prepared from aqueous standards having a sensitivity of 0.150 ± 0.0006 A/ng. An absolute detection limit of $(3\sigma)14$ pg is achieved. The precision of determination at 100 pg/L is 4% relative standard deviation.

The application of this technique is also discussed in Section 3.70 (Table 3.4B).

3.28.4 ANODIC STRIPPING VOLTAMMETRY

Clem [392] described an electrochemical cell in which rapid deoxygenation of the sample solution was achieved by allowing a jet of nitrogen to impinge on the surface of the liquid, while the cell was rotated to maintain the solution as a thin film on the cell wall; 15 mL of solution can be deoxygenated in 1–1.5 min. The cell has been used for determining, by anodic stripping voltammetry, 11.2 and 4.1 parts per 10^9 of lead in acidified seawater (pH 2), and for the amperometric titration (with lead) of organic ligands in non-acidified seawater.

Acebal et al. [393] discussed the quantitative behaviour of lead (and copper) when voltammetric determinations are done at mercury film electrodes and hanging mercury drop electrodes. The samples were collected in polyethylene bottles and, generally, were not acidified immediately after collection. This might place some doubt on the results reported.

Automated chemical stripping has been used for the determination of lead in seawater [394].

Quentel et al. [395] studied the influence of dissolved organic matter in the determination of lead in seawater by anodic stripping voltammetry.

Svensmark [396] gave details of equipment and procedure for the rapid determination of lead by the modification of anodic stripping voltammetry, using staircase voltammetry at high scan rates to strip the lead plated on a rotating mercury film electrode. This allowed rapid determination of lead without the need for deoxygenation, rest periods, electrode rotation, or stirring. At a concentration as low as 0.1 µg/L, lead could be determined in less than 4 min. Results obtained on a sample of seawater are presented.

To determine down to 6 ppt of lead in seawater, Wu and Batley [397] used absorptive stripping voltammetry with ligand competition using xylenol orange.

Garcia-Monco Carrá et al. [398] discussed the use of 'hybrid' mercury film electrode for the voltammetric analysis of lead (and copper) in acidified seawater.

The application of this technique is also discussed in Section 3.70 (Table 3.4G).

3.28.5   MASS SPECTROMETRY

Flegal and Stukas [399] described the special sampling and processing techniques necessary for the prevention of lead contamination of seawater samples, prior to stable lead isotopic ratio measurements by thermal ionisation mass spectrometry. Techniques are also required to compensate for the absence of an internal standard and the presence of refractory organic compounds. The precision of the analyses is 0.1%–0.4%, and a detection limit of 0.02 ng/kg allows the tracing of lead inputs and biogeochemical cycles.

3.28.6   MISCELLANEOUS

Ultraviolet spectroscopy has been applied for determining lead and lead speciation studies [400]. Scaule and Patterson [401] used isotope dilution mass spectrometry to determine the lead profile in the open North Pacific Ocean.

Capodaglio et al. [624] studied the evolution of lead, cadmium, and copper by organic ligands along the water column during the 1990/1991 summer in the Gerlache Inlet (Antarctica). The complexation was estimated by the determination of the total dissolved concentration of metals, in labile concentration, ligand concentrations, and relative conditional stability constants. The main total dissolved concentration for lead ranged from 0.083 nM until December to 0.030 nM in the upper 100 m at the end of the summer. The labile fraction did not change during the season and represented 39% of the total dissolved concentration. Lead was complexed by one class of ligands; its mean concentration ranged between 0.60 nM at the beginning of summer and 1.2 nM at the end of the season. The results are compared to those obtained for surface samples collected in the whole of Terra Nova Bay between the 1987/1988 summer and the 1989/1990 summer.

Frache et al. [625] studied the copper and cadmium profiles in the coastal waters of Gerlache Inlet (Terra Nova Bay, Western Ross Sea) during the 1997/1998 Austral Summer. In order to relate the distributions of trace metals with the physical and biological processes, a series of temperature and salinity measures were made, and water samples were collected to determine nutrients and chlorophyll. Samples of pack ice and marine microlayers (50–150 μm) were also collected and analysed.

With respect to the surface layer, it was found that metal concentrations were mainly affected by the dynamics of the pack ice melting and phytoplankton activity. The first process influences both the input of metals from meltwaters and the covering of the seawater surface, allowing atmospheric dust input only when all ice has been melt or removed.

Surface distribution is further complicated by the effect of phytoplankton activity, which removes Cu and Cd from water, incorporating them into organic particulate. Finally, in the absence of pack ice, there is evidence of inputs of lead and cadmium due to the atmospheric dust brought into the column water through marine microlayer.

In intermediate and deep waters, the vertical distribution of lead and cadmium was characterised by substantially constant profiles, while copper showed, during the end of the summer and in the absence of a well-defined water column stratification, a 'scavenging-type' distribution which overlaps its 'nutrient-type' behaviour.

3.28.7   FURTHER TECHNIQUES

Other techniques employed to determine lead in seawater, discussed in Section 3.70, include Zeeman atomic absorption spectrometry (Table 3.4C), inductively coupled plasma atomic emission spectrometry (Table 3.4E), inductively coupled plasma spectrometry (Table 3.4F), cathodic stripping voltammetry (Table 3.4L), differential pulse anodic stripping voltammetry (Table 3.4H), potentiometric stripping voltammetry (Table 3.4M), plasma emission spectrometry (Table 3.4J), isotope chelation method (Table 3.4K), X-ray fluorescence spectrometry (Table 3.4N), neutron activation analysis (Table 3.4O), and high-performance liquid chromatography (Table 3.4P).

3.29   LITHIUM

3.29.1   ATOMIC ABSORPTION SPECTROMETRY

Benzwi et al. [403] determined lithium in Dead Sea water using atomic absorption spectrometry. The sample was passed through a 0.45 μm filter, and lithium was then determined by the method of standard additions. Solutions of lithium in hexanol and 2-ethylhexanol gave greatly enhanced sensitivity.

The application of this technique is also discussed in Section 3.70 (Table 3.4A).

3.29.2   Isotope Dilution Methods

Isotope dilution mass spectrometry has been used to determine traces of lithium in seawater [402,404].

3.29.3   Neutron Activation Analysis

Wiernik and Amiel [405] used neutron activation analysis to measure lithium and its isotopic composition in Dead Sea brines.

3.29.4   Gel Permeation Chromatography

Rona and Schumuckler [406] used gel permeation chromatograph to separate lithium from Dead Sea brine. The elements emerged from the column in the order potassium, sodium, lithium, magnesium, and calcium, and it was possible to separate a lithium-rich fraction also containing some potassium and sodium but no calcium and magnesium.

3.30   LUTECIUM

3.30.1   Isotope Dilution Methods

The application of this technique is discussed in Section 3.70 (Table 3.4K).

3.31   MAGNESIUM

3.31.1   Titration Methods

The application of this technique is discussed in Section 3.70.

3.31.2   Spectrophotometric Methods

The application of this technique is discussed in Section 3.70.

3.31.3   Neutron Activation Analysis

The application of this technique is discussed in Section 3.70 (Table 3.4O).

3.31.4   Miscellaneous

Das et al. [407] carried out a direct gravimetric determination of magnesium in seawater by precipitation with N-benzoyl-N-phenylhydroxylamine. The precipitate is weighed as $Mg(C_{13}H_{10}O_2N)_2$. The coefficient of variation for 5 mg of magnesium was 0.55%. Magnesium could be determined in the presence of barium or strontium; coprecipitation of calcium was avoided by adding tartrate; nickel, cobalt, copper, mercury, and zinc were masked with cyanide and tartrate. Phosphate, oxalate, fluoride, and EDTA interfered. Tin, iron, aluminium, and beryllium were separated by prior precipitation with N-benzoyl-N-phenylhydroxylamine at pH 0.5 to 1.0, 3.5, 4.0, and 5.5, respectively.

3.32   MANGANESE

The natural water chemistry of manganese, which is an important trace element both biologically and geologically, is complicated by nonequilibrium behaviour. From thermodynamic considerations, manganese dioxide (manganese(IV)) is expected to be the stable form of manganese in seawater [408]. However, seawater contains a significant quantity of dissolved manganese(II), which only oxidises slowly [409]. Investigations of the oxidation of manganese(II) have shown that the process is autocatalytic, the product being a solid manganese oxide phase whose composition depends on the reaction conditions [409–412]. The heterogenous nature of the oxidation process can thus account for the extremely slow oxidation of manganese(II) in seawater where concentrations of particulate matter can be relatively small [409].

Manganese is of particular interest because of its central role in many marine geochemical processes and involvement in biological systems.

Manganese and many other trace metals are present in open ocean waters at concentrations in the order of nmol/L or less.

The dissolved manganese is easily precipitated by oxidation to manganese(IV) oxide, which acts as a powerful scavenger for trace elements. The solidified manganese in sediments is reduced to manganese(II) and is regenerated into the water column under mild reducing conditions, for example in the oxygen minimum zone and the near-shore anoxic sediments. In recent years, it has also been found that a copious amount of manganese is injected into deep waters by hydrothermal emanations through active ocean crusts. Therefore, it is very important to clarify the distribution of manganese in seawater to understand marine geochemistry. Furthermore, manganese is thought to be a promising element as a chemical tracer for probing the hydrothermal activities if it can be analysed easily and quickly onboard ship.

3.32.1   SPECTROPHOTOMETRIC METHODS

Olaffson [413] described a semiautomated determination of manganese in seawater using leucomalachite green. The autoanalyser had a 620 nm interference filter and 50 min flow cells. Findings indicated initial poor precision was related to pH, temperature, and time variations. With strict controls on sample acidity and reaction conditions, the semiautomated method had high precision, at least as good as that achieved by preconcentration and atomic absorption procedures and provided precise, rapid, shipboard information on the continental distribution of manganese and on anomalies associated with geothermal seafloor activity. The method was not suitable for estuarine samples or quite sensitive enough for the study of the open ocean manganese distribution.

Brewer and Spencer [414] described a method for the determination of manganese in anoxic seawaters based on the formulation of a chromophore with formaldoxime to produce a complex with an adsorption maximum at 450 nm. Sulphide (50 µg/L), iron, phosphate (8 µg/L), and silicate (100 µg/L) do not interfere in this procedure. The detection limit is 10 µg/L manganese.

3.32.2 SPECTROFLUOROMETRIC METHODS

Biddle and Wehry [415] carried out fluorometric determination of manganese(II) in seawater via catalysed enzymatic oxidation of 2,3-diketogulonate. The detection limit was 8 µmol/L Mn(II).

3.32.3 FLOW INJECTION ANALYSIS

Resing and Matti [416] described a method for determining manganese(II) and total dissolvable manganese in seawater based on flow injection analysis with online preconcentration and spectrometric detection. The method involved online concentration of Mn(II) onto δ-hydroxyquinoline immobilised onto a vinyl polymer gel. Mn(II) is then eluted from the gel by acid, and its concentration is determined by spectrophotometric detection of the malachite green formed from the reaction of leucomachite green and potassium peliodate with Mn(II) added as a catalyst. The limit of detection was 36 pmol/L when concentrating 15 mL of seawater. The accuracy of the method was proved using MASS-2 standard seawater. The method had a precision of approximately 5% for Mn(II) concentrations of 400 pmol/L and analysis time is about 5 min per sample.

3.32.4 ATOMIC ABSORPTION SPECTROMETRY

Burton [417] described an atomic absorption method for the determination of down to 0.3 nmol/L manganese in seawater.

Samples for the analysis of manganese were pressure filtered through 0.4 µm nucleopore filters. To the mL filtrate, 20 mL of an aqueous solution of 2% w/v in both ammonium and diethyl ammonium diethyl dithiocarbamate was added, and the solution was extracted first with freon for 6 min and then with 100 µL of concentrated nitric acid for 30 s.

The aqueous solutions were returned to the shore laboratory, and manganese was determined by electrothermal atomic absorption spectrometry.

The application of this technique is also discussed in Section 3.70 (Table 3.4A).

3.32.5 GRAPHITE FURNACE ATOMIC ABSORPTION SPECTROMETRY

Graphite furnace atomic absorption spectrometry, although element selective and highly sensitive, is currently unable to directly determine manganese at the lower end of their reported concentration ranges in open ocean waters. Techniques that have been successfully employed in recent environmental investigations have thus used a preliminary step to concentrate the analyte and separate it from the salt matrix prior to determination by atomic absorption spectrometry.

The determination of manganese in seawater using graphite furnace atomic absorption spectrometry has been investigated by many researchers [418–426]. If the seawater matrix is atomised with the analyte, the result is a large background signal, which is often beyond the correcting capabilities of current instrumentation. The presence of large amounts of chlorides has also been shown to provide interferences [378,427] usually making direct analysis difficult.

To reduce these problems, most researchers either have used matrix modification [481,422–426,433] or have extracted the metal from the seawater matrix [420,424]. Few researchers have been successful with the direct determination in seawater after volatilisation of the matrix during the char program step [418,419,423,425,428]. Slavin and Manning [429–431] showed that by using a furnace at steady-state temperature (the L'vov platform), the interference of chloride in manganese determination was greatly reduced as long as the background signal was within the limits that the deuterium arc background corrector could handle.

Segar and Gonzales [419] attributed the reduced sensitivity for manganese in a seawater matrix to co-volatilisation of some manganese within the salt matrix. It has been claimed that this reduced sensitivity is a vapour-phase binding of a portion of the manganese by chloride.

Ediger et al. [375] showed that it was necessary to char away as much as possible of the seawater matrix to get maximum sensitivity for manganese and to be free of interference.

Segar and Cantillo [425] developed a direct method for manganese in seawater with a detection limit for manganese of about 0.3 µg/L. They found that, as the graphite tube aged, the analytical signal fell linearly at the rate of 50% per 100 firings. Since variations in salinity produced relatively large changes in signal, the methods of standard additions were required.

McArthur [421] preferred to use ammonium nitrate matrix modification to determine manganese in seawater. Most of his paper discussed the charring process. He found considerable salinity, which was dependent on the age of the tubes.

Kingston et al. [424] resorted to extraction on Chelex–100 followed by stripping into nitric acid. The ammonium nitrate matrix modification technique was used by Montgomery and Peterson [426] for the determination of manganese in seawater. They showed that the pyrolitically coated tubes they used deteriorated very rapidly using the combination of ammonium nitrate and seawater. Manganese was determined in seawater (with copper and cobalt) by Hydes [422] after adding 1% ascorbic acid to the sample. He used the Perkin-Elmer HGA–2100 furnace and found significant loss of manganese from seawater between 600°C and 900°C.

The direct furnace method of Sturgeon et al. [418] for manganese was very similar to the method of Segar and Cantillo [425,428]. The Sturgeon detection limit was 0.22 µg/L for manganese in seawater using 20 µg/L samples in the furnace and pyrolitically coated graphite tubes. They found a loss in sensitivity during the life of the tubes. They had to use the method of additions to accommodate small residual interference.

Statham [432] optimised a procedure based on chelation with ammonium pyrrolidine dithiocarbamate and diethylammonium diethyldithiocarbamate for the preconcentration and separation of dissolved manganese from seawater prior to extraction with freon–TF and determination of manganese in the aqueous nitric acid back-extract by graphite furnace atomic absorption spectrometry. The procedure concentrates trace metals in the seawater 67-fold and has a recovery of 100% and a detection limit of 0.1 ng nmol/L.

Klinkhammer [434] described a method for determining manganese in a seawater matrix for concentrations ranging from about 30 to 5500 ng/L. The samples were extracted with 4 nmol/L 8-hydroxyquinoline in chloroform, and manganese in the organic phase was then back-extracted into 3 M nitric acid. The manganese concentrations were determined by graphite furnace atomic absorption spectrophotometry. The blank of the method was about 3.0 ng/L, and the precision from duplicate analysis is ± 9% (1 SD).

Bender et al. [435] and Klinkhammer and Bender [436] determined the total and soluble manganese in seawater. The samples were collected into 500 mL polyethylene bottles. All samples were brought to pH 2 with nitric acid free of trace metals.

The pH was adjusted to approximately pH 8 using concentrated ammonia (Ultrapure, G. Frederick Smith). Chelating cation-exchange resin in the ammonia from (Chelex–100, 100–200 mesh, Bio-Rad) was added to the samples, and they were batch extracted on a shaker table for 36 h. The resin was decanted into columns and manganese was eluted using 2 N nitric acid [424]. The eluent was then analysed by graphite furnace atomic absorption spectrophotometry. Replicate analyses of samples indicate a precision of about 5%.

Lan and Alfassi [437] determined manganese in seawater in amounts down to 50 ppt using 50 μL sample graphite furnace atomic absorption spectrometer.

To determine manganese [408,438], factors had to be controlled carefully to obtain results against simple standards that were independent of salinity and variations in matrix composition. Use of the L'vov platform and integration of the absorbance signal reduced the sensitivity to the matrix composition [438]. Pyrolitically coated graphite reduced variations that depended on the life of the tubes. The tubes appeared to fail by intercalation of the sodium or sodium chloride matrix. The char temperature must not vary outside the range of 1100°C–1300°C. A Zeeman background correction permitted the use of large seawater samples. The detection limit of the procedure using 20 μL samples was 0.1 μg/L (2 pg) manganese. By the use of the Zeeman background corrector, less than 0.02 μg/L manganese was detected in seawater using 75 μL sample.

Carnrick et al. [439] used problems in the determination of manganese in seawater as a model to study the graphite furnace system at steady-state temperature. Several factors had to be controlled carefully to obtain reliable results against simple standards that were independent of salinity and variations in matrix composition. Use of the L'vov platform and integration of the absorbance signal reduced the sensitivity to matrix composition. Pyrolytically coated graphite reduced variations that depend on the life of the tubes. The tubes appeared to fall by intercalation of the Na or NaCI matrix. The char temperature must not vary outside the range of 1100°C–1300°C. Zeeman background correction permitted the use of larger seawater samples. The detection limit of the procedure using 20 μL samples was 0.1 μg/L

(2 pg). Manganese was detected in seawater using a 75 μL sample. Seawater samples can be processed in less than 30 min per sample.

The application of this technique is also discussed in Section 3.70 (Table 3.4B).

3.32.6 Differential Pulse Anodic Stripping Voltammetry

O'Halloran [440] discussed the determination of manganese in seawater at a mercury film electrode. This technique is suitable for ultra-trace determinations of manganese(II) in seawater. Samples can be preserved by acidification and then buffered with sodium tetraborate prior to measurement, with precautions to avoid calomel formation on the electrode. Interference effects of other trace metals are negligible for open ocean water, partly because zinc interacts with copper to minimise the formation of a copper–manganese intermetallic compound. Rapid determinations of manganese(II) at concentrations down to 0.01 μg/L are possible. Manganese levels in confined of Port Phillip Bay were found to be an order of magnitude greater than open ocean levels in the Tasman Sea. Results for the ocean water were in close accord with those found elsewhere by an extraction–radiotracer method. Manganese concentrations in Port Phillip Bay, Tasmania, were between 0.01 and 0.1 μg/L.

3.32.7 Polarography

Knox and Tuner [441] described a polarographic method for manganese(II) in estuarine and seawaters, which covers the lower concentration range of 100–300 μg/L. The method, which is specific to manganese(II) and its labile complexes, was used in conjunction with a colorimetric technique to compare the levels of manganese(II) and total dissolved manganese in an estuarine system. They showed that polarographically determined manganese(II) can vary widely from 100% to less than 10% by the total dissolved manganese, determined spectrophotometrically at 450 nm by the formaldoxime method [442] calibrated in saline medium to overcome any salt effects. It has been suggested that the manganese not measured by the polarographic method is in a colloidal form.

3.32.8 Neutron Activation Analysis

Neutron activation analysis has been used to determine total manganese in seawater [443,444]. The application of this technique is also discussed in Section 3.70 (Table 3.4O).

3.32.9 Further Techniques

Further techniques for the determination of manganese in seawater, discussed in Section 3.70, include inductively coupled plasma atomic emission spectrometry (Table 3.4E), inductively coupled plasma mass spectrometry (Table 3.4F), cathodic stripping voltammetry (Table 3.4L), high-performance liquid chromatography (Table 3.4P), X-ray fluorescence spectroscopy (Table 3.4N), and anodic stripping voltammetry (Table 3.4G).

3.33 MERCURY

3.33.1 ATOMIC ABSORPTION SPECTROMETRY

Atomic absorption and atomic fluorescence techniques using closed system reduction-aeration have been applied widely to determine mercury concentrations in natural samples [445–459].

A typical method is that of Topping and Pirie [455] in which mercury is concentrated by drawing air for 5 h (500 mL/min) through a mixture of the sample (4 L) and a 20% stannous chloride solution in 5 M of hydrochloric acid (45 mL) and absorbing mercury vapour from the air stream in a 20 mL 2% $KMnO_4$ solution: 50% (v/v) sulphuric acid (1:1). To the absorption mixture was added 15 mL of the stannous chloride solution, and this mixture was aerated at 21 mL/min. The air and mercury vapour are passed through at 15 cm gas cell (with silica windows) in an atomic absorption spectrophotometer for the measurement at 253.65 nm. Samples containing down to 2 ng Hg/L could be analysed by this procedure.

Olafsson [459] described a similar procedure in which the sample (450 mL) was acidified with nitric acid, aqueous stannous chloride was added, and mercury was entrained in a stream or argon into a silica tube wound externally with resistance wire and containing pieces of gold foil, on which the mercury was retained. The tube and its contents were then heated electrically to about 320°C, and the vaporised mercury was swept by argon into a 10 cm silica absorption cell in an atomic absorption spectrophotometer equipped with a recorder. The absorption (measured at 253.7 nm) was directly proportional to the amount of mercury in the range of 0–24 ng per sample.

Voyce and Zeitlin [460] used adsorption colloid flotation to determine mercury in seawater. The sample (500 mL) is treated with concentrated hydrochloric acid; an aqueous solution of cadmium sulphate and a fresh aqueous solution of sodium sulphate are added. The pH is adjusted to pH 1.0 and then poured into a flotation cell with a nitrogen flow of 10 mL/min. Ethanolic octadecyltrimethylammonium chloride is injected and the froth is dissolved in aqua regia in a flameless atomic absorption cell. Following reduction of mercury with stannous chloride, the mercury vapour is flushed from the system.

Fitzgerald et al. [461] have described a method based on cold trap preconcentration prior to gas-phase atomic absorption spectrometry for the determination of down to 2 ng/L of mercury in seawater.

Fitzgerald et al. [461] also showed that the most significant quantities of mercury occurred in the waters of the Atlantic Ocean's continental shelf, and a slope of 21–78 ng/L was found compared with open ocean samples (2–11 ng/L). These researchers distinguished between inorganic mercury obtained by direct analysis on the sample as received and organic mercury (the difference between total mercury, obtained upon ultraviolet irradiation of the sample) and inorganic mercury.

To preserve natural water samples for mercury analysis, much care must be exercised to prevent the loss of mercury during storage [462–464].

Coyne and Collins [462] recommended pre-acidification of the sample bottle with concentrated nitric acid to yield a final pH 1 in the sample solution. However, when this procedure was used for the storage of seawater, freshwater, and distilled deionised water in low-density polyethylene storage containers, abnormally high absorption was observed. This absorption at the mercury wavelength (253.7 nm) is due to the presence of volatile organic plasticiser material and any polyethylene residue leached by the concentrated nitric acid and the acid solution at pH 1. Those procedures employing acidified storage of samples in polyethylene bottles and gas-phase mercury detection may be subject to artificial mercury absorption due to the presence of organic material.

Reported mercury values in the oceans determined since 1971 have spanned three orders of magnitude caused in part at least due to errors caused by incorrect sampling [465]. Olafsson [466] attempted to establish reliable data on mercury concentrations obtained in cruises in North Atlantic waters.

Olafsson [467] reported the results obtained in an international intercalibration for mercury in seawater. Sixteen countries participated in this exercise, which involved the analysis of a seawater and seawater spiked with 15.4 and 143 ng/L mercury. The results show good accuracy and precision in the recovery of spikes for the majority of calibrations but serious errors in the low-level determinations on seawater.

Since the intercalibration exercise had been ultraviolet irradiated, the majority of precipitants in this exercise preferred to analyse the samples without any pretreatment.

Blake [468] described a method for the determination of trace amounts of mercury, with a limit of detection of less than 2 ng/L in fresh and saline waters. It was based on generating mercury vapour from the sample by reduction, together with trapping on gold mesh, subsequent desorption and measurement by cold vapour atomic absorption spectrometry. Checks on the precision on the recovery of the method with respect to inorganic mercury were described, and the recovery of methylmercury was also investigated. The performance of the method was within the limits implied in the requirements of the harmonised monitoring scheme and the application of EC Directives concerned with water quality monitoring.

Gill and Fitzgerald [469] determined picomolar quantities of mercury in seawater using stannous chloride reduction and two-stage amalgamation with gas-phase detection. The gas flow system used two gold-coated bead columns (the collection and the analytical columns) to transfer mercury into the gas cell of an atomic absorption spectrometer. By careful control and estimation of the blank, a detection limit of 0.21 pM was achieved using 2 L of seawater. The accuracy and precision of this method were checked by comparison with aqueous laboratory and the National Bureau of Standards (NBS) reference materials spiked into acidified natural water samples at picomolar levels. Further studies showed that at least 88% of mercury in the open ocean and coastal seawater consisted of labile species, which could be reduced by stannous chloride under acidic conditions.

3.33.2 Graphite Furnace Atomic Absorption Spectrometry

Filippelli [470] determined mercury at the sub-nanogram level in seawater using graphite furnace atomic absorption spectrometry. Mercury(II) was concentrated using the ammonium tetramethylenedithiocarbamate (ammonium pyrrolidine dithiocarbamate, APDC)–chloroform system, and the chloroform extract was introduced into the graphite tube. A linear calibration graph was obtained for 5–1500 ng of mercury in 2.5 mL chloroform extract. Because of the high stability of the Hg^{II}–APDC complexes, the extract may be evaporated to obtain a crystalline powder to be dissolved with a few microlitres of chloroform.

About 84% of mercury was recovered in a single extract (97% in two extractions). The coefficient of variation of this method was about 2.6% at the 1 µg/L mercury level. The calibration graph is linear over the range of 5–1500 µg mercury.

The application of this technique is also discussed in Section 3.70 (Table 3.4B).

3.33.3   ZEEMAN ATOMIC ABSORPTION SPECTROMETRY

Hedeishi and McLaughlin [471] reported the application of the Zeeman effect for the determination of mercury.

3.33.4   INDUCTIVELY COUPLED PLASMA ATOMIC EMISSION SPECTROMETRY

Watling [472] described an analytical technique for the accurate determination of mercury at picogram per litre levels in freshwater and seawater. Mercury, released by tin(II) chloride reduction of water samples, is amalgamated onto silver wool contained in quartz amalgamation tubes. The wool is then heated, and the mercury thus released is flushed by argon into a plasma, where it is excited. The emission signal thus produced results in a detection limit of 3.10^{-17} g and an analytical range of 1.10^{-14} g to 1.10^{-7} g.

3.33.5   INDUCTIVELY COUPLED PLASMA MASS SPECTROMETRY

Bloxam et al. [473] used liquid chromatography with an inductively coupled plasma mass spectrometric detector in speciation studies on ppt levels of mercury in seawater.

Debrak and Denoyer [474] determined ppt levels of mercury in seawater by first converting mercury salts to elemental mercury using stannous chloride. The mercury was then trapped on gold deposited on platinum gauze and released by heating prior to determination by inductively coupled plasma mass spectrometry. The application of this technique is also discussed in multimetal analysis (Table 3.4F).

3.33.6   ANODIC STRIPPING VOLTAMMETRY

Turyan and Mandler [475] determined mercury in seawater by anodic stripping voltammetry using a glassy carbon electrode spin coated with 4,7,13,16,21,42-hexaoxa-1,10-diazabicyclo[8.8.8] hexaconane.

3.33.7   ATOMIC EMISSION SPECTROMETRY

Wrembel [476] gave details of a procedure for the determination of mercury in seawater by low-pressure ring-discharge atomic emission spectrometry with electrolytic preconcentration on copper and platinum mesh electrodes. Between 40 ± 5 (open sea) and 50 ± 8 (shore area) µg/L mercury was found in Baltic Sea waters.

Wrembel and Pajak [477] evaporated mercury from natural water samples with argon and amalgamated the mercury with a gold foil. The mercury was excited in a ring-discharge plasma and determined by atomic emission spectroscopy. The method was applied for determining mercury in seawater in the range of 0.01–1.0 μg/L.

3.33.8 MISCELLANEOUS

Other techniques used include subtractive differential pulse voltammetry at twin gold electrodes [478], anodic stripping voltammetry using glassy carbon electrodes [479], X-ray fluorescence analysis [480], and neutron activation analysis [481,482].

Agemian and Da Silva [483] described an automated method for total mercury in saline waters.

Bioassay methods have been used to obtain estimates of low mercury concentrations (5–20 μg/L) in seawater [484,485]. This method is useful for detecting comparatively small enhancements over background mercury concentrations in estuarine and seawater.

This method consists of suspending 70 mussels (*Mytilus edulis*), each of standard weight, for a standard time in a plastic-coated wire cage 2 m below the surface. Mercury in the mussels was determined by cold vapour atomic absorption spectrometry [486]. The procedure is calibrated by plotting determined mercury content of mussels against the mercury content of the seawater in the same area.

Quareshi et al. [487] presented pseudo first-order rate constants for gross photoreduction and gross photooxidation of mercury in surface seawater in the Atlantic Ocean. They showed that reduction and oxidation of mercury in oceans did not follow a simple two-species reversible reaction pathway and suggested two possible redox pathways that reproduce the pattern of dissolved gaseous mercury concentrations. On both the aforementioned pathways, HgCo, the major constituent dissolved gaseous mercury, is converted to an unidentified oxidised species that is different from the reducible form that was originally present.

Pseudo first-order rate constants for reductions were determined to be in the range of 0.15–0.93 h^{-1} and pseudo first-order rate constants for the oxidation of Hg(0) to be in the range of 0.4–1.9 h^{-1}.

3.33.9 FURTHER TECHNIQUES

Other techniques for determining mercury in seawater, discussed in Section 3.70, include X-ray fluorescence spectroscopy (Table 3.4N), neutron activation analysis (Table 3.4O), and differential pulse anodic stripping voltammetry (Table 3.4H).

3.34 MOLYBDENUM

3.34.1 SPECTROPHOTOMETRIC METHODS

Shriadah et al. [488] determined molybdenum(VI) in seawater by densitometry after enrichment as the Tiron complex on a thin layer of anion-exchange resin. There were no interferences from trace elements or major constituents of seawater, except for chromium and vanadium. These were reduced by the addition of ascorbic acid. The concentration of dissolved molybdenum(VI) determined in Japanese seawater was 11.5 µg/L with a relative standard deviation of 1.1%.

In a method described by Kiriyama and Kuroda [489], molybdenum was sorbed strongly on Amberlite CG 400 (Cl form) at pH 3 from seawater containing ascorbic acid. This is easily eluted with 6 M nitric acid. Molybdenum in the effluent can be determined spectrophotometrically with potassium thiocyanate and stannous chloride. The combined method allows selective and sensitive determination of traces of molybdenum in seawater. The precision of the method is 2% at a molybdenum level of 10 µg/L, and recoveries were satisfactory.

An adsorbing colloid formation method has been used to separate molybdenum from seawater prior to its spectrophotometric determination by the thiocyanate procedure [490].

Kuroda and Tarui [491] developed a spectrophotometric method for molybdenum based on the fact that molybdenum(VI) catalyses the reduction of ferric iron by divalent tin ions. The plot of initial reaction rate constant versus molybdenum concentration is rectilinear in the range of 0.01–0.03 mg/L molybdenum. Several elements interfere, for example titanium, rhenium, palladium, platinum, gold, arsenic, selenium, and tellurium.

Thorium hydroxide has been used as a collector of molybdenum from seawater [492].

Anion-exchange resins have been used to preconcentrate molybdenum in seawater prior to its spectrophotometric determination as the Tiron complex [488,493,494].

Kuroda and Kawabuchi concentrated molybdenum by anion exchange from seawater containing acid and thiocyanate [494] or hydrogen peroxide [491,494] and determined it spectrophotometrically.

Korkisch et al. [495] concentrated molybdenum from natural waters on Dowex 1–X8 in the presence of thiocyanate and ascorbic acid. A sodium citrate and ascorbic acid system has also been worked out for the concentration of molybdenum on Dowex 1–X8 (citrate form) as a citrate complex.

Nucatsuka et al. [496] determined molybdenum in seawater by the formation of its phenylfluorone complex, which was then extracted onto a membrane filter, and the absorbance of the filter was then measured.

3.34.2 ATOMIC ABSORPTION SPECTROMETRY

Chau and Lum-Shui-Chan [497] investigated the use of atomic absorption in conjunction with solvent extraction using 1% oxime in methyl isobutyl ketone for preconcentration. The detection limit is 3 µg/L, in which a preconcentration factor of 20 is employed. The disadvantages of the system are that it requires a 100 mL sample and there are interferences, although some of these can be eliminated [498].

The application of atomic absorption spectrometry in the determination of molybdenum in seawater is discussed in Section 3.70 (Table 3.4A).

3.34.3 GRAPHITE FURNACE ATOMIC ABSORPTION SPECTROMETRY

A limited amount of work has been carried out on the determination of molybdenum in seawater by graphite furnace atomic absorption spectrometry [175,524].

In a recommended procedure [500], a 50 mL sample of seawater at pH 2.5 is passed through a column of 0.5 g p-aminobenzylcellulose. Then the column is left in contact with 1 M ammonium carbonate for 3 h, after which three 5 mL fractions are collected. Finally, molybdenum is determined by atomic absorption at 313.2 nm with the use of the hot-graphite-rod technique. At the 10 mg/L level, the standard deviation was 0.13 μg.

Emerick [501] showed that sulphate interfered with the graphite furnace atomic absorption determination of molybdenum in aqueous solutions with concentrations of only 0.1% (w/v) sodium sulphate causing complete elimination of the molybdenum absorbance peak in solutions free of other salts. Matrix modification with 0.5% (w/v) $CaCl_2 \cdot 2H_2O$, in a volume equal to that sample, facilitates the determination of molybdenum in the presence of solutions containing as high as 0.4% (w/v) sodium sulphate. The need for matrix modification for molybdenum determination in natural waters appears to exist when sulphate greatly exceeds the equivalent calcium content. The routine use of a volume of 0.5% $CaCl_2 \cdot 2H_2O$ equal to sample volume is recommended in the determination of molybdenum in seawater.

Nakahara and Chakrabarti [176] showed that the seawater salt matrix can be removed from the sample by selective volatilisation at 1700°C–1850°C. But the presence of sodium chloride, sodium sulphate, and potassium chloride causes a considerable decrease in molybdenum absorbance, and magnesium chloride and calcium chloride cause pronounced enhancement. The presence of magnesium chloride prevents the depressive effects. Samples of less than 50 μL can be analysed directly without using a background corrector with a precision of 10%.

These researchers conclude that the selective volatilisation technique is highly suitable for the determination of traces of molybdenum in seawater samples. The sensitivity achieved allows seawater samples to be analysed for molybdenum, because the concentration of molybdenum in seawater is usually 2.1–18.1 μg/L. The selected temperature of 1700°C–1850°C during the charging stage permits separation of the seawater matrix from the analyte prior to atomisation with the Perkin-Elmer Model 603 atomic absorption spectrometer equipped with a heated graphite atomiser (HGA–2100).

Kuroda et al. [502] determined traces of molybdenum in seawater by combined anion-exchange graphite furnace atomic absorption spectrometry.

In this method, trace amounts of molybdenum were concentrated from acidified seawater on a strongly basic anion-exchange resin (Bio-Rad AG1, X-8 in the chloride form) by treating the water with sodium azide. Molybdenum(VI) complexes with azide were stripped from the resin by elution with ammonium chloride/ammonium hydroxide solution.

The application of this technique is also discussed in Section 3.70 (Table 3.4B).

3.34.4 INDUCTIVELY COUPLED PLASMA MASS SPECTROMETRY

It is widely recognised that the natural isotopic variation of molybdenum can provide crucial information about the geochemical circulation of molybdenum, and the ocean is an important reservoir of molybdenum. To obtain isotopic data on molybdenum in seawater samples using multiple collector inductively coupled plasma mass spectrometry (MC-ICPMS), Nakagawa et al. [503] developed a preconcentration technique using 8-hydroxyquinoline bonded covalently to a vinyl polymer resin (TSK-8HQ). Molybdenum in seawater could be effectively separated from matrix elements such as alkali, alkaline earth, and transition metals. With this technique, even with a 50-fold enrichment factor, the changes in the $^{98}Mo/^{95}Mo$ ratio during preconcentration were smaller than twice the standard deviation n(SD) in this study. Mass discrimination of molybdenum isotopes during the measurement was externally corrected for by normalising 85Sr/88Sr to 0.1194 using an exponential law. Nakagawa et al. [503] evaluated $\delta^{98/95}Mo$ to a precision of ±0.08% (±2 SD); this value was found to be less than one-third of previous reported values.

The application of this technique in the determination of molybdenum is also discussed in Section 3.70 (Table 3.4F).

3.34.5   DIFFERENTIAL PULSE AND LINEAR SWEEP VOLTAMMETRY

Van der Berg [504] carried out direct determinations of molybdenum in seawater by adsorption voltammetry. The method is based on complex formation of molybdenum(VI) with 8-hydroxyquinoline (oxine) on a hanging mercury drop electrode. The reduction current of adsorbed complex ions was measured by differential pulse adsorption voltammetry. The method was accurate up to 300 nmol/L. The detection limit was 0.1 nmol/L.

Willie et al. [505] used linear sweep voltammetry for the determination of molybdenum. Molybdenum was adsorbed by Eriochrome Blue Black R complex on a static mercury electrode. The method was reported to have a limit of detection of 0.50 µg/L, and the results agreed well with certified values for two reference seawater samples.

Hua et al. [506] described an automated method for the determination of molybdenum in seawater by means of constant current reduction of the adsorbed 8-quinolinol complex in a computerised flow potentiometric stripping analyser. The complex was adsorbed onto a molybdenum film electrode at −0.2 V and stripped at −0.42 V. A concentration at 8.9 ± 1.3 µg/L was obtained in reference seawater NASS–1 with a certified value of 11.5 ± 1.9 µg/L.

3.34.6   POLAROGRAPHY

Hidalgo et al. [507] reported a method for the determination of molybdenum(VI) in natural waters based on differential pulse polarography. The catalytic wave caused by molybdenum(VI) in nitrite medium following preconcentration by co-flotation on ferric hydroxide was measured. For seawater samples, hexadecyltrimethylammonium bromide with octadecylamine was used as the surfactant. The method was applied to molybdenum in the range of 0.7–5.7 µg/L.

3.34.7   X-RAY FLUORESCENCE SPECTROSCOPY

Monien et al. [508] compared the results obtained in the determination of molybdenum in seawater by three methods based on inverse voltammetry, atomic absorption spectrometry, and X-ray fluorescence spectroscopy. Only the inverse voltammetric method can be applied without prior concentration of molybdenum in the sample, and a sample volume of only 10 mL is adequate. Results of determinations by all three methods on water samples from the Baltic Sea are reported, indicating their relative advantages with respect to reliability.

X-ray fluorescence was used for the determination of molybdenum in seawater in a method described by Kimura et al. [509]. Molybdenum is coprecipitated with sodium diethyldithiocarbamate, which is measured by X-ray fluorescence. They reported a detection limit of 0.3 μg/L and a relative standard deviation of 2.9%.

The application of this technique is also discussed in Section 3.70 (Table 3.4N).

3.34.8   Neutron Activation Analysis

Neutron activation of molybdenum in seawater has been carried out on the β-naphthoin oxime [510] complex and the pyrrolidine dithiocarbamate and diethyl-dithiocarbamate complex [511]. The neutron activation analysis method was capable of determining down to 0.32 μg/L molybdenum in seawater.

The application of this technique is also discussed in Section 3.70 (Table 3.4N).

3.34.9   High-Performance Liquid Chromatography

Nagaosa and Kahayashi [512] separated chelates of Mo(VI) and Mn(II) with 8-hydroxyquinoline on C_{18} column by reversed-phase high-performance liquid chromatography. The mobile phase was 1:1 acetonitrile/0.02 mol/dm³ acetate buffer (pH 4.1) containing the ligand at 10^{-3} mol/dm³. The calibration curves were linear over the concentration range from 0.5 to 200 μg/dm³ for both metals using spectrophotometric detection (390 nm). The detection limits were 0.2 μg for Mn(II) and 0.4 μg/dm³ for Mo(VI) at a signal-to-noise ratio of 3, when the metal chelates were formed prior to injection onto the C_{18} column (pre-column chelation method).

3.34.10   Further Techniques

Further techniques for the determination of molybdenum in seawater, discussed in Section 3.70, include cathodic stripping voltammetry (Table 3.4L), Zeeman atomic absorption spectrometry (Table 3.4C), inductively coupled plasma atomic emission spectrometry (Table 3.4E), and inductively coupled plasma mass spectrometry (Table 3.4F).

3.35   NEODYMIUM

3.35.1   Isotope Dilution Analysis

The application of this technique is discussed in Section 3.70 (Table 3.4K).

3.36 NEPTUNIUM

3.36.1 RADIONUCLIDES

The determination of radioneptunium is discussed in Section 3.70 (Table 3.4O).

3.37 NICKEL

The concentration of nickel in natural waters is so low that enrichment steps are necessary before instrumental analysis. The most common method is graphite furnace atomic absorption after preconcentration by solvent extraction [513] or coprecipitation [514]. This technique has been used successfully for nickel analyses of seawater [515,516].

3.37.1 SPECTROPHOTOMETRIC METHODS

Nickel has been determined spectrophotometrically in seawater in amounts down to 0.5 µg/L as the dimethylglyoxime complex [517,518]. In one procedure [517], the dimethylglyoxime complex of nickel is added to the sample and the pH is adjusted to 9–10. The nickel complex is extracted into chloroform. Then 1 M hydrochloric acid is oxidised with aqueous bromine and adjusted to pH 10.4. The extinction of the nickel complex is measured at 442 nm. There is no serious interference from iron, cobalt, copper, or zinc, but manganese may cause low results. In another procedure [378], the sample of seawater (0.5–3 L) is filtered through a membrane filter (pore size 0.7 µm), which is then wet ashed. Nickel is separated from the resulting solution by extraction as the dimethylglyoxime complex and is then determined by its catalysis of the reaction of Tiron and diphenylcarbazone with hydrogen peroxide with spectrophotometric measurement at 413 nm.

3.37.2 ATOMIC ABSORPTION SPECTROMETRY

Rampon and Cavelier [519] used atomic absorption spectrometry to determine down to 0.5 µg/L nickel in seawater. Nickel is extracted into chloroform from seawater (500 mL) at pH 9–10 as its dimethylglyoxime complex. Several extractions and a final washing of the aqueous phase with carbon tetrachloride are required for 100% recovery. The combined organic phases are evaporated to dryness, and the residue is dissolved in 5 mL of acid for atomic absorption analysis.

Lee [520] described a method for the determination of nanogram or sub-nanogram amounts of nickel in seawater. Dissolved nickel is reduced by sodium borohydride to its elemental form, which combines with carbon monoxide to form nickel carbonyl. The nickel carbonyl is stripped from solution by a helium–carbon monoxide mixed gas stream, collected in a liquid nitrogen trap, and atomised in a quartz tube burner of an atomic absorption spectrophotometer. The sensitivity of the method is 0.05 ng of nickel. The precision for 3 ng nickel is about 4%. No interference by other elements is encountered in this technique.

Nishoika et al. [521] coprecipitated nickel from seawater with sodium diethyldithiocarbamate, filtered, and re-dissolved the precipitate in nitric acid, followed by electrothermal atomic absorption spectrophotometric determination of nickel. The detection limit was 0.5 µg/L, and the relative standard deviation was 13.2% at the 2 µg/L level.

The application of this technique is also discussed in Section 3.70 (Table 3.4A).

3.37.3 CATHODIC STRIPPING VOLTAMMETRY

Van den Berg and Nimmo [522] studied the complexation of nickel with dimethylglyoxime to determine nickel complexing capacities in seawater. Seawater containing 0.01 M borate buffer and 0.0001 M dimethylglyoxime was pipetted into 10–15 separate Teflon voltammetric cells. Nickel was then added to give a concentration range between 1 and 20 nM. After equilibrating, cathodic stripping voltammetry was used to determine the labile nickel concentration by measuring the reduction current of nickel–dimethylglyoxime complex absorbed on the hanging mercury drop electrode. Initial concentrations of total dissolved nickel were measured by cathodic stripping voltammetry with 0.0001 M dimethylglyoxime after UV irradiation for 2 h. Values for nickel complexing capacities with dimethylglyoxime were determined for seawater of several salinities by ligand competition with EDTA.

The application of this technique is also discussed in Section 3.70 (Table 3.4L).

3.37.4 FURTHER TECHNIQUES

Other procedures discussed in Section 3.70 for the determination of nickel in seawater include graphite furnace atomic absorption spectrometry (Table 3.4C), inductively coupled plasma atomic emission spectrometry (Table 3.4E), inductively coupled plasma mass spectrometry (Table 3.4F), anodic stripping voltammetry (Table 3.4G), chromopotentiometric analysis (Table 3.4I), plasma emission spectrometry (Table 3.4J), isotope chelation methods (Table 3.4K), X-ray fluorescence spectroscopy (Table 3.4N), neutron activation analysis (Table 3.4O), and high-performance liquid chromatography (Table 3.4P).

3.38 OSMIUM

3.38.1 MASS SPECTROMETRY

To determine osmium in seawater, Koide et al. [523] first separated the osmium on an anion-exchange column and distilled off the osmium tetroxide formed followed by resonance ionisation mass spectrometry.

Chen and Sharman [524] presented a chemical separation and mass spectrometric procedure, which permits precise and accurate determination of osmium concentration and isotopic composition of osmium in seawater.

Water samples with added ^{190}Os tracer are heated at 300°C at 128 bar in quartz carius tubes with $Cr^{vi}O_3$. This allows tracer-sample equilibration via complete oxidation of all osmium species to OsO_4, which is then separated and purified using distillation. Samples are measured using negative thermal ionisation mass spectrometry. Ionisation efficiency is about 5% for 200 fg of osmium (1 fg of Os = 3×10^6 atoms). The total yield of the chemical procedure is ~90%, and the total procedure blank is 3.6 fg. Chen and Shaman [526] applied this procedure to seawater depth profiles from the Atlantic and Pacific Oceans. They found that the procedure gave osmium concentrations distinctly higher than those obtained by alternative methods.

3.39 PALLADIUM

3.39.1 MISCELLANEOUS

Wang et al. [768] used a liquid membrane containing tri-N-octylamine to separate palladium from seawater.

Chen et al. [769] described a technique to quantify ultra-trace ^{231}Pa (50–2000 ag; 1 ag = 10^{-18} g) concentrations in seawater at the microgram level using isotope dilution thermal ionisation mass spectrometry. The method is a modification of a process developed by Pickett et al. [770] and extends the technique to very low levels. The procedural blank is 16 ± 15 ag (2σ), and the ionisation efficiency (ion generated/atom loaded) approached 0.5%, measurement time is less than 0.1 L. Replicate measurements made on known standards and seawater samples demonstrate that the analytical precision approximates that expected from counting statistics in a minimum sample size of surface water of ~2 L for suspended particulate matter and less than 0.1 L for filtered (<0.4 μm) seawater, respectively. Replicate measurements made on known standards and seawater samples demonstrate that the analytical precision approximates to be expected from counting statistics in a minimum sample size of surface seawater of ~2 L for suspended particulate matter and <0.1 L for filtered (<0.4 μm) seawater, respectively. For a dissolved fraction, 0.5–1 L seawater yields ^{231}Pa measurements with a precision of 1%–10%. Sample size requirements are orders of magnitude less than traditional decay-counting techniques and significantly less than previously reported ICP-MS techniques.

3.40 PLATINUM

3.40.1 CATHODIC STRIPPING VOLTAMMETRY

Van den Berg and Jacinto [525] determined platinum by adsorptive cathodic stripping voltammetry at 0.9251 V on a hanging mercury drop electrode. The detection limit was 0.014 pM platinum.

3.41 PLUTONIUM

Delle Site et al. [527] used extraction chromatography to determine plutonium in seawater, sediments, and marine organisms. They used double extraction chromatography with microthene-210 (microporous polyethylene) supporting tri-n-octyl-phosphate oxide (TOPO). 236-plutonium and 242-plutonium were used as the internal standards to determine the overall plutonium recovery. The 5.42 MeV α-lines of 228-thorium interfere with those of 238-plutonium (5.50 MeV), and so complete purification from thorium isotopes is required.

Plutonium sources were counted by an α-spectrometer with good resolution, background, and counting yield. The counting apparatus used had a resolution of 40 keV. The mean (±SD) background value was 0.0004 ± 0.0003 cpm in the 239- and 240-plutonium energy range and 0.0001 ± 0.0001 cpm in the 238-plutonium energy range. The mean (±SD) counting yield, obtained with 239-plutonium, 240-plutonium reference sources counted in the same geometry, was found to be 25.08 ± 0.72%.

3.42 POLONIUM

3.42.1 RADIONUCLIDES

Skwarzec and Bojanoski [528], in a study of the accumulation of 210-polonium in Southern Baltic seawater, showed that this mean concentration of 210-polonium was 0.19 m Bq per dm^3 of which 80% was dissolved and the remainder adsorbed onto phytoplankton and 300 plankton.

Tsunogai and Nozaki [529] analysed Pacific Ocean surface waters by consequetive coprecipitations of polonium with calcium carbonate and bismuth oxychloride after the addition of lead and bismuth carriers to acidified seawater samples. After concentrations, polonium was spontaneously deposited onto silver planchets.

Shannon et al. [530] determined 216-polonium and 210-lead in seawater by the adjustment of seawater to pH 2 and extraction with a solution of ammonium pyrollidine dithiocarbamate in isobutylmethyl ketone. The two elements were then back-extracted into hydrochloric acid and plated out of solution by the technique of Flynn [531].

Cowen et al. [532] showed that polonium can be electrodeposited onto carbon rods directly from acidified seawater, stripped from the rods and autoplated directly onto silver counting discs with overall recovery of 85% for an electrodeposition time of 16 h.

Biggin et al. [771] developed a time-efficient method for the determination of ^{210}Pb, ^{210}Bi, and ^{210}Po in seawater using liquid scintillation spectrometry.

3.43 POTASSIUM

3.43.1 TITRATION METHODS

Torbjoern and Jaguer [533,534] used a potassium selective valinomycin electrode and a computerised semiautomatic potentiometric titrator to determine potassium. Samples were titrated with standard additions of aqueous potassium so that the potassium to sodium ion ratio increased on the addition of the titrant and the contribution from sodium ions to the membrane potential could be neglected. The initial concentration of potassium ions was then derived by an extrapolation procedure.

Marquis and Lebel [535] precipitated potassium from seawater or marine sediment pore water using sodium tetraphenylborate, after first removing halogen ions with silver nitrate. Excess tetraphenylborate was then determined by silver nitrate titration using a silver electrode for end-point detection.

The content of the potassium in the sample is obtained from the difference between the amount of tetraphenylboron measured and the amount initially added.

The amount of potassium recovered varied from 98% to 102%. This confirms that the recovery is quantitative and that there is no systematic variation related to the amount of potassium added.

3.43.2 ION-SELECTIVE ELECTRODE

Ward [536] evaluated various types of potassium ion-selective electrodes for the analysis of seawater. Three types of potassium ion-selective electrodes were evaluated for their suitability for continuous monitoring and *in situ* measurement applications in water of varying salinities and at temperatures of 10°C and 25°C. The three types include glass membrane single electrode, glass membrane combination electrode, and liquid ion-exchange electrode. Although all the three electrode systems performed well in freshwater, the results obtained with the liquid ion-exchange electrode in seawater were significantly better than those with glass membranes. An accuracy of 5% could be achieved under certain conditions but response times generally exceeded 10 min and glass membrane electrodes were sensitive to external motion flow variations.

3.43.3 POLAROGRAPHY

Polarography has also been applied in the determination of potassium in seawater [537]. The sample (1 mL) is heated to 70°C and treated with 1 mL 0.1 M sodium tetraphenylborate. The precipitated potassium tetraphenylborate is filtered off, washed with 1% acetic acid, and dissolved in 5 mL acetone. This solution is treated with 3 mL 0.01 M thallium nitrate and 1.25 mL 2 M sodium hydroxide, and the precipitate of thallium tetraphenylborate is filtered off. The filtrate is made up to 25 mL and, after de-aeration with nitrogen, unconsumed thallium is determined polarographically. There is no interference from 60 mg sodium, 0.2 mg calcium or magnesium, 20 µg barium, or 2.5 µg strontium. Standard deviations at concentrations of 375, 750, and 1125 µg potassium per mL were 26.4, 26.9, and 30.5, respectively.

3.43.4 FURTHER TECHNIQUES

Other techniques have been used to determine potassium in seawater, which are discussed in Section 3.70, "Multimetal Analysis" (Table 3.4A). These include atomic absorption spectrophotometry (Table 3.4A), X-ray fluorescence spectroscopy (Table 3.4O), and neutron activation analysis (Table 3.4O).

3.44 PRASEODYMIUM

3.44.1 ISOTOPE DILUTION METHODS

The application of this technique is discussed in Section 3.70 (Table 3.4K).

3.45   PROMETHIUM

3.45.1   ISOTOPE DILUTION METHODS

The application of this technique is discussed in Section 3.70 (Table 3.4K).

3.46   RADIUM

3.46.1   RADIONUCLIDES

Perkins [538] carried out radium and radio barium measurements in seawater using sorption and direct multidimensional gamma-ray spectrometry. 265-radium, 230-thorium, and 210-lead in large volumes of seawater have been collected on manganese oxyhydroxide-impregnated cartridges prior to determination by radio analytical methods [539].

3.47   RHENIUM

Rhenium is one of the last stable elements discovered, one of the least abundant metals in the earth's crust, and one of the most important sentinels of reducing aqueous environments through its abundance in sediments. Although its chemistry is fairly well described, its marine chemistry is poorly developed. In addition, the understanding of rhenium's marine chemistry will provide an entry to the understanding of the marine chemistry of technetium, an element which is just above rhenium in Group VIIA (Group 7 in the 1985 notation of the periodic table). Technetium only has unstable isotopes whose origins are primarily in nuclear weapon detonations and in nuclear reactor wastes. These two elements have remarkably similar chemistries. Rhenium's solution chemistry primarily involves anionic species in the IV, V, and VIII oxidation states. The oxo-anion perrhenate is especially stable.

3.47.1   GRAPHITE FURNACE ATOMIC ABSORPTION SPECTROMETRY

Koide et al. [540] described a graphite furnace atomic absorption method for the determination of rhenium at picomolar levels in seawater and parts-per-billion levels in marine sediments based on the isolation of heptavalent rhenium species upon anion-exchange resins. All steps are followed with 186-rhenium as a yield tracer. A crucial part of the procedure is the separation of rhenium from molybdenum, which significantly interferes with the graphite furnace detection when the Mo/Re ratio is 2 or greater. The separation is accomplished through an extraction of tetraphenylarsonium perrhenate into chloroform in which the molybdenum remains in the aqueous phase.

It was observed by these researchers that the rhenium signal was attenuated by as little as 10 ng or less of molybdenum in the isolate. Thus, importance is placed upon molybdenum decontamination steps. In seawaters as well as in many marine sediments the Mo/Re varies about 1000-fold. In addition, a clean separation of rhenium from other elements (the salt effect) is required. Otherwise, false peaks result upon atomisation due to the high background generated by impurities.

The seawater concentration of rhenium is in the range under 3–11 ng/L compared to iridium, platinum, and gold, whose concentrations usually do not exceed 0.3 ng/L.

3.47.2 Neutron Activation Analysis

Matthews and Riley [541] described the following procedure for determining down to 0.06 µg/L rhenium in seawater. From 6 to 8 µg/L rhenium was found in Atlantic seawater. The rhenium in a 15 L sample of seawater, acidified with hydrochloric acid, is concentrated by adsorption on a column of De-acidite FF anion-exchange resin (Cl-form), followed by elution with 4 M nitric acid and evaporation of the eluate. The residue (0.2 mL), together with standards and blanks, is irradiated in a thermal neutron flux of at least 3×10^{12} neutrons/cm^2/s for at least 50 h. After a decay period of 2 days, the sample solution and blank are treated with potassium perrhenate as carrier and evaporated by dryness with a slight excess of sodium hydroxide. Each residue is dissolved in 5 M sodium hydroxide. Hydroxylammonium chloride is added (to reduce technetium(VII)), which arises from 99 m-technetium from the activation of molybdenum present in the samples, and rhenium(VII) is extracted selectively with ethyl methyl ketone. The extracts are evaporated, and the residue is dissolved in formic acid–hydrochloric acid (19:1). Rhenium is adsorbed on a column of Dowex I, and the column is washed with the same acid mixture followed by water and 0.5 M hydrochloric acid. Rhenium is eluted at 0°C with acetone–hydrochloric acid (19:1) and is finally isolated by precipitation and tetraphenylarsonium perrhenate. The precipitate is weighed to determine the chemical yield, and the 186-rhenium activity is counted with an end-window Geiger–Muller tube. The irradiated standards are dissolved in water together with potassium permanganate. At a level of 0.057 µg/L rhenium, the coefficient of variation was ±7%.

3.48 RUBIDIUM

3.48.1 Atomic Absorption Spectrometry

Shen and Li [542] extracted rubidium (and caesium) from brine samples with 4-tert-butyl-2-(α-methyl-benzyl) phenol prior to the atomic absorption determination of the metal.

The application of this technique is also discussed in Section 3.70 (Table 3.4A).

3.48.2 Isotope Dilution Mass Spectrometry

Isotope dilution mass spectrometry has been used to determine the traces of rubidium in seawater [543].

3.48.3 Spectrochemical Methods

Schoenfold and Held [544] used a spectrochemical method to determine rubidium in seawater. Concentrations of rubidium found were in the range of 8–50 µg/L.

3.48.4 FURTHER TECHNIQUES

Other procedures for the determination of rubidium in seawater, discussed in Section 3.70, "Multimetal Analysis", include inductively coupled plasma mass spectrometry (Table 3.4F), X-ray fluorescence spectroscopy (Table 3.4N), and high-performance liquid chromatography (Table 3.4P).

3.49 SAMARIUM

3.49.1 ISOTOPE DILUTION METHOD

The application of this technique is discussed in Section 3.70, "Multimetal Analysis" (Table 3.4K).

3.50 SCANDIUM

3.50.1 NEUTRON ACTIVATION ANALYSIS

The application of this technique is discussed in Section 3.70, "Multimetal Analysis" (Table 3.4O).

3.51 SELENIUM

In recent years, the physiological role of selenium as a trace element has created considerable speculation and some controversy. Selenium has been reported as having carcinogenic and toxic properties; other authorities have presented evidence that selenium is highly beneficial as an essential nutrient [545,546]. Its significance and involvement in marine biosphere are now known. A review of the marine literature indicates that selenium occurs in seawater as selenite ions (SeO_3^{2-}) with a reported average of 0.2 µg/L [547].

Various techniques have been applied for determining selenium in seawater, including flameless atomic absorption spectrometry [548,549], gas chromatography [550–552], adsorption colloid flotation [553], spectrophotometry [554–557], and neutron activation analysis [558,559].

3.51.1 ATOMIC ABSORPTION SPECTROMETRY

Neve et al. [549] digested the sample with nitric acid. After digestion the sample is reacted selectively with an aromatic o-diamine, and the reaction product is detected by flameless atomic absorption spectrometry, after the addition of nickel(III) ions. The detection limit is 20 mg/L, and both selenium(IV) and total selenium can be determined. There was no significant interference in a saline environment with three times the salinity of seawater.

3.51.2 GRAPHITE FURNACE ATOMIC ABSORPTION SPECTROMETRY

Sturgeon et al. [560] preconcentrated selenium(IV) by the absorption of their ammonium pyrrolidine diethyl dithiocarbamate chelates onto C_{18} bonded silica prior to desorption and determination by graphite furnace atomic adsorption spectrometry. The detection limit was 7 ng/L selenium(IV), based on a 300 mL water sample.

The application of this technique is also discussed in Section 3.70 (Table 3.4B).

3.51.3 HYDRIDE GENERATION ATOMIC ADSORPTION SPECTROMETRY

Cutter [561] surveyed the application of this technique to the determination of selenium in seawater.

3.51.4 CATHODIC STRIPPING VOLTAMMETRY

Certain trace substances such as selenium(IV) can be determined by differential cathodic stripping voltammetry (DPCSV). For selenium, a rather positive preconcentration potential of -0.2 V is adjusted. Selenium(IV) is reduced to Se^{2-}, and Hg originating from the electrode is oxidised to Hg^{2+} at this potential. It forms, with Se^{2-} on the electrode, a layer of insoluble HgSe, and in this manner, the preconcentration is achieved. Subsequently, the potential is altered in the cathodic direction in the differential pulse mode. The mercury produced by the Hg^{2+} reduction in the analyte.

3.51.5 GAS CHROMATOGRAPHY

Shimoishi [550] determined selenium by gas chromatography with electron capture detection. To 50–100 mL of seawater was added 5 mL concentrated hydrochloric acid and 2 mL 1% 4-nitro-*o*-phenylenediamine, and after 2 h, the product formed was extracted into 1 mL of toluene. The extract was washed with 7.5 M hydrochloric acid. Then 5 µL is injected into a glass gas–liquid chromatography column (1 × 4 mm) packed with 15% of SE-30 on Chromosorb W (60–80 mesh) and operated at 200°C with nitrogen (53 min^{-1}) as the carrier gas. There is no interference from other substances present in seawater.

Measures and Burton [551] used gas chromatography to determine selenite and total selenium in seawater. Siu and Berman [552] determined selenium(IV) in seawater by gas chromatography after coprecipitation by hydrous ferric oxide.

After coprecipitation, selenium is derivatised to 5-nitropiaz-selenol, extracted into toluene, and quantified by electron capture detection. The detection limit is 5 ng/L with 200 mL sample, and the precision at the 0.025 µg Se per litre level is 6%.

Ferric hydroxide coprecipitation techniques are lengthy, 2 days being needed for a complete precipitation. To speed up this analysis, Tzeng and Zeitlin [553] studied the applicability of an intrinsically rapid technique, namely adsorption colloid flotation. This separation procedure uses a surfactant–collector–inert gas system, in which a charged surface-inactive species is adsorbed on a hydrophobic colloid collector of opposite charge; the colloid with the absorbed species is floated to the surface with a suitable surfactant and inert gas, and the foam layer is removed manually for analysis by a methylene blue spectrometric procedure.

The advantages of the method include a rapid separation, simple equipment, and excellent recoveries.

Using this method, Tzeng and Zeitlin [553] found 0.40 ± 0.12 µg/L selenium in seawater.

3.51.6 FURTHER TECHNIQUES

These include inductively coupled plasma hydride X-ray fluorescence spectroscopy (Table 3.4N) and neutron activation analysis (Table 3.4O).

3.52 SILVER

3.52.1 ATOMIC ABSORPTION SPECTROMETRY

Bermejo-Barrera et al. [562] described an electrothermal atomic absorption spectrometric method for the determination of silver at the ppb level in seawater.

The application of this technique is also discussed in Section 3.70 (Table 3.4A).

3.52.2 GRAPHITE FURNACE ATOMIC ABSORPTION SPECTROMETRY

Miller and Bruland [563] described an equilibration/solvent extraction method based on competition for silver between sample ligands and added diethyldithiocarbamate for the determination of µg/L levels of silver in seawater. Detection was achieved by graphite furnace atomic absorption spectrometry.

3.52.3 NEUTRON ACTIVATION ANALYSIS

Kawabuchi and Riley [564] used neutron activation analysis to determine silver in seawater. Silver in a 4 L sample of seawater was concentrated by ion exchange on a column containing De-acidite FF–IP resin, previously treated with 0.1 M hydrochloric acid. Silver was eluted with 0.4 M aqueous thiourea, and the eluate was evaporated to dryness, transferred to a silica irradiation capsule, heated at 200°C, and ashed at 500°C. After sealing, the capsule was irradiated for 24 h in a thermal neutron flux of 3.5×10^{12} neutrons/cm^2/s, and after a decay period of 2–3 days, the 100 m silver arising from the reaction $^{199m}Ag(n, \gamma)$ ^{110m}Ag was separated by a conventional radiochemical procedure. The activity of the 100 m silver was counted with an end-window Geiger–Muller tube, and the purity of the final precipitate was checked with a GE(Li) detector coupled to a 400-channel analyser. The method gave a coefficient of variation of ±10% at a level of 40 ng/L silver.

3.52.4 X-RAY FLUORESCENCE SPECTROSCOPY

The application of this technique is discussed in Section 3.70 (Table 3.4N).

3.53 SODIUM

3.53.1 POLARIMETRY

In the indirect polarimetry method [565], sodium is precipitated as the zinc uranyl acetate salt and the uranium present in the precipitate is determined polarometrically after reaction with +-tartaric acid.

3.53.2 AMPEROMETRIC METHODS

In the indirect amperometric method [566], saturated uranyl zinc acetate solution is added to the sample containing 0.1–10 mg sodium. The solution is heated for 30 min at 100°C to complete precipitation. The solution is filtered, and the precipitate is washed several times with saturated uranyl acetate and then with 99% ethanol saturated with sodium uranyl zinc acetate. The precipitate is dissolved and diluted to a known volume. To an aliquot containing up to 1.7 mg zinc, 1 M tartaric acid 2–3 mL and 3 M ammonium acetate 8–10 mL are added and pH is adjusted to 7.5–8.0 with 2 M aqueous ammonia. The solution is diluted to 25 mL, and an equal volume of ethanol is added. It is titrated amperometrically with 0.01 M $K_4Fe(CN)_6$ using a platinum electrode. Uranium does not interfere with the determination of sodium.

3.53.3 NEUTRON ACTIVATION ANALYSIS

The application of this technique is discussed in Section 3.70 (Table 3.4O).

3.54 STRONTIUM

3.54.1 ATOMIC ABSORPTION SPECTROMETRY

Carr [567] studied the effects of salinity on the determination of strontium in seawater by atomic absorption spectrometry using air–acetylene flame. Using solutions containing 7.5 mg/L strontium and between 15% and 14% sodium chloride, he demonstrated a decrease in absorption with increasing sodium chloride concentration. To overcome this effect, a standard additions procedure is recommended.

3.54.2 MISCELLANEOUS

Flame photometry [568] and inductively coupled plasma atomic emission spectrometry [569] have been applied for determining strontium in seawater.

3.54.3 FURTHER TECHNIQUES

As discussed in Section 3.70, the following techniques have also been applied for determining strontium in seawater: Zeeman atomic absorption spectrometry (Table 3.4C), X-ray fluorescence spectroscopy (Table 3.4N), and neutron activation analysis (Table 3.4N).

3.55 TECHNETIUM

3.55.1 PRECONCENTRATION

Chen et al. [570] preconcentrated 99-technetium in seawater on an anion-exchange column to determine it in amounts down to 3 mBq/m^3.

3.56 TELLURIUM

3.56.1 GRAPHITE FURNACE ATOMIC ABSORPTION SPECTROMETRY

Andreae [571] coprecipitated tellurium(V) and tellurium(VI) from seawater and other natural waters with magnesium hydroxide. After dissolution of the precipitate with hydrochloric acid, tellurium(IV) was reduced to tellurium hydride in 3 M hydrochloric acid. The hydride was trapped inside the graphite tube of a graphite furnace atomic absorption spectrometer heated to 300°C and tellurium(IV) was determined. Tellurium(VI) was reduced to tellurium(IV) by boiling with hydrochloric acid and total tellurium was determined. Tellurium(VI) was then calculated. The limit of detection was 0.5 pmol/L and precision was 10%–20%.

The application of this technique is also discussed in Section 3.70 (Table 3.4B).

3.56.2 HYDRIDE GENERATION ATOMIC ABSORPTION SPECTROMETRY

Petit [572] described a method for the determination of tellurium in seawater at picomolar concentrations. Tellurium(VI) was reduced to tellurium(IV) by boiling in 3 M hydrochloric acid. After preconcentration by coprecipitation with magnesium hydroxide, tellurium was reduced to the hydride by sodium borohydrate at 300°C for 120 s then 257°C for 12 s. The hydride was then analysed by atomic absorption spectroscopy. Recovery was 90%–95%, and the detection limit was 0.5 pmol/L.

The application of this technique is also discussed in Section 3.70 (Table 3.4D).

3.57 TERBIUM

3.57.1 ISOTOPE DILUTION METHODS

The application of this technique is discussed in Section 3.70 (Table 3.4K).

3.58 THALLIUM

As discussed in Section 3.70, various techniques have been applied for determining thallium in seawater, including graphite furnace atomic absorption spectrometry (Table 3.4B), inductively coupled plasma atomic emission spectrometry (Table 3.4E), isotope dilution methods (Table 3.4K), anodic scanning voltammetry (Table 3.4G), and chronopotentiometry (Table 3.4I).

3.59 THORIUM

3.59.1 NEUTRON ACTIVATION ANALYSIS

Huh and Bacon [573] used neutron activation analysis to determine 232-thorium in seawater. Seawater samples were subjected to pre- and post-irradiation procedures. Separation and purification of the isotopes, using ion-exchange chromatography and solvent extraction, were performed during pre-irradiation. After irradiation, protactinium-233 was extracted and counted. Yields were monitored with thorium-230 and protactinium-231 tracers. Thorium-232 concentrations were 27×10^{-7} dpm/kg for deep-water samples from below 500 m.

The application of this technique is also discussed in Section 3.70 (Table 3.4O).

3.60 THULIUM

3.60.1 ISOTOPE DILUTION METHODS

The application of this technique is discussed in Section 3.70 (Table 3.4K).

3.61 TIN

3.61.1 SPECTROPHOTOMETRIC METHODS

In an early method, Kodama and Tsubota [574] determined tin in seawater by anion-exchange chromatography and spectrophotometry with catechol violet.

The sample is absorbed on a column of Dower I–X resin (CI— form), and elution is then effected with 2 M nitric acid. Tin is determined in the eluate by the spectrophotometric catechol violet method. There is no interference from 0.1 mg of each of aluminium, manganese, nickel, copper, zinc, arsenic, cadmium, bismuth, and uranium; any titanium, zirconium, and antimony are removed by the ion exchange.

3.61.2 ATOMIC ABSORPTION SPECTROMETRY

Electrothermal atomic absorption spectrophotometry was used for the determination of inorganic and butyl-tin in seawater in a method described by Burns et al. [575]. Butyl-Sn is extracted into toluene, and inorganic tin is extracted, as its Sn (IV) 8-hydroxyquinoline chelate, into chloroform. The detection limit was 0.7 ng of tin.

3.61.3 GRAPHITE FURNACE ATOMIC ABSORPTION SPECTROMETRY

Dogan and Haerdi [576] and Bergerioux and Haerdi [577] determined total tin in seawater by graphite furnace atomic absorption spectrometer. These researchers added 0.1–1.0 mL of 0.25 M, 10-phenanthroline and 0.1–1.0 mL 0.2 M tetraphenylboron to 50–1000 mL of water sample. The pH of this solution was adjusted to 5.0 before the addition of coprecipitating reagents. The precipitant thus obtained was either filtered or centrifuged and dissolved in a 1–5 mL aliquot of ammoniacal alcohol (methanol, ethanol, or isopropanol) solution, pH = 8–9 or Lumatom® prior to atomic absorption spectrometry.

The application of this technique is also discussed in Section 3.70 (Table 3.4B).

3.61.4   ANODIC STRIPPING VOLTAMMETRY

A method described by Florence and Farrer [578] separated tin from its associated lead by distillation from an aqueous sulphuric acid medium into which the vapour from boiling 50% hydrobromic acid is passed. The distillate provides an ideal supporting electrolyte for the determination of tin(II) (produced by reduction with hydrazinium hydroxide) by anodic stripping at a rotating vitreous–carbon electrode in the presence of co-deposited mercury [579,580]. The tin is deposited at −0.70 V versus the SCE for 5 min and then stripped at −0.50 V during a sweep from −0.70 V to −0.45 V at 5 V per min. Tin in seawater is coprecipitated on ferric hydroxide, and the precipitate is then dissolved in the aqueous sulphuric acid and subjected to the aforementioned procedure. The average content for Pacific coastal waters was found to be 0.58 μg/L.

3.61.5   FURTHER TECHNIQUES

Other techniques for the determination of tin in seawater, discussed in Section 3.70, include neutron activation analysis (Table 3.4O) and hydride generation atomic absorption spectrometry (Table 3.4D).

3.62   TITANIUM

3.62.1   SPECTROPHOTOMETRIC METHODS

Yang et al. [581] described a spectrophotometric method for the determination of dissolved titanium in seawater after preconcentration using sodium diethyldithiocarbamate.

3.62.2   X-RAY FLUORESCENCE SPECTROSCOPY

The application of this technique is discussed in Section 3.70 (Table 3.4N).

3.62.3   CATALYTIC CATHODIC STRIPPING VOLTAMMETRY

Croot [582] described a cathodic stripping voltammetric method for the determination of titanium in seawater at extremely low concentrations, viz down to 380×110^{-12} nmol/L.

This method is capable of being used directly at sea. This method utilises the catalytic enhancement of the reduction of the complex formed between Cupferron

(*N*-nitrosophenylhydroxylamine) and Ti(IV). While Cupferron itself acts as both a complexing agent and an oxidising agent, it was found that the optimal sensitivity was with bromate as an auxiliary oxidant. An advantage of this method is that it is useable over the pH range of 5.5–8. Under the conditions employed in this work, detection limits ranged from 5 to 12 pM. This method has been extensively tested at sea in the Atlantic and Southern Oceans.

3.63 TUNGSTEN

3.63.1 MISCELLANEOUS

Van der Sloot and Das [583] described a method for the determination of tungsten in seawater.

3.64 URANIUM

3.64.1 SPECTROPHOTOMETRIC METHODS

Agrawal et al. [584] determined down to 1 ppm of uranium in seawater by liquid extraction with *N*-phenyl-3-styrylacrylohydroxamine acid followed by a spectrophotometric finish.

Chang et al. [585] described a simple and effective method for extractive preconcentration of uranium(VI) in seawater, in which the ionic liquid, 1-octyl-3-methylimidazolium hexafluorophosphate (DMPASF), was employed to form a neutral uranium(VI)-DMPASF complex. DMPASF reacted with uranium(VI) rapidly to form a stable red complex. The complex was then extracted into $[C_8mim][PF_6]$ phase, and the uranium(VI) in $[C_8mim][PF_6]$ was back-extracted into aqueous phase again with 3.0 mol/L hydrochloric acid. The extraction efficiency, back-extraction efficiency, and preconcentration factor were 98.9%, 96.5%, and 200 times for 100 μg of standard uranium(VI) in 1000 mL of water samples, respectively. The preconcentration coupled with spectrophotometry with chlorophosphonazo III (CPAIII) was developed for the determination of ultra-trace uranium(VI) in natural water. The apparent molar absorbance of the uranium(VI)–CPAIII complex and the detection limit were 3.17×10^6 1 mol^{-1} cm^{-1} and 0.22 ng/mL of uranium(VI), respectively. The absorbance of the uranium(VI)–CAPIII complex at 670 nm increases linearly with the concentration of uranium(VI) up to 45 μg of uranium(VI) in 1000 mL of aqueous solution. An interference study showed that the determination of uranium(VI) is free from the interference of almost all positive and negative ions found in water samples.

3.64.2 CATHODIC STRIPPING VOLTAMMETRY

Van der Berg and Zhang [586] determined uranium(VI) in seawater by cathodic stripping voltammetry at pH 6.8 of uranium(VI)–cathodic ions. A hanging mercury drop electrode was used. The detection limit was 0.3 nmol^{-1} after a collection period of 25 min. Interference by high concentrations of iron(III) was overcome by selective adsorption of the uranium ions at a collection potential near the reduction potential of iron(III). Organic surfactants reduced the peak heights for uranium by up to 75%

at high concentrations. EDTA was used to eliminate competition by high concentration of copper(II) for space on the surface of the drop.

Hua et al. [587] described an automated flow system for the constant current reduction of uranium(VI) onto a mercury film–coated fibre electrode. Interference from iron(III) was eliminated by the addition of sulphite. The results obtained for uranium(VI) in two reference seawater samples, NASS-1 and CASS-1, were 2.90 and 2.68 µg/L with standard deviations of 0.57 and 0.75 µg/L, respectively.

Van der Berg and Nimmo [588], in their determination or uranium in seawater, added sample aliquots to the voltammetric cell of a polarograph together with buffer comprising piperazine-N,-N'-bis (2-ethanesulphonic acid) monosodium salt and sodium hydroxide to give a final buffer concentration of 0.01 M. Oxine solution at 20 µM was also added. The best sensitivity for uranium and lack of interference occurred with 8-hydroxyquinoline.

Economon et al. [589] determined down to 0.1 µg/L of uranium (VI) in seawater by square-wave absorptive stripping voltammetry.

Square-wave cathodic stripping voltammetry has been used to determine down to 10 nmol/L of uranium in seawater [590].

The application of this technique is also discussed in Section 3.70 (Table 3.4L).

3.64.3 MISCELLANEOUS

Spencer [591] reviewed methods for the determination of uranium in seawater. Quinoline–chloroform extraction has been used to separate uranium from seawater but with limited success [592].

Billon et al. [593] studied the distribution coefficient and redox behaviour of uranium in Authie Bay, Northern France.

3.64.4 FURTHER TECHNIQUES

Further techniques discussed in Section 3.70 for the determination of uranium include inductively coupled plasma mass spectrometry (Table 3.4F), isotope dilution methods (Table 3.4K), X-ray fluorescence spectroscopy (Table 3.4N), and neutron activation analysis (Table 3.4O).

3.65 VANADIUM

3.65.1 SPECTROPHOTOMETRIC METHODS

Nishimura et al. [594] described a spectrophotometric method using 2-pyridyl azoresorcinol for the determination of down to 0.025 µg/L vanadium in seawater. Vanadium was determined as its complex with 4-(2-pyridylazo) resorcinol formed in the presence of 1,2-diaminocyclohexane-N,N,N',N'-tetraacetic acid. The complex was extracted into chloroform by coupling with zephiramine. The extinction of the chloroform layer was measured at 560 nm against water as was that of a blank prepared with vanadium-free artificial seawater.

Kirkyama and Kuroda [595] combined ion-exchange preconcentration with spectrophotometry using 2-pyridyl azoresorcinol in the determination of vanadium in seawater.

The sample as a couplex with ammonium thiocyanate was passed down a column of Dowex 1–X8 in its thiocyanate form on which vanadium is retained. A hydrochloric acid eluate of this column is reacted with diaminocyclohexane-N,N,N',N'-tetraacetic acid, and vanadium is determined spectrophotometrically at 545 nm with 4-(2-pyridylazo) resorcinol. Vanadium was determined in seawater at levels of 1.65 μg/L.

3.65.2 ATOMIC ABSORPTION SPECTROMETRY

In the atomic absorption spectrometric procedure [596], potassium thiocyanate and ascorbic acid (to reduce to vanadium(VI)) were dissolved in 1 L of seawater.

These samples were passed through a Dowex 1–X8 ion-exchange column, which was then washed with 20 mL distilled water and vanadium eluted with 150 mL eluent solution prior to atomic absorption spectrometry. The average concentration and standard deviation of the Pacific Ocean waters (μg/L) was 1.86 ± 0.12 μg/L. For the Adriatic water, the values were about 1.7 μg/L.

Machackova and Zemberyova [597] compared various modifiers (ascorbic acid, NH_4NO_3, EDTA, NH_4SCN, and a mixture of $Pd/Mg(NO_3)_2$) for the accurate determination of vanadium in natural waters by electrothermal atomic absorption spectrometry. The interferences of compounds commonly present in natural waters, such as NACl, $CaCl_2$, $MgCl_2$, and $FeCl_3$, were studied. Matrix interferences were effectively eliminated by ascorbic acid or ammonium nitrate. For comparison, the standard addition method was applied without a modifier, which provided satisfactory results. The accuracy of the method was confirmed by the analysis of certified waters. The limit of detection for vanadium and characteristic masses for ascorbic acid and ammonium nitrate as modifiers were 1.71 and 1.56 μg/L and 70 and 67 pg, respectively. Recovery was in the range of 98%–105%, and RSD was less than 5%.

3.65.3 GRAPHITE FURNACE ATOMIC ABSORPTION SPECTROMETRY

Monien and Stangel [598] studied the performance of a number of chelating agents, namely 4(2-pyridyldazo) resorcinol in conjunction with tetraphenyl arsonium chloride and tetramethyldithiocarbamate, both in chloroform solution, for the extraction of vanadium from seawater.

The extracts were injected directly into a graphite furnace atomic absorption spectrometer coated with lanthanum carbide.

For both reagents, a linear concentration dependence was obtained between 0.5 and 7 μg/L after the extraction of a 100 mL sample.

The application of this technique is also discussed in Section 3.70 (Table 3.4B).

3.65.4   INDUCTIVELY COUPLED PLASMA MASS SPECTROMETRY

Hastings et al. [599] described a method for the determination of picogram quantities of vanadium in seawater by isotope dilution inductively coupled plasma mass spectrometry with electrothermal vaporisation to introduce the sample into the plasma. A ^{50}V isotope spike enriched to 44 atom% was equilibrated with samples, followed by chemical purification by cation-exchange chromatography. Samples were introduced into the electrothermal vaporisation unit with palladium modifier and heated to 1000°C. This quantitatively eliminates the ClO$^+$ isobaric interference with vanadium at m/z 51 for solutions up to 0.5 N hydrochloric acid. Corrections for 50-titanium and 50-chromium, which interfere with the vanadium signal, were made by the measurement of 49-titanium and 53-chromium. These isobaric interferences and variable ArC levels were the limiting sources of error in the ID measurement and diminished the detection limit of 6 pg of vanadium. The detection limit for non-isotope dilution applications was 0.3 pg of vanadium in seawater.

The application of this technique is also discussed in Section 3.70 (Table 3.4F).

3.65.5   CATHODIC STRIPPING VOLTAMMETRY

Van der Berg and Huang [600] carried out direct electrochemical stripping of dissolved vanadium in seawater using cathodic stripping voltammetry performed with a hanging mercury drop electrode. The detection limit was 0.3 nmol/L after a collection period of 2 min.

The application of this technique is also discussed in Section 3.70 (Table 3.4L).

3.65.6   FURTHER TECHNIQUES

Other procedures discussed in Section 3.70 for the determination of vanadium in seawater include inductively coupled plasma atomic emission spectrometry (Table 3.4E), X-ray fluorescence spectroscopy (Table 3.4N), neutron activation analysis (Table 3.4O), and high-performance liquid chromatography (Table 3.4P).

3.66   YTTERBIUM

3.66.1   ISOTOPE DILUTION METHOD

The application of this technique is discussed in Section 3.70 (Table 3.4K).

3.67   YTTRIUM

3.67.1   X-RAY FLUORESCENCE SPECTROSCOPY

The application of this technique is discussed in Section 3.70 (Table 3.4N).

3.68 ZINC

Interest in the zinc concentration in oceans stems from its dual role as a required mononutrient and as a potential toxicant due to its widespread industrial and marine usage [601]. The major inputs of zinc to surface seawater include atmospheric deposition (both natural and anthropogenic in origin), fluvial runoff, and upwelled waters. Zinc exists at natural levels in North Pacific surface water at a total concentration of approximately 0.1 NM, increasing to 3 nM at 500 m and reaching a maximum of ~9 nM at depths greater than 2000 m [602,603].

3.68.1 SPECTROFLUOROMETRIC METHODS

Nowicki et al. [604] described a sensitive technique for the shipboard determination of zinc in seawater. The technique couples flow injection analysis with fluorometric detection. A cation-exchange column was used to separate zinc from interfering alkali and alkaline earth ions and to concentrate zinc from seawater. The organic indicator ligand, p-tosyl-8-aminoquinoline, was used to form a complex with zinc, the fluorescence of which was determined with a flow-through fluorimeter. The detection limit (defined as three times the standard deviation of the blank, $n = 4$) was 0.1 nM for 4.4 mL sample. The precision based on the replicate analysis of samples containing 4.3 nM Zn was ±6% ($n = 5$).

3.68.2 ATOMIC ABSORPTION SPECTROMETRY

A method described by Hirata and Honda [605] used a flow injection analysis manifold for the pH adjustment of a seawater sample followed by the concentration of zinc on a column packed with Chelex–100 resin. Zinc was eluted with nitric acid and determined by atomic absorption spectrophotometry. The detection limit is 0.5 µg/L, and the relative standard deviation is 2.7% at the 10 µg/L level.

The application of this technique is also discussed in Section 3.70 (Table 3.4K).

3.68.3 GRAPHITE FURNACE ATOMIC ADSORPTION SPECTROMETRY

Graphite furnace atomic adsorption spectrometry has also been used to determine zinc [606–608] in seawater with a detection limit of 2 µg [605].

Guevremont [607] discussed the use of organic matrix modifiers for the direct determination of zinc.

Huang and Shih [609] determined down to 24 ppt of zinc in seawater by graphite furnace atomic absorption spectrometry using a stabilised temperature platform furnace technique. Atomisation from the graphite furnace pretreated with vanadium gave improved detection limits.

Akatsuka et al. [610] preconcentrated zinc from seawater on a column comprising methyl-tricapryl ammonium chloride coated on C_{18} resin. The final determination was carried out by graphite furnace atomic absorption spectrometry. Down to 2.4 ng dm^3 of zinc could be determined by this procedure by using a 500 mL sampler.

The application of this technique is also discussed in Section 3.70 (Table 3.4B).

3.68.4 ANODIC STRIPPING VOLTAMMETRY

Anodic stripping voltammetry has been used [611] using a tubular mercury–graphite electrode to determine zinc in seawater. Zinc concentrations of 1×10^{-9} M can be detected within 5 min using this system.

Muzzarrelli and Sipos [612] showed that a column of chitosan (15 × 10 mm) can be used to concentrate zinc from 3 L of seawater before determination by anodic stripping voltammetry with a composite mercury–graphite electrode. Zinc (also lead) is eluted from the column by 50 mL 2 M ammonium acetate, copper by 10 mL of 0.01 M EDTA, and cadmium by 3 mL 0.1 M potassium cyanide.

Andruzzi and Trazza [613] described a new kind of semi-stationary mercury electrode, the long-lasting sessile-drop electrode, and gave diagrams of the electrolysis cell and details of the electrode. It allows a longer electrolysis time with more vigorous stirring, a longer rest time, a larger potential range at slower scan rates, and a higher current response. The sensitivity and reproducibility of the electrode were demonstrated by the determination of zinc in seawater samples by differential pulse anodic stripping voltammetry.

Lieberman and Zirino [614] used a tubular mercury and graphite electrode to determine zinc in seawater.

The application of this technique is also discussed in Section 3.70 (Table 3.4G).

3.68.5 CATHODIC STRIPPING VOLTAMMETRY

Van der Berg [615] determined zinc complexing capacity in seawater by cathodic stripping voltammetry of zinc–ammonium pyrrolidine dithiocarbamate complex ions.

The successful application of cathodic stripping voltammetry, preceded by absorptive collection of complexes with ammonium pyrrolidine dithiocarbamate, for the determination of zinc complexing capability in seawater is described. The reduction peak of zinc was depressed as a result of ligand competition by natural organic material in the sample.

Investigations of electrochemically reversible and irreversible complexes in seawater of several salinities are detailed, together with experimental measurements of ligand concentrations and conditional stability constants for complexing ligands. Results obtained were comparable with those obtained by other equilibrium techniques, but the aforementioned methods had a greater sensitivity.

Van der Berg [616] also reported a direct determination of sub-nanomolar levels of zinc in seawater by cathodic stripping voltammetry. The ability of ammonium pyrrolidine dithiocarbamate to produce a significant reduction peak in the presence of low concentrations of zinc was used to develop a method for analysis at levels two orders of magnitude below those achieved with anodic stripping voltammetry. Interference from nickel and cobalt ions could be overcome by using a collection potential of 1.3 V and interference by organic complexing material by ultraviolet irradiation. Zinc could be determined in seawater and freshwater. Zinc and nickel could be determined simultaneously by using dimethylgloxime at a collection potential of −0.7 V, followed by ammonium pyrrolidine dithiocarbamate at −1.3 V. The sensitivity for this determination was 3 pmol/L.

Zima and Van der Berg [617] determined zinc in seawater in amounts down to 3 nmol by cathodic stripping voltammetry.

The application of this technique is also discussed in Section 3.70 (Table 3.4L).

3.68.6 MISCELLANEOUS

Adsorption colloid flotation using didecylamine as surfactants has been used to separate zinc with 95% efficiency from seawater [608].

Guevremont [607] discussed the use of organic matrix modifiers for the direct determination of zinc in seawater.

3.68.7 FURTHER TECHNIQUES

Further techniques, discussed in Section 3.70, for the determination of zinc in seawater include potentiometric stripping analysis (Table 3.4M), plasma emission spectrometry (Table 3.4J), isotope dilution methods (Table 3.4K), X-ray fluorescence spectrometry (Table 3.4N), neutron activation analysis (Table 3.4O), inductively coupled plasma atomic emission spectrometry (Table 3.4E), and inductively coupled plasma mass spectrometry (Table 3.4F).

3.69 ZIRCONIUM

3.69.1 X-RAY FLUORESCENCE SPECTROSCOPY

The application of this technique in the determination of zirconium is discussed in Section 3.70 (Table 3.4O).

3.70 MULTIMETAL ANALYSIS

It is often the case that a method of analysis covers a range of elements, not a single element as discussed in Sections 3.1 through 3.70. Some more recent examples of the type of analysis are reviewed as follows. These cover the following transition metals: cadmium, copper, nickel, and lead.

Soylak et al. [764] used a preconcentration/separation method for the atomic absorption spectrometric determinations of Cu(II), NI(II), and Pb(II) ions in seawater. This was achieved using a chromatographic column filled with Amberlite XAD-1180. Recoveries of Cu(II), NI(II), and Pb(II) ions were quantitative (>95%) at pH 8–9. The influences of the various analytical parameters, including the amount of calmagite, amount of Amberlite XAD-1180, eluent type, and so on, were investigated. The influence of the seawater matrix was also examined. Simultaneous enrichment and determination of copper, nickel, and lead ions in seawater are possible with recoveries >95%, RSD <9%.

Dakova et al. [765] determined total dissolved and labile concentrations of Cd(II), Cu(II), Ni(II), and Pb(II) of the Bourgas Gulf of the Bulgarian Black Sea coast. Solid-phase extraction procedure based on monodisperse, sub-micrometer silica spheres modified with 3-amino minopropyltrimethoxysilane followed by the electrothermal atomic absorption spectrometry was applied to quantify the total dissolved metal concentrations in seawater. Quantitative sorption of Cd, Cu, Ni, and Pb was achieved in the pH range of 7.5–8 for 30 min, and the adsorbed elements were easily eluted with 2 mL 2 mol/L nitric acid. Detection limits achieved for total dissolved metal quantification were as follows: Cd 0.002 μg/L, Cu 0.005 μg/L, and Pb 0.02 μg/L. Detection limits achieved for total dissolved metal quantification were as follows: Cd 0.002 μg/L, Cu 0.005 μg/L, Ni 0.03 μg/L, and Pb 0.02 μg/L.

Escaleira et al. [766] determined Cd, Pb, and Ni in high-salinity waters by inductively coupled plasma optical emission spectrometry. This method is based on cloud-point extraction of these metals as complexes of diethyldithiocarbamate in micellar media of nonionic surfactant octylphenoxypolyethoxethanol (Triton X-114). The effect of interference from residual salinity in surfactant-rich phase was also investigated. The developed procedure achieved enhancement factors of 20.0, 20.4, 19.5, and 20.6, along with limits of detection ($3\sigma_B$) of 0.030, 2.1, 06.2, and 0.27 μg/L, and precision expressed as relative standard deviation (%RSD, $n = 10$) of 3.7 (40.0 μg/L), 6.6 (20.0 μg/L), and 3.1% (10.0 μg/L) for Cd, Pb, Cu, and Ni, respectively. The accuracy was evaluated by spike tests on the seawater (salinity of 35%). It was obtained by recoveries between 70% and 105%.

Wen et al. [767] described a procedure for the determination of a wider range of elements in seawater. 8-Hydroxyquinoline immobilised on polyacrylonitrile hollow fibre membrane was used for the preconcentration of cadmium, lead, copper, manganese, bismuth, indium, beryllium, and silver in seawater prior to their determination by inductively coupled plasma mass spectrometry. The optimum experimental conditions such as pH, sample flow rate, and volume of eluents were investigated. The concentration factor of at least 300 for analytes of interest in seawater and separation of matrix components can be achieved. The method has been applied for the determination of trace elements in coastal seawater. The results indicated that the recovery ranged from 91% to 107%, and the relative standard deviations were found to be less than 5% for the trace elements at ng/L level.

Table 3.4 presents a summary of methods used to carry out multimetal analysis, Wherever available, detection limits for the methods are reported. The methods are cross-referenced with the method for determining individual single elements described in Sections 3.1 through 3.69.

TABLE 3.4
Multimetal Analysis

Elements Determined	LD	References
3.4A Atomic Absorption Spectrometry		
An, Zn, Pb, Fe,		[618]
Mn, Ne, Co, Ag		
Cd, Pb, C, Zn,	1.0 ppb	[619]
Cu, Pb, Zn		
Cu, Pb, Zn	Cu 0.07 µg/L	[620]
	Pb 0.03 µg/L	
	Cd 0.05 µg/L	
Cu, Cd, Co	–	[621]
K, Li, Rb	–	[622]
Co, Cu, Mn	–	[623]
Ca, Pb, Cu	nM	[624]
Cu, Cd, Pb	nM	[625]
3.4B Graphite Furnace Atomic Absorption Spectrometry		
Cd, Cu, Pb, Ni, Zn, Co	0.02 µg/L	[385]
Cd, Cu, Pb, Ni, Zn	Cd 0.01 µg/L	[626]
	Ni 50 µg/L	
Cu, NI, Cd	–	[627]
Cu, Cd, Pb, Ni, Zn	–	[628]
Cu, Cd, Zn, Ni	–	[629]
Cu, Fe, Mn	Fe 0.09 µmol/L	[630]
Za	Mn 0.2 µmol/L	
Cd, Co, Cu, Fe, Mn, Ni, Pb, Zn	–	[424]
Cd, Pb, Ca, Cu, Mn, Ni	Cd 0.1 µg/L	[631,632]
	Pb 4 µg/L	
	Cr 0.2 µg/L	
Cd, Pb, Cu, Mn, Ni, Cr	–	[632]
Cd, Pb, Cu, Mn, Ni, Zn	–	[424]
Fe, Mn, Zn	Fe 0.2 µg/L	[418,425]
	Mn 0.2 µg/L	
	Zn 0.4 µg/L	
Fe, Cr, Mn	–	[633]
Cd, Cu, Ag, Pb	µg/L	[634]
Ag, Cd, Pb	–	[635]
Pb, Mn, V, Mo	–	[636–638]
Hg, Pb, Cd	–	[656]
Cu, Fe, Mn, Co, Ni	–	[419]
Ni, Cu, Mo, Mn	–	[639]
As, Bi, In, Pb, Sb, Se,	–	[640–642]
Sn, Te, Te		
Pt, Ir	–	[643]

(Continued)

TABLE 3.4 (*Continued*)
Multimetal Analysis

Elements Determined	LD	References
3.4C Zeeman Atomic Absorption Spectrometry		
Miscellaneous	–	[644–659]
Cu, Pb, Cd, Co, Ni, Sr	–	[660,661]
Cr, Ni, Mn, Cd, As, Mo	–	[662–665]
3.4D Hydride Generation Atomic Absorption Spectrometry		
As, Sb, Ge, Sn	40 pg	[666–676]
3.4E Inductively Coupled Plasma Atomic Emission Spectrometry		
Zo elements	–	[624]
Cd, Cr, Cu, Pb, Ni, Zn	–	[675–677]
Fe, Mn, Zn, Cu, Ni	–	[678]
Mn, Zn, Fe, Ni, Cu, Pb, Cd, Cr, Co	–	[678,679]
Cd, Zn, Pb, Fe, Mn, Cu, Cr, Ni, Co		[165]
Cr, Mn, Co, Ni, Cu, Cd, Pb	1 μ/L	[679]
Pb, Zn, Cd, Ni, Mn, Fe, V, Cu	Mn 0.053 μg/L Zn 0.63 μg/L Cd 0.25 μg/L Fe 0.25 μg/L V 0.38 μg/L Ni 0.5 μg/L Cu 0.5 μg/L Pb 2.5 μg/L	[680]
Bi, Cd, Cu, Co, In, Ni, Pb, Tl, Zn	–	[681]
Cd, Pb, Zn, Fe, Cu, Ni, Mo, V	–	[677]
Cd, Cu, Fe, Mn, Zn	–	[682]
3.4F Inductively Coupled Plasma Mass Spectrometry		
Zn, Mn, Co, Cu, Cr, Ni, Fe Cd, Pb, Hg,	Fe 1.9 μg/L, Cd 50 Mn, Ni, Hg 10–100 ppt	[683]
Cu, Co, Mn, Ni, V, Mo, Cd, Pb, U	–	[684]
Cu, Co, Mn, Ni, V	3–40 ppt	[685]
Ni, As, V	–	[686]
Ca, Be, Al, Zn, Rb, In, Pb	–	[687]
Sb, As, Hg	0.6 ppt	[688]
2o Metals	–	[689–691]

(Continued)

TABLE 3.4 (*Continued*)
Multimetal Analysis

Elements Determined	LD	References
3.4G Anodic Stripping Voltammetry		
Zn, Cd, Pb, Cu	1–10 mg/L	[694]
Zn, Cd, Pb, Cu		
	Pb Zn 0.01–0.1 mg/L	[695]
	Cu	
Cd	0.001–0.1 mg/L	[696]
Cd, Pb, Cd	0.18 µg/L	[696]
Pb	0.21 µg/L	
Bi, Cu, Pb, Cd, Zn	–	[695]
Cu, Pb, Cd, Zn	–	[692]
TI	0.2–1 µmol/L	[695]
Pb, Cd	1 µg/L	[702]
Pb	4 µg/L	[697]
Cd, Cu, Pb, Cd	0.1 µg/L	
Cu	0.3 µg/L	[698]
Pb	0.7 µg/L	[699]
Zn, Cd, Pb	–	[700,701]
Co, Ni, Cr	–	[218,385]
Pb, Cd	–	[171]
Cd, Pb, Cu	–	[698]
Cd, Pb, Cu	–	[703]
Cd, Cu, Pb, Ni, Zn, Cd, Pb	–	[704]
Pb, Cu	–	[705]
Cu, Pb, Cd	Ni 0.45 mm	[706]
Ni, Cu, Zn	–	[707]
Cd, Cu, Zn, Mn, Fe	Cd 0.6 ng/L	[708]
	Pb 8 ng/L	
	Cu 5 ng/L	
3.4H Differential Pulse Anodic Stripping Voltammetry		
Zn, Ca, Pb, Cu	–	[709]
Cd, Pb, Cu,	–	[710,711]
Ni, Cu	–	[712]
Cd, Cu, Zn	–	[713]
Cu, Pb, Cd	–	[714]
Zn, Cd, Pb, Cu	–	[715]
Cu, Cd	–	[716]
Zn, Cd, Pb	–	[717]
Cd, Pb, Cu	10 µg/L	[711,718]
Cu, Pb	–	[719]
Cu, Hg	–	[720]

(*Continued*)

TABLE 3.4 (*Continued*)
Multimetal Analysis

Elements Determined	LD		References
Zn, Cd, Pb, Cu, Sb, Bi	Zn	0.1 μg/kg	[721]
	Cu	0.1 μg/kg	
	Cd	0.01 μ	
	Pb	0.01 μ	
	Sb	0.05 μg/kg	
	Bi	0.05 μg/kg	
3.4I Chronopotentiometry			
Co, Ni	–		[722]
Cu, Cd, Tl, Pb	–		[723]
3.4J Plasma Emission Spectrometry			
Cd, Co, Cu, Pb, Ni, Zn	–		[724]
3.4K Isotope Dilution Mass Spectrometry			
Cu, Cd, Tl, Pb	–		[723]
Te, Cd, Zn, Cu, NI, Pb, U	–		[249]
Cu, Cd, Pb, Zn, Te	–		[725]
Lanthanides	–		[726]
3.4L Cathodic Stripping Voltammetry			
Cu, Ni	–		[727,728]
Ni, Co	Ni	2 nm	[729]
	Co	50 pm	
Cd, Cu, Fe, Mn, Ni, Zn	–		[730]
Pb, Cd, Cu, Zn, U, V, Mo, Ni, Co	–		[731]
3.4M Potentiometric Stripping Analysis			
Pb, Cu, Zn, Cd	–		[732–734]
Pb, Cd, Zn	–		[735]
3.4N X-ray Fluorescence Spectroscopy			
Cu, Ni, Co, Zn, Mn	–		[736]
Cr, Mn, Fe, Co, Ni, Cu, Zn	–		[737]
Fe, Co, Ni, Cu, Zn, Pb	Mn	15 μg/L	[738]
	Fe	16 μg/L	
	Co	8 μg/L	
	Ni	8 μg/L	
	Cu	13 μg/L	
	Zn	13 μg/L	
	Pb	40 μg/L	
Mn, Fe, Co, Ni, Ci, Zn, Pb, Cd, Se, V, Mo, Hg, U K, Cu, Ti, V, Cr, Mn	–		[739]

(*Continued*)

TABLE 3.4 (*Continued*)
Multimetal Analysis

Elements Determined	LD	References
Fe, Co, Ni, Cu, Zn, Ga	–	[739]
Zr, Mo, Ag, Cd, Sb		
Ba		
Heavy metals	–	[740]
3.4O Neutron Activation Analysis		
Al, Cd, Co, Cr, Cu, Eu, Fe	–	[741,742]
Mn, Mo, Ni, Sc, Sn, Th,		[743,744]
U, V, An		
Co, Cr, Ca, Fe, Pb, Se, Sr	–	[745]
As, Cu, Sb	–	[746]
Ai, V, Cu, Mo, Zn, U	–	
As, Cd, Ce, Co, Cr, Eu	–	[747]
Fe, Hg, La, Mo, Sc, Se, U,		
Zn, As, Sb		
Ba, Ca, Cd, Le, Cr, Cu, Se,	–	[748]
Fe, La, Ma, So, U, V, Zn		
Hg, Au, Cu	–	[749]
Transition metals	–	[750]
As, Mo, U, V	–	[751]
Co, Cu, Hg	–	[752]
Ca, Cr, Cu, Fe, Mi	–	[753]
Mo, Ni, Sc, Th, Zn, Ag,	–	[754]
Ag, Cd, Cr, Cu, Mn, Th, U, Zr		
Cd, Co, Cr, Cu, Fe, Mn,	–	[756]
Ni, Sc, Sn, Th, U, V, Zn		
Np	–	[772]
Sc, Cr, Fe, Co, Cu, Zn, Ag, Sb, Hg	–	[755]
3.4P Performance Liquid Chromatography		
Al, Fe, Mn	–	[756]
Cu, Ni, V	–	[757]
Transition metals	–	[758]
Ni, Co, Cu, Zn, Pb	Co 0.5 µg/L	[759,760]
	Ni 15 µg/L	
Zu, Fe, Mn, Cu, Ni	Zn 0.5 µg/L	[761]
Ce^{III}, Cr^V	µg/L	
3.4Q Molecular Photoluminescence Spectrometry		
Sb, As	–	[762,763,766,768,772]

REFERENCES

1. R.M. Moore, In *Proceedings of a NATO Advanced Research Institute on Trace Metals in Seawater* (eds. C.S. Wong et al.), Sicily, Italy, 30 March–3 April 1981, Plenum, New York, p. 198.
2. W.F. Fitzgerald, *Analytical Methods in Oceanography* (ed. T.R.P Gibb), American Chemical Society, Washington, DC, 1975.
3. K.K.S. Pillay, C.C. Thomas, I.A. Sondel, and C.M. Hyche, *Analytical Chemistry*, 1971, 41, 1419.
4. R.E. Sturgeon, S.S. Berman, S.N. Willie, and J.A.H. Desaulniers, *Analytical Chemistry*, 1981, 53, 2337.
5. D. Jagner, M. Josefson, and S. Westerlund, *Analytica Chimica Acta*, 1981, 129, 153.
6. L.G. Danielson, R. Magnusson, and S. Westerlund, *Analytica Chimica Acta*, 1978, 98, 47.
7. D. Jagner, M. Josefson, and S. Westerlund, *Analytica Chimica Acta*, 1981, 129, 153.
8. A. Mujazaki, A. Kimura, K. Bansho, and Y. Amezaki, *Analytica Chimica Acta*, 1981, 144, 213.
9. B. Magnusson and A. Westerlund, In *Proceedings of a NATO Advanced Research Institute on Trace Metals in Seawater* (eds. C.S. Wong et al.), Sicily, Italy, 30 March–3 April 1981, Plenum, New York, 1981, p. 116.
10. L. Brugmann, L.G. Danielsson, B. Magnusson, and S. Westerlund, *Marine Chemistry*, 1983, 13, 327.
11. E.A. Boyle and J.M. Edmond, *Analytica Chimica Acta*, 1977, 91, 189.
12. B. Magnusson and S. Westerlund, *Marine Chemistry*, 1980, 8, 231.
13. V.J. Stukas and C.S. Wong, In *Proceedings of a NATO Advanced Research Institute on Trace Metals in Seawater* (eds. C.S. Wong et al.), 30 March–3 April 1981, Sicily, Italy, Plenum Press, New York, 1981, p. 120.
14. H. Armannsson, *Analytica Chimica Acta*, 1979, 110, 71.
15. W.C. Campbell and J.M. Ottaway, *Analyst*, 1977, 102, 495.
16. D. Grimaud and G. Michard, *Marine Chemistry*, 1974, 2, 229.
17. T. Kuwamoto and S. Murai, Preliminary Report of the Hukuho-Maru Cruise K11–68–4, Ocean Research Institute University, Tokyo, Japan, 1970, p. 72.
18. R. Tukai, *Nature*, 1967, 213, 901.
19. S.N. Willie, R.E. Sturgeon, and S.S. Berman, *Analytical Chemistry*, 1983, 55, 981.
20. G.E. Batley and J.P. Matousek, *Analytical Chemistry*, 1980, 52, 1570.
21. S. Motomizu, *Analytica Chimica Acta*, 1973, 64, 217.
22. R.R. Greenberg and H.M. Kingston, *Journal of Radioanalytical Chemistry*, 1982, 71, 147.
23. N. Sheffrin and E.E. Williams, *Analytical Proceedings*, 1982, 19, 483.
24. L. Mort, H.W. Nurnberg, and D. Dyrssen, In *Proceedings, of a NATO Advanced Research Institute on Trace Metals in Seawater* (eds. C.S. Wong et al.), Sicily, Italy, 30 March–3 April 1981, Plenum Press, New York, 1981.
25. A. Miyazaki, A. Kimura, K. Bansho, and Y.Z. Umezaki, *Coastal Analytica Chimica Acta*, 1981, 144, 213.
26. B.K. Schaube and C.C. Patterson, In *Proceedings of a NATO Advanced Research Institute on Trace Metals in Seawater* (eds. C.S. Wong et al.), Sicily, Italy, 30 March to 3 April 1981, Plenum Press, New York, 1981, p. 148.
27. E. Leyte and S.W. Huested, In *Proceedings of a NATO Advanced Research Institute on Trace Metals in Seawater* (eds. C.S. Wong et al.), Sicily, Italy, 30 March–3 April 1981, Plenum Press, New York, 1981.
28. W.F. Fitzgerald, W.B. Lyons, and C.D. Hunt, *Analytical Chemistry*, 1974, 46, 1882.

29. J. Olafsson, In *Proceedings of a NATO Advanced Research Institute on Trace Metals in Seawater* (eds. C.S. Wong et al.), Sicily, Italy, 30 March–3 April 1981, Plenum Press, New York, 1981, p. 170.
30. G.E. Batley and J.P. Matousek, *Analytical Chemistry*, 1977, 49, 2031.
31. H.M.A. Sheriadah, M. Katoaka, and K. Ohzeki, *Analyst*, 1985, 110, 125.
32. B. Pihlar, P. Valenta, and H.W. Nurnberg, *Fresenius Zeitschrift für Analytische Chemie*, 1981, 307, 337.
33. V.J. Stuckas and C.S. Wong, In *Proceedings of a NATO Advanced Research Institute on Trace Metals in Seawater* (eds. C.S. Wong et al.), Sicily, Italy, 30 March–3 April 1981, Plenum Press, New York, 1981, p. 177.
34. A.P. Mykytiuk, D.S. Russell, and R.R. Sturgeon, *Analytical Chemistry*, 1980, 52, 1281.
35. H. Elderfield and H.J. Greaves, In *Proceedings of a NATO Advanced Research Institute on Trace Metals in Seawater* (eds. C.S. Wong et al.), Sicily, Italy, 30 March–3 April 1981, Plenum Press, New York, 1981, p. 203.
36. A.D. Matthews and J.P. Riley, *Analytica Chimica Acta*, 1970, 51, 445.
37. S.R. Willie, R.E. Sturgeon, and S.S. Berman, *Analytical Chemistry*, 1986, 58, 1140.
38. F.E. Brinkmann, In *Proceedings of a NATO Advanced Research Institute on Trace Meals in Seawater* (eds. C.S. Wong et al.), Sicily, Italy, 30 March–3 April 1981, Plenum Press, New York, 1981, p. 220.
39. M.B. Collela, S. Siggia, and R.M. Barnes, *Analytical Chemistry*, 1980, 52, 2347.
40. C.M.G. Van Der Berg and O. Huang, *Analytical Chemistry*, 1984, 56, 2383.
41. M. Ishibashi and T. Kawai, *Nippon Kagaku Zasshi*, 1952, 73, 380.
42. M. Ishibashi and K. Motojima, *Nippon Zasshi*, 1952, 23, 491.
43. K. Motojimi, *Nippon Kagaku Zasshi*, 1955, 76, 902.
44. Japan Industrial Standard (JIS), 1974, K0102.
45. M.O. Andreae, *Analytical Chemistry*, 1977, 49, 820.
46. C.C. Foreback, Some studies on the detection of determination of mercury arsenic and antimony in gas discharges, Thesis, University of South Florida, Tampa, FL, 1973.
47. L.H. Simons, P.H. Monaghan, and M.S. Taggart, *Analytical Chemistry*, 1953, 25, 989.
48. W. Sackett and G. Arrhenius, *Geochimica Cosmochimica Acta*, 1962, 26, 955.
49. Y. Nishikawa, K. Hiraki, K. Morishige, A. Tsuchiyama, and T. Shigematsu, *Bunseki Kagaku*, 1968, 17, 1092.
50. T. Shigematus, Y. Nishikawa, K. Hiraki, and N. Nagano, *Bunseki Kagaku*, 1970, 19, 551.
51. D.J. Hydes and P.S. Liss, *Analyst*, 1976, 101, 922.
52. D.Y. Hsu and W.O. Pipes, *Environmental Science and Technology*, 1972, 6, 645.
53. H.L. Kahn, *Environmental Analytical Chemistry*, 1973, 3, 121.
54. R.A. Barnes, *Chemical Geology*, 1975, 15, 177.
55. M.L. Lee and D.C. Burrell, *Analytica Chimica Acta*, 1973, 66, 245.
56. T.A. Gosink *Analytical Chemistry*, 1975, 47, 165.
57. L.I. Pavlenko and N.S. Safronova, *Zhurnal Analiticheskoi Khimii*, 1975, 30, 775.
58. T. Fujinaga, R. Kusaka, M. Koyama et al., *Journal of Radioanalytical Chemistry*, 1973, 13, 301.
59. J.A. Reising and C.I. Measures, *Analytical Chemistry*, 1994, 66, 4105.
60. T. Korenaga, S. Motomizu, and K. Toei, *Analyst*, 1980, 105, 328.
61. A.G. Howard, A.J. Coxhead, A. Potter, and A.P. Watt, *Analyst*, 1986, 111, 1379.
62. M. Salgado Ordonez, A. Garcia de Torres, and J.M. Cano Pavon, *Talanta*, 1985, 32, 887.
63. R. Suarez, B. Horskoffe, C.M. Duerl-Verda, *Analytical Chemistry*, 2012, 84, 9462.
64. C.H.G. Van Der Berg, K. Murphy, and J.P. Riley, *Analytica Chimica Acta*, 1986, 188, 177.
65. D. Spencer and P.L. Sachs, *Atomic Absorption Newsletter*, 1969.

66. B.S. Newell, *Journal of the Marine Biological Association UK*, 1967, 47, 271.
67. K. Matsunaga and M. Nishimura, *Japan Analyst*, 1971, 29, 993.
68. R.T. Emmet, Naval Ship Research and Development Centre, Report 2570, 1968.
69. F. Koroleff, *Information on Techniques and Methods for Seawater Analysis, Interlaboratory Report*, 1970, 3, 19.
70. L. Solorzano, *Limnology and Oceanography*, 1969, 14, 799.
71. P.C. Head, *Deep Sea Research*, 1971, 18, 531.
72. K. Grasshoff and H. Johansson, *Journal of Conservation of International Mercantile Exploration*, 1972, 34, 516.
73. M.I. Liddicoat, S. Tibbits, and E.I. Butler, *Limnology and Oceanography*, 1975, 20, 131.
74. G. Slawyk and J.J. MacIsaac, *Deep Sea Research*, 1972, 19, 521.
75. R. Benesch and P. Mangelsdorf, *Helgolander wiss Meeresunters*, 1972, 23, 365.
76. J.E. Harwood and D.J. Huyser, *Water Research*, 1970, 4, 695.
77. J.E. Harwood and A.L. Kuhn, *Water Research*, 1970, 4, 805.
78. D. Degobbis, *Limnology and Oceanography*, 1973, 18, 146.
79. G. Dal Pont, M. Hogan, and B. Newell, Laboratory Techniques in Marine Chemistry II, Determination of Ammonia in Seawater and the Preservation of Samples for Nitrate Analysis. Commonwealth Scientific and Industrial Research Organisation (Australia) Division of Fisheries and Oceanography Report No. 55, Marine Laboratory, Cromella, Sydney, New South Wales, Australia, 1974.
80. B.R. Berg and M.I. Abdullah, *Water Research*, 1977, 11, 637.
81. B.L. Hampson, *Water Research*, 1977, 11, 305.
82. K. Grasshoff and H. Johanssen, *Journal of Conservation and International Mercantile Exploration*, 1974, 36, 90.
83. R.F.C. Mantoura and E.M.S. Woodward, *Estuarine Coastal and Shelf Science*, 1983, 17, 219.
84. L. Prochazkova, *Analytical Chemistry*, 1964, 36, 865.
85. R. Johnston, *International Conference on Exploration of the Sea*, Palermo, cm 1966, N:11, 1966.
86. J.D.H. Strickland and K.H. Austin, *Journal of Conservation of International Mercantile Exploration*, 1959, 24, 446.
87. F.A. Richards and R.A. Kletsch, In *Recent Researches in the Fields of Hydrosphere Atmosphere and Nuclear Geochemistry* (eds. Y. Maysho and T. Kryama), Maruzen, Tokyo, Japan, 1964, pp. 65, 81.
88. K. Matsunava and M. Nishimura, *Analytica Chimica Acta*, 1974, 73, 204.
89. P. Le Corre and P. Treguer, *Journal de Consei*, 1978, 38, 147.
90. M.A. Brzezinska, *Marine Chemistry*, 1987, 20, 277.
91. N. Amornthammorory and Jia-zhang Zhang, *Analytical Chemistry*, 2008, 80, 1019.
92. S.W. Willason and K.S. Johnson, *Marine Biology*, 1986, 91, 285.
93. T.R. Gilbert and A.M. Clay, *Analytical Chemistry*, 1973, 45, 1757.
94. J.D. McLean, V.A. Stenger, R.E. Reim, and M.W. Long, *Analytical Chemistry*, 1978, 50, 1309.
95. W.S. Gardner and P.A. St John, *Analytical Chemistry*, 1991, 63, 537.
96. W.S. Gardner, L.R. Herche, P.A. St John, and S.P. Seitzinger, *Analytical Chemistry*, 1991, 63, 1838.
97. Y.A. Afanasev, A.I. Ryabinin, L.T. Azhipa, and A.S. Romanov, Kuban State University, Krasndar and State Oceanographic Institute USSR, 1970, p. 1832.
98. K.K. Bertine and D.S. Lee, In *Proceedings of a NATO Advanced Research Institute on Trace Metals in Seawater* (eds. C.C. Wong et al.), Sicily, Italy, 30 March–30 April 1981, Plenum Press, New York, 1981, p. 222.
99. R. Sturgeon, S.N. Willie, and S.S. Berman, *Analytical Chemistry*, 1985, 57, 2311.

100. Y.A. Afansev, A. Ryabinin, and L. Azhipa, *Zhurnal Analiticheskoi Khimii*, 1975, 30, 1830.
101. Arsenic in potable and seawater by spectrophotometry 1978 tentative method, Department of the Environment/National Water Council Standing Committee and Analysts, HMSO, London, UK, 1980, 20pp (RP22A ENV).
102. A.G. Howard and S.D.W. Comber, *Mikrochimica Acta*, 1992, 109, 27.
103. J.D. Burton, In *Proceedings of a NATO Advanced Research Institute on Trace Metals in Seawater* (eds. C.S. Wong, et al.), Sicily, Italy, 30 March–30 April 1981, Plenum Press, New York, 1981, p. 145.
104. J.T. Creed, M.L. Magnuson, C.A. Brockhoff, I. Chamberlain, and M. Sivaganesan, *Journal of Analytical Atomic Spectroscopy*, 1996, 11, 505.
105. S. Jaya, T.P. Rao, and Q.D. Rao, *Talanta*, 1987, 34, 574.
106. C. Hua, D. Jagner, and L. Renman, *Analytica Chimica Acta*, 1987, 220, 263.
107. H. Huiliang, D. Jagner, and L. Renman, *Analytica Chimica Acta*, 1988, 207, 37.
108. N.S.C. Becker, V.M. McRae, and J.D, Smith, *Analytica Chimica Acta*, 1985, 173, 361.
109. A.I. Ryabinin, A.S. Romanov, S.L. Khatawov, A.A. Kist, and R. Khamidova, *Zhurnal Analiticheskoi Khimii*, 1972, 27, 94.
110. A.N. Yusov, Z.B. Ishsan, and A.K.H. Wood, *Journal of Radioanalytical and Nuclear Chemistry*, 1994, 179, 277.
111. M.G. Haywood and J.P. Riley, *Analytica Chimica Acta*, 1975, 85, 219.
112. T.J. Chow and E.P. Goldberg, *Geochimica Cosmochimica Acta*, 1961, 20, 192.
113. M.G. Epstein and A.T. Zander, *Analytical Chemistry*, 1979, 57, 915.
114. K.K. Roe and P.N. Froelich, *Analytical Chemistry*, 1984, 56, 2724.
115. F. Dehairs, M. De Bondt, W. Baeyens, P. Van Den Winkel, and M. Hoenig, *Analytica Chimica Acta*, 1987, 196, 33.
116. J.K.B. Bishop, *Analytical Chemistry*, 1990, 62, 553.
117. T. Okutani, Y. Tsuruta, and A. Sakuragawa, *Analytical Chemistry*, 1993, 65, 1273.
118. C. Matsubara, *Bunseki Kagaku*, 1974, 23, 878.
119. N. Yamahuchi, T. Nishida, and H. Nishida, *Bunseki Kagaku*, 1989, 38, 48.
120. H. Nishida, *Bunseki Kagaku*, 1990, 39, 805.
121. H. Tao, T. Imagawa, A. Miyazaki, and K. Bansho, *Bunseki Kagaku*, 1987, 36, 447.
122. B. Fleet, K.V. Liberty, and T.S. West, *Talanta*, 1970, 17, 203.
123. S. Terashima, *Bunseki Kagaku*, 1973, 22, 1317.
124. K. Matsuzaki, *Bunseki Kagaku*, 1975, 24, 442.
125. S. Terashima, *Bunseki Kagaku*, 1982, 31, 727.
126. T. Asami and F. Fukuzawa, *Soil Science and Plant Nutrition*, 1985, 31, 43.
127. S. Shimomura, H. Morita, and M. Kubota, *Bunseki Kagaku*, 1986, 25, 539.
128. R. Nakajima, *Bunseki Kagaku*, 1978, 27, 185.
129. D.C. Paschal and G.G. Bailey, *Atomic Spectroscopy*, 1986, 7, 1.
130. W.F. Schmidt and F. Diotl, *Fresenius Zeitschrift für Analytische Chemie*, 1987, 326, 40.
131. W.G. Ellis, V.R. Hodge, D.A. Darby, and C.L. Jones, *Atomic Spectroscopy*, 1988, 9, 181.
132. X.Q. Shan, Z. Yian, and Z.M. Ni, *Analytica Chimica Acta*, 1989, 217, 271.
133. Y. Shijo, M. Mitsuhashi, T. Shimizu, and S. Sakurai, *Analyst*, 1992, 117, 1929.
134. D.S. Lee, *Analytical Chemistry*, 1982, 54, 1182.
135. T.R. Gilbert and D.N. Hume, *Analytica Chimica Acta*, 1973, 65, 451.
136. T.M. Florence, *Journal of Electroanalytical Chemistry*, 1974, 49, 255.
137. D. Jaguer and K. Aren, *Analytica Chimica Acta*, 1978, 100, 375.
138. L.G. Danielsson, D. Jagner, M. Josefson, and S. Westerlund, *Analytica Chimica Acta*, 1981, 127, 147.
139. D. Jagner, M. Josefson, and S. Westerlund, *Analytica Chimica Acta*, 1981, 128, 155.
140. J.F. Portman and J.P. Riley, *Analytica Chimica Acta*, 1966, 34, 201.

141. G. Gillaim, A. Duychaerts, and A. Disteche, *Analytica Chimica Acta*, 1979, 106, 23.
142. F. Koroleff, *Acta Chimica Scandinavia*, 1947, 1, 503.
143. F. Koroleff (ed.), *Proceedings of the 12th Conference of the Baltic Oceanographers*, Leningrad, Russia, 1980.
144. G. Kormorsky-Lovric, *Analytica Chimica Acta*, 1988, 204, 161.
145. L.R. Uppstroem, *Analytica Chimica Acta*, 1968, 43, 475.
146. P. Hulthe, L.R. Uppstroem, and G. Oestling, *Analytica Chimica Acta*, 1970, 51, 31.
147. R.A. Nicolson, *Analytica Chimica Acta*, 1971, 56, 147.
148. M. Oshima, S. Motomizu, and K. Toei, *Analytical Chemistry*, 1974, 56, 948.
149. M. Marcantoncetos, G. Gamba, and D. Marnier, *Analytica Chimica Acta*, 1973, 67, 220.
150. T.M.T.C. Horta and A.J. Curtins, *Analytica Chimica Acta*, 1977, 95, 207.
151. T.M.T.C. Horta and A.J. Curtins, *Analytica Chimica Acta*, 1978, 96, 209.
152. S.P. Tsaikov, *Comptes Rendus de l'Academic Bulgare de Science*, 1982, 35, 61.
153. G.E. Bailey and Y.J. Farrah, *Analytica Chimica Acta*, 1978, 99, 283.
154. J. Gardner and J. Yates, *International Conference Management and Control of Heavy Metals in the Environment*, Water Research Centre, Stevenage Laboratory, Stevenage, CEP Consultants, Edinburgh, UK, 1979, pp. 427–430.
155. H. Han, X. Le, and Z. Ni, *Huanjing Huaxue*, 1986, 5, 34.
156. W.C. Campbell and J.H. Ottaway, *Analyst*, 1977, 102, 495.
157. R. Guevremont, *Analytical Chemistry*, 1980, 52, 1574.
158. R. Guevremont, R.E. Sturgeon, and S.S. Berman, *Analytica Chimica Acta*, 1980, 115, 163.
159. L.G. Danielson, B. Magnusson, and S. Westerlund, *Analytica Chimica Acta*, 1978, 98, 47.
160. L.G. Danielsson, B. Magnusson, and S. Westerlund, *Analytica Chimica Acta*, 1978, 98, 48.
161. R.E. Sturgeon, S.S. Berman, A. Desaulniers, and D.S. Russell, *Talanta*, 1980, 27, 85.
162. K.W. Bruland, R.P. Franks, G.A. Knauer, and J.H. Martin, *Analytica Chimica Acta*, 1979, 105, 233.
163. R.G. Smith and H.L. Windom, *Analytica Chimica Acta*, 1980, 113, 39.
164. L. Rasmussen, *Analytica Chimica Acta*, 1981, 125, 117.
165. R.E. Sturgeon, S.S. Berman, J.A.H. Desaulniers, A.P. Mykytink, J.W. McLaren, and D.M. Russell, *Analytical Chemistry*, 1980, 52, 1585.
166. K.R. Sperling, *Analytical Chemistry*, 1978, 292, 113.
167. M. Bengtsson, L.G. Danielsson, and B. Magnusson, *Analytical Letters*, 1979, 12, 1367.
168. K.R. Sperling, *Analytical Chemistry*, 1980, 301, 294.
169. H.M. Kingston, I.L. Barnes, T.J. Brady et al. *Analytical Chemistry*, 1978, 50, 2084.
170. W. Lund, and B.V. Larsen, *Analytica Chimica Acta*, 1974, 72, 57.
171. G.E. Batley, *Analytica Chimica Acta*, 1981, 124, 121.
172. G. Lundgren, L. Lundmark, and G. Johansson, *Analytical Chemistry*, 1974, 46, 1028.
173. K.R. Sperling, *Zeitschrift für Analytische Chemie*, 1977, 287, 23.
174. K.R. Sperling, *Zeitschrift für Analytische Chemie*, 1980, 301, 294.
175. E. Pruszkowska, C.R. Carnrick, and W. Slavin, *Analytical Chemistry*, 1983, 55, 182.
176. T. Nakahara and C.L. Chakrabarti, *Analytica Chimica Acta*, 1979, 104, 99.
177. S.W. Brewer, *Analytical Chemistry*, 1985, 57, 724.
178. M. Knowles, Methods for the determination of cadmium in seawater with Zeeman Background Correction, Varian Atomic Absorption No. AA 71, Agilent Technologies, 1987.
179. K.R. Lum and M. Callaghan, *Analytica Chimica Acta*, 1986, 187, 157.
180. Z. Zhang, J. Liu, R. Lin, Z Yang, and H. He, *Zhongshan Daxue Zuebao, Ziran Kexueban*, 1988, 3, 109.

181. E. Margul, C. Fontas, K. Van Meel, R. Van Lirinken, I. Queract, and M. Hidalgo, *Analytical Chemistry*, 2008, 80, 2357.
182. R. Guevremont, *Analytical Chemistry*, 1980, 52, 1574.
183. S.P. Kounaves and A. Zirino, *Analytica Chimica Acta*, 1979, 109, 327.
184. D.R. Turner, S.G. Robinson, and M. Whitfield, *Analytical Chemistry*, 1984, 56, 2387.
185. R.J. Stolzberg, *Analytica Chimica Acta*, 1977, 92, 193.
186. W. Yoshimura and Z. Uzawa, *Bunseki Kagaku*, 1987, 36, 367.
187. P. Frigieri, R. Trucco, I. Ciaccolini, and G. Pampurini, *Analyst (London)*, 1980, 105, 651.
188. V.M.T. Ganzerli, R. Stella, L. Maggi, and G. Ciceri, *Journal of Radioanalytical and Nuclear Chemistry*, 1987, 114, 105.
189. Z. Shen and P. Li, *Fenxi, Huaxue*, 1986, 14, 55.
190. D. Jagner, *Analytica Chimica Acta*, 1974, 68, 83.
191. R.W. Schmid and C.N. Reilly, *Analytical Chemistry*, 1957, 29, 264.
192. A.G. Ringbom, G Pensar, and E. Wanninen, *Analytica Chimica Acta*, 1958, 19, 525.
193. F.S. Sadek, R.W. Schmid, and C.N. Reilley, *Talanta*, 1959, 2, 38.
194. E. Wanninen, *Talanta*, 1961, 8, 355.
195. Y. Date and K. Toei, *Bulletin of the Chemical Society of Japan*, 1963, 36, 518.
196. F. Culkin and R.A. Cox, *Deep-Sea Research*, 1966, 13, 789.
197. S. Tsunogai, M. Nishimura, and S. Nakaya, *Talanta*, 1968, 15, 835.
198. G. Schwarzenbach and H. Flashka, *Complexometric Titrations*, 2nd English edn., Methuen, London, UK. 1969.
199. Y. Horibe, K. Endo, and H. Tsuboto, *Earth Planet Science Letters*, 1974, 23, 136.
200. J. Lebel and A Poisson, *Marine Chemistry*, 1976, 4, 321.
201. D. Krumgalz and R. Holzer, *Limnology and Oceanography Society Japan*, 1980, 25, 367.
202. S. Kanamori and H. Ikegami, *Journal of the Oceanographic Society Japan*, 1980, 36, 177.
203. E.J. Olson and C.T.A. Chen, *Limnology and Oceanography*, 1982, 27, 375.
204. B. Van't Riet and J.E. Wynn, *Analytical Chemistry*, 1969, 41, 158.
205. U. Ezat, *Analusis*, 1988, 16, 168.
206. W.E. Blake, M.W.R. Brvant, and A. Waters, *Analyst*, 1969, 94, 49.
207. T.J. Chow and T.S. Thompson, *Analytical Chemistry*, 1955, 27, 910.
208. M. Whitfield, J.V. Leyendekkers, and J.D. Kerr, *Analytical Chemistry*, 1969, 45, 399.
209. J.B. Brenner, H. Eldad, S. Erlich, and N. Dalman, *Analytica Chimica Acta*, 1984, 166, 51.
210. M.R. Smith and H.B. Cockran, *Atomic Spectroscopy*, 1981, 2, 97.
211. J. Pybus, *Clinica Chimica Acta*, 1969, 23, 309.
212. T. Shigematsu, Y. Nishikawa, K. Hiraki, S. Goda, and Y. Tsujinaaatu, *Japan Analyst*, 1971, 20, 575.
213. E. Nakayama, T. Kuwamoto, H. Tokoro, and Fujinaca, *Analytica Chimica Acta*, 1981, 131, 247.
214. R.E. Cranston and J.W. Murray, *Analytica Chimica Acta*, 1978, 99, 275.
215. G.J. Dejong and U.A.Th. Brinkman, *Analytica Chimica Acta*, 1978, 98, 243.
216. T.R. Gilbert and A.M. Clay, *Analytica Chimica Acta*, 1973, 67, 289.
217. S.T. Crosmun and T.R. Mueller, *Analytica Chimica Acta*, 1975, 75, 199.
218. G.E. Batley and J.P. Matousek, *Analytical Chemistry*, 1980, 52, 1570.
219. J.F. Pankow and G.E. Junauer, *Analytica Chimica Acta*, 1974, 69, 97.
220. A. Miyazaki and R.M. Barnes, *Analytical Chemistry*, 1981, 53, 364.
221. E.B. Sandell, *Colorimetric Determination of Traces of Metals*, Interscience, New York, 1959.
222. H. Marchart, *Analytica Chimica Acta*, 1984, 30, 11.

223. G.J. Willems and C.J. de Ranter, *Analytica Chimica Acta*, 1974, 68, 111.
224. G.V. Myasoedova and S.G. Savvin, *Zhurnal Analiticheskoi Khimii*, 1982, 37, 499.
225. D.E. Leyden and W. Wegscheider, *Analytical Chemistry*, 1981, 63, 1059A.
226. S.M. Willie, R.E. Sturgeon, and S.S. Berman, *Analytical Chemistry*, 1983, 55, 981.
227. C.A. Chang, H.H. Patterson, L.M. Mayer, and D.E. Bause, *Analytical Chemistry*, 1980, 52, 1264.
228. L.I. Dubovenko, Q.A. Zaporozhets, and I.I. Pyatnitskii, *Khimiya i Tekhnologiya Vody*, 1986, 8, 50.
229. G.J. De Jong and U.A.T. Brinkman, *Analytica Chimica Acta*, 1978, 98, 243.
230. R.E. Cranston and J.W. Murray, *Limnology and Oceanography*, 1980, 25, 1104.
231. E. Nayakama, T. Kuwamoto, H. Tokoro, and T. Fiyinaka, *Analytica Chimica Acta*, 1981, 130, 289.
232. D. Grimaud and G. Michard, *Marine Chemistry*, 1974, 2, 229.
233. E. Nayakama, T. Kuwamoto, H. Tokoro, and T. Fiyinaka, *Analytica Chimica Acta*, 1981, 130, 401.
234. E. Nakayama, T. Kuwamoto, H. Tokoro, and T. Fiyinaka, *Analytica Chimica Acta*, 1981, 131, 247.
235. E. Takayama, T. Kuwamoto, H. Tokoro, and T. Fujinaka, *Analytica Chimica Acta*, 1981, 130, 289.
236. M. Ishibashi and T. Shigematsu, *Bulletin of the Institute of Chemical Research, Kyoto University*, 1950, 23, 59.
237. L. Chuecas and J.P. Riley, *Analytica Chimica Acta*, 1966, 35, 240.
238. R. Fukai, *Nature*, 1967, 213, 901.
239. R. Fukai and D. Vas, *Journal of the Oceanographic Society Japan*, 1967, 23, 298.
240. T. Kuwamoto and S. Murai, Preliminary Report of the Hakuho-maru Cruise KH–68–4, Ocean Research Institute University, Tokyo, Japan, 1970, p. 72.
241. T. Yawamoto, S. Kadowski, and J.H. Carpenter, *Geochemical Journal*, 1974, 8, 123.
242. R.L. Mullins, *Analytica Chimica Acta*, 1984, 165, 97.
243. J.H. Moffett, Varian Atomic Absorption Bo AA-69, Measurement of Varian Techtron Pty Ltd, Mulgrave, Victoria, Australia, Chromium in Environmental Waters by Zeeman Corrected Graphite Tube Atomisation, January 1987.
244. R.K. Mugo and K.J. Orians, *Analytica Chimica Acta*, 1992, 271, 1.
245. J. Posta, A. Alimonti, E. Petrucci, and S. Caroli, *Analytica Chimica Acta*, 1996, 325, 185.
246. W.M. Shi, H. Bednas, and S.S. Berman, *Analytical Chemistry*, 1983, 55, 473.
247. K.G. Heumann, *Toxicological Environmental Analytical Review*, 1980, 3, 11.
248. B.N. Colby, A.E. Rosecrance, and M.E. Colby, *Analytical Chemistry*, 1981, 53, 107.
249. A.P. Mykytiuk, D.S. Russell, and R.E. Sturgeon, *Analytical Chemistry*, 1980, 52, 1281.
250. R.E. Sturgeon, S.S. Berman, S.N. Willie, and J.A.H. Desaulniers, *Analytical Chemistry*, 1981, 53, 2337.
251. M. Boussemart and E.M.G. Van den Berg, *Analyst*, 1994, 119, 1349.
252. I. Ahern, J.M. Eckert, N.C. Payne, and K.L. Williams, *Analytica Chimica Acta*, 1985, 175, 147.
253. E. Kentner and H. Zeitlin, *Analytica Chimica Acta*, 1970, 49, 587.
254. T.A. Kouimtzis, C. Apostolopoulou, and I. Staphiliakis, *Analytica Chimica Acta*, 1980, 113, 185.
255. G. Vasilikiotis, Th. Kouimtzis, C. Apostolopoulou, and A. Voulgaropoulos, *Analytica Chimica Acta*, 1974, 70, 319.
256. C.P. Spencer, *Journal of General Microbiology*, 1957, 16, 282.
257. R. Johnston, *Journal of the Marine Biological Association UK*, 1962, 43, 427.
258. R. Johnson, *Journal of the Marine Biological Association UK*, 1964, 44, 87.
259. R.T. Barber and J.H. Ryther, *Journal of Exploratory Marine Biology and Ecology*, 1969, 3, 191.

260. S. Motomizu, *Analytica Chimica Acta*, 1973, 64, 217.
261. A. Yuzefovsky, R. Honardo, M. Wang, and R.G. Michel, *Journal of Analytical Atomic Spectroscopy*, 1994, 9, 1195.
262. N. Sakamoto – Arnold, Private Communication.
263. A. Malaholff, Y. Kolotrykiris, and L.K. Shipgun, *Analyst*, 1996, 121, 1037.
264. M. Carole, V.S. Sakamoto, and S. Kenneth, *Analytical Chemistry*, 1987, 59, 1789.
265. M. Vega and C.M.G. Van den Berg, *Analytical Chemistry*, 1997, 69, 874.
266. B.R. Harvey and J.W.R. Dutton, *Analytica Chimica Acta*, 1973, 67, 377.
267. G.K. Pagenkopf, R.C. Russo, and R.V. Thurston, *Journal of the Fisheries Research Board Canada*, 1974, 31, 462.
268. R.N. Sylva, *Water Research*, 1976,10, 789.
269. W.G. Sunda and A.K. Hanson, *Limnology and Oceanography*, 1987, 32, 537.
270. W.G. Sunda and J.M. Lewis, *Limnology and Oceanography*, 1978, 23, 870.
271. W.G. Sunda and P.A. Gillespie, *Journal of Marine Research*, 1979, 37, 761.
272. D.R. Turner, M. Whitfield, and D. Dickson, *Geochimica Cosmochimica Acta*, 1981, 45, 855.
273. A. Zirino and S. Yamamoto, *Limnology and Oceanography*, 1972, 17, 661.
274. A. Zirino and S. Yamamoto, *Limnology and Oceanography*, 1972, 17, 663.
275. W. Stumm and J.J. Morgan, *An Introduction Emphasising Chemical Equilibria in Natural Waters*, 2nd edn., Wiley Interscience, New York, 1981.
276. T.H. Sibley and J.H. Morgan, Equilibrium speciation of trace metals in freshwater; Seawater mixtures, In *Proceedings of International Conference on Heavy Metals in the Environment* (ed. H.C. Hutchinson), University of Toronto, Toronto, Ontario, Canada, 1975, pp. 310–338.
277. E.W. Davey, M.J. Morgan, and S.J. Ericksen, *Limnology and Oceanography*, 1973, 18, 993.
278. M.S. Shuman and G.F. Woodward, *Analytical Chemistry*, 1973, 45, 2032.
279. R. Chau and K. Lum-Shue-Chan, *Research*, 1974, 8, 383.
280. P.C. Campbell, M. Blisson, R. Gagne, and A. Tessior, *Analytical Chemistry*, 1977, 49, 2358.
281. M.J. Stiff, *Water Research*, 1971, 5, 585.
282. G.E. Batley and T.M. Florence, *Analytical Letters (London)*, 1982, 9, 379.
283. M. Plasvic, D. Krznacic, and M. Branica, *Marine Chemistry*, 1982, 11, 17.
284. C.M.G. Van den Berg, *Marine Chemistry*, 1982, 11, 323.
285. T.M. Florence and G.E. Batley, *Talanta*, 1976, 23, 179.
286. P. Figura and B. McDuffie, *Analytical Chemistry*, 1979, 51, 120.
287. I. Ruzic, *Analytica Chimica Acta*, 1982, 140, 99.
288. T.D. Waite and F.M.M. Morel, *Analytical Chemistry*, 1983, 55, 1268.
289. J. Abraham, M. Winpe, and D.E. Ryan, *Analytica Chimica Acta*, 1974, 69, 35.
290. H.H. Zeng, R.B. Thompson, B.P. Miliwal, G.R. Fones, J.W. Moffatt, and C.A. Flerke, *Analytical Chemistry*, 2003, 75, 6807.
291. C. Fairless and A.J. Bard, *Analytical Chemistry*, 1973, 45, 2289.
292. R.A.A. Muzzaralli and R. Rocchetti, *Analytica Chimica Acta*, 1974, 69, 35.
293. A. Zlatkis, W. Bruening, and E. Bayer, *Analytical Chemistry*, 1970, 42, 1201.
294. J.Y. Cabon and A. Le Bihan, *Analytica Chimica Acta*, 1987, 198, 87.
295. J.Y. Cabon, *Analytica Chimica Acta*, 1987, 198, 103.
296. R. Tasinski, I. Trachtenberg, and D. Andrychuk, *Analytical Chemistry*, 1974, 46, 364.
297. S.L. Belli and A. Zirino, *Analytica Chimica Acta*, 1993, 65, 2583.
298. R. De Marco, *Analytical Chemistry*, 1994, 66, 3202.
299. J.C. Westall, F.M.M. Morel, and D.X. Hume, *Analytical Chemistry*, 1979, 51, 2016.
300. A. Leivenstam, A. Hulanicki, and E. Ghali, In *Contemporary Electron Chemistry*, (eds. I. Ivaska et al.), Plenum Press, New York, 1990.

301. M.S. Shuman and L.C. Michael, *Environmental Science and Technology*, 1978, 12, 1069.
302. G. Scorano, E. Marelli, A. Seritti, and A. Zirino, *Analytical Chemistry*, 1990, 62, 943.
303. F. Quentel, C. Ellouet, and C. Made, *Electroanalysis*, 1994, 6, 683.
304. J. Wang, N. Foster, S. Armalis, D. Larsen, A. Sirino, and K. Olsen, *Analytica Chimica Acta*, 1995, 310, 223.
305. R. Garcia-Monco Carrá, A. Sanchez-Misiego, and A. Zirino, *Analytical Chemistry*, 1995, 67, 4484.
306. G. Capodaglio, G. Tascan, and G. Scarppon Plescon, *International Journal of Environmental Analytical Chemistry*, 1994, 55, 129.
307. H.V. Weiss, P.R. Kenis, J. Korkisch, and I. Steffan, *Analytica Chimica Acta*, 1979, 104, 337.
308. D.R. Turner, S.G. Robinson, and M. Whitfield, *Analytical Chemistry*, 1984, 56, 2387.
309. K.T. Marvin, R.R. Proctor, and R.A. Neal, *Limnology and Oceanography*, 1970, 15, 320.
310. Y.P. Virmani and E.J. Zeller, *Analytical Chemistry*, 1974, 46, 324.
311. W.G. Sunda and A.K. Hanson, *Limnology and Oceanography*, 1987, 325, 37.
312. D. Buerge-Weirich and B. Suztzberger, *Environmental Science and Technology*, 2004, 38, 1843.
313. D. Li, J. Haskins, K. Sukola, and G.T. Wallace, *Environmental Science and Technology*, 2011, 45, 5660.
314. M.S. Shuman and G.P. Woodward, *Analytical Chemistry*, 1973, 45, 2032.
315. R. Chan, K. Lum, and S. Chau, *Water Research*, 1974, 8, 383.
316. T.M. Florence and G.B. Bailey, *Journal of Electroanalytical Chemistry*, 1977, 75, 791.
317. A. Zirino and S. Yamamoto, *Limnology and Oceanography*, 1972, 17, 661.
318. W. Stumm and H. Bilinksi, In *Sixth International Conference on Advances in Water Pollution Research*, Jerusalem, Israel, 8–23 June 1972, Pergamon, Oxford, UK. 1973, p. 39.
319. A. Zirino and M.L. Healy, *Limnology and Oceanography*, 1970, 15, 956.
320. W.L. Bradford, *Limnology and Oceanography*, 1978, 18, 757.
321. H. Bilinski, R. Huston, and W. Stumm, *Analytica Chimica Acta*, 1976, 84, 157.
322. A.M. Wood, D.W. Evans, and J.J. Alberts, *Marine Chemistry*, 1983, 13, 305.
323. N.G. Zorkin, E.V. Grill, and A.G. Lewis, *Analytica Chimica Acta*, 1986, 183, 163.
324. A. Zirino and S.P. Kounaves, *Analytica Chimica Acta*, 1980, 13, 29.
325. N. Shefrin and F.E. Williams, *Analytical Proceedings*, 1982, 79, 487.
326. M.S. Shuman and L.C. Michael, *Proceedings of the International Conference on Heavy Metals in the Environment*, Vol. 1, John Wiley, New York, 1975, p. 227.
327. M. Odier and V. Plichon, *Analytica Chimica Acta*, 1971, 55, 209.
328. C.M.G. Van Der Berg and Z.Q. Huang, *Analytica Chimica Acta*, 1984, 164, 209.
329. C.M.G. Van der Berg, *Analytica Chimica Acta*, 1984, 164, 195.
330. M.V. Andreae and P.N. Froelich, *Analytical Chemistry*, 1981, 53, 287.
331. K.K. Falkner and J.M. Edmond, *Analytical Chemistry*, 1990, 62, 1477.
332. A.T. Pilipenko and U.R. Pavlova, *Zhurnal Analiticheskoi Khimii*, 1972, 27, 1253.
333. F. Ogata and N. Kawasaki, *Journal of Chemical engineering Data*, 2014, 59, 412.
334. A. Matthews and J.P. Riley, *Analytica Chimica Acta*, 1970, 51, 287.
335. J.H. Martin and S.E. Fitzwater, *Nature (London)*, 1988, 331, 342.
336. J.H. Martin, R.M. Gordon, S.E. Fitzwater, and W.W. Broenkow, *Deep Sea Research*, 1989, 36, 649.
337. J.H. Martin, R.M. Gordon, and S.E. Fitzwater, *Nature (London)*, 1990, 345, 156.
338. J.H. Martin, S.E. Fitzwater, and R.M. Gordon, *Global Biochemical Cycles*, 1990, 4, 5.
339. J.H. Martin, *Nature, (London)*, 1991, 353, 123.
340. P.R. Betzer, K.L. Carder, R.A. Duce et al., *Nature (London)*, 1988, 336, 568.
341. R.W. Young, K. Karter, P.R. Betzer et al. *Global Biochemical Cycles*, 1991, 5, 119.

342. W.S. Broecker, *Global Biochemical Cycles*, 1990, 4, 13.
343. R.C. Dugdale and F.P. Wilkerson, *Global Biochemical Cycles*, 1990, 4, 13.
344. G. Zhuang, Z. Yi, R.A. Duce, and P.R. Brown, *Nature (London)*, 1992, 355, 537.
345. V.L. Rigin and A.I. Blokhin, *Zhurnal Analiticheskoi Khimii*, 1977, 32, 312.
346. A.T. Pilipenko, V.A. Baravskii, and I.E. Kalinichenko, *Zhurnal Analiticheskoi Khimii*, 1978, 33, 1880.
347. A. Obata, H. Karatano, and E. Nakayama, *Analytical Chemistry*, 1993, 65, 1524.
348. M.M.A. Shirdar and K. Odzeki, *Analyst*, 1986, 111, 55.
349. S. Blair and P. Tregeur, *Analytica Chimica Acta*, 1995, 308, 425.
350. N.A. Eirod, K.S. Johnson, and K.H. Coale, *Analytical Chemistry*, 1991, 63, 893.
351. D.W. O'Sullivan, A.K. Hanson, and D.R. Lester, *Marine Chemistry*, 1995, 49, 65.
352. K. Hiiro, T. Tanaka, and T. Sawada, *Japan Analyst*, 1972, 21, 635.
353. R.M. Moore, In *Proceedings of a NATO Advanced Research Institute on Trace Metals in Seawater* (eds. C.S. Wong et al.), Sicily, Italy, 30 March–30 April 1981, Plenum Press, New York, 1981, p. 301.
354. L. Lacan, A Roche, M. Lebatat, C. Jeandil, F. Poit Rasson, G. Sarthon, C. Pradoux, J. Chemeleff, and R. Freydier, *Analytical Chemistry*, 2010, 82, 7103.
355. I.K.S. Cathy, C.C. Lanz, D. Timothy, J. Jickets, I. David, J. Richardson, and D.A. Russell, *Analytical Chemistry*, 2006, 78, 5040.
356. G.M. Sherma and H.R. Du Bois, *Analytical Chemistry*, 1978, 50, 516.
357. L.M. Laglera and C.M. Van der Berg, *Environmental Science and Technology*, 2007, 41, 2296.
358. B.E. Johnson, R.S. Addleman, M. Douglas, R.D. Rutledge, W. Chouyyek, J.D. Davidson, G.E. Fryxell, and J.M. Schumantes, *Environmental Science and Technology*, 2012, 46, 11251.
359. M.A. Bolshov, A.V. Zybin, V.G. Koloshnikov, and M.V. Vesnetsov, *Spectrochimica Acta*, 1981, 36B, 345.
360. V. Cheam, J. Lechner, I. Sekerka, and R. Desrosiers, *Journal of Analytical Atomic Spectroscopy*, 1994, 9, 315.
361. K. Ohta and M. Suzuki, *Fresenius Zeitschrift für Analytische Chemie*, 1979, 298, 140.
362. W. Frech and A. Cedergreu, *Analytica Chimica Acta*, 1977, 88, 57.
363. M.C. Halliday, C. Houghton, and J.M. Ottaway, *Analytica Chimica Acta*, 1986, 119, 67.
364. B.K. Schaule and C.C. Patterson, *Earth and Planetary Science Letters*, 1981, 54, 97.
365. Y. Hirao, K. Fukomoto, H. Sugisaki, and K. Kimura, *Analytical Chemistry*, 1979, 51, 651.
366. G. Torsi, E. Desimoni, F. Palmisano, and L. Sabbantini, *Analytica Chimica Acta*, 1981, 124, 143.
367. J.W. McLaren and R.C. Wheeler, *Analyst*, 1977, 102, 542.
368. M.J. Adams, G.F. Kirkbright, and P. Rienvatana, *Atomic Absorption Newsletter*, 1977, 14, 105.
369. B.R. Culver and T. Surles, *Analytical Chemistry*, 1975, 47, 920.
370. M.W. Pritchard and R.D. Reeves, *Analytica Chimica Acta*, 1976, 82, 103.
371. S. Yasuda and H. Kakihama, *Analytica Chimica Acta*, 1976, 84, 291.
372. G. Tessari and G. Torsi, *Analytical Chemistry*, 1975, 47, 842.
373. R.E. Sturgeon, C.L. Chakrabarti, and C.N. Langford, *Analytical Chemistry*, 1976, 48, 1792.
374. F.J. Fernandez and D.C. Manning, *Atomic Absorption Newsletter*, 1971, 10, 3.
375. R.D. Ediger, G.E. Peterson, and J.D. Kerner, *Atomic Absorption Newsletter*, 1974, 13, 61.
376. R.D. Ediger, *Atomic Absorption Newsletter*, 1975, 14, 127.
377. J.G.T. Regan and J. Warren, *Analyst (London)*, 1976, 101, 220.
378. D.C. Manning and W. Slavin, *Analytical Chemistry*, 1978, 50, 1234.

379. H. Branderberger and H. Bader, *Atomic Absorption Newsletter*, 1967, 6, 101.
380. W. Lund and B.V. Larsen, *Analytica Chimica Acta*, 1974, 70, 299.
381. W. Lund and B.V. Larsen, *Analytica Chimica Acta*, 1974, 72, 57.
382. W. Lund, B.V. Larsen, and N. Gundersen, *Analytica Chimica Acta*, 1976, 81, 319.
383. W. Lund, Y. Thomassen, and P. Doyle, *Analytica Chimica Acta*, 1977, 93, 53.
384. L. Fairless and A.J. bard, *Analytical Letters*, 1972, 5, 433.
385. G.B. Batley and J.P. Matousek, *Analytical Chemistry*, 1977, 49, 2031.
386. F.O. Jensen, J. Dolezal, and F.J. Langmyhr, *Analytica Chimica Acta*, 1975, 72, 245.
387. M.P. Newton, J.V. Chauvin, and D.G. Davis, *Analytical Letters (London)*, 1973, 6, 89.
388. M.P. Newton and D.G. Davis, *Analytical Chemistry*, 1975, 47, 2003.
389. J.B. Dawson, D.J. Ellis, T.F. Hartley, M.E.A. Evans, and K.W. Metcalf, *Analyst*, 1974, 99, 602.
390. G. Torsi, *Analytica Chimica, Rome*, 1977, 67, 557.
391. R.E. Sturgeon, S.N. Willie, and S.S. Berman, *Analytical Chemistry*, 1989, 61, 1867.
392. R.G. Clem, MPI application, University of California Berkley Laboratory, Notes 8, 1 1973.
393. S.A. Acebal, DeLuca, and A. Rebello, *Analytica Chimica Acta*, 1983, 148, 71.
394. D.R. Turner, S.G. Robinson, and W. Whitfield, *Analytical Chemistry*, 1984, 56, 2387.
395. F. Quentel, C. Madoc, and J. Courtot-Coupez, *Water Research*, 1986, 20, 325.
396. B. Svensmark, *Analytica Chimica Acta*, 1987, 197, 239.
397. Q. Wu and G.E. Batley, *Analytica Chimica Acta*, 1995, 309, 95.
398. R. Garcia-Monco Carrá, A. Sanchez-Misiego, and A. Zirino, *Analytical Chemistry*, 1995, 67, 4484.
399. A. Flegal and V.J. Stukas, *Marine Chemistry*, 1987, 22, 163.
400. R.H. Byrne, *Nature (London)*, 1981, 290, 487.
401. D.M. Scaule and C.C. Patterson, In *Proceedings of a NATO Advanced Research Institute on Trace Metals in Seawater* (eds. C.S. Wong et al.), Sicily, Italy, 30 March–30 April 1981, Plenum Press, New York, 1981, p. 488.
402. R.M. Thomas, Private communication.
403. N. Benzwi, *Israel Journal of Chemistry*, 1972, 10, 967.
404. T.J. Chow and E.D. Goldberg, *Journal of Marine Research*, 1979, 20, 163.
405. M. Wiernik and S. Amiel, *Journal of Radioanalytical Chemistry*, 1970, 5, 123.
406. M. Rona and G. Schumuckler, *Talanta*, 1973, 20, 237.
407. D.K. Das, M. Muzumdan, and S.C. Schome, *Analytica Chimica Acta*, 1972, 60, 439.
408. W. Stumm and J.J. Morgan, *Aquatic Chemistry*, Wiley Interscience, New York, 1970.
409. J.W. Murray and P.G. Brewer, *Marine Manganese Deposits* (ed. G.P. Glasby), Elsevier, Amsterdam, the Netherlands, 1977, pp. 291–325.
410. J.J. Morgan, *Principles and Applications of Water Chemistry* (eds. S.D. Faust and J.V. Hunter), Wiley, New York, 1967, pp. 561–624.
411. M.A. Kessick and J. J. Morgan, *Environmental Science and Technology*, 1975, 9, 157.
412. W. Stumm and R. Gioyanoli, *Chimica*, 1976, 30, 423.
413. J. Olaffson, *Science of the Total Environment*, 1986, 49, 101.
414. P.G. Brewer and D.W. Spencer, *Limnology and Oceanography*, 1971, 16, 107.
415. V.L. Biddle and E.L. Wehry, *Analytical Chemistry*, 1978, 50, 867.
416. J.A. Resing and M.J. Matti, *Analytical Chemistry*, 1997, 64, 2682.
417. J.D. Burton, In *Proceedings of a NATO Advance Research Institute on Trace Metals in Seawater* (eds. C.S. Wong et al.), Sicily, Italy, 30 March–30 April 1981, Plenum Press, New York, 1981, p. 419.
418. R.E. Sturgeon, S.S. Berman, J.A.H. Desaulniers, and D.G. Russel, *Analytical Chemistry*, 1979, 51, 2364.
419. D.A. Segar and J.G. Gonzales, *Analytica Chimica Acta*, 1972, 58, 7.
420. G.P. Klinkhammer, *Analytical Chemistry*, 1980, 52, 117.

421. J.M. McArthur, *Analytica Chimica Acta*, 1977, 93, 77.
422. D.J. Hydes, *Analytical Chemistry*, 1980, 52, 959.
423. D.A. Segar, *Advances in Chemistry Services*, 1975, 147, 56.
424. H.M. Kingston, I.L. Barnes, T.J. Brady, T.C. Rains, and M.A. Champ, *Analytical Chemistry*, 1978, 50, 2064.
425. D.A. Segar and A.Y. Cantillo, *Analytical Chemistry*, 1980, 52, 1766.
426. J.R. Montgomery and G.N. Peterson, *Analytica Chimica Acta*, 1980, 117, 397.
427. B.V. L'vov, *Spectrochimica Acta Part B*, 1978, 33B, 153.
428. D.A. Segar and A.Y. Cantillo, *Spectrochimica Symposium American Society of Limnology and Oceanography*, 1976, 2, 171.
429. W. Slavin and D.C. Manning, *Analytical Chemistry*, 1979, 51, 261.
430. D.C. Manning and W. Slaving, *Analytica Chimica Acta*, 1980, 118, 301.
431. W. Slavin and D.C. Manning, *Spectrochimica Acta Part B*, 1980, 35B, 701.
432. P.J. Statham, *Analytica Chimica Acta*, 1985, 169, 149.
433. International Council for Exploration of the Sea, Marine Environment Quality Committee Report (CM1983/E24), Charlottenhind, Denmark, 1983.
434. G.P. Klinkhammer, *Analytical Chemistry*, 1980, 42, 117.
435. M.I. Bender, G.P. Klinkhammer, and D.W. Spencer, *Deep Sea Research*, 1977, 24, 799.
436. G.P. Klinkhammer and M.I. Bender, *Earth Planet Science Letters*, 1980, 46, 361.
437. C.R. Lan and Z.B. Alfassi, *Analyst*, 1994, 119, 1033.
438. G.R. Carnrick, W. Slavin, and D.C. Manning, *Analytical Chemistry*, 1981, 53, 1866.
439. S.R. Carnrick, W. Slavin, and D.C. Manning, *Analytical Chemistry*, 1981, 53, 1866.
440. R.J. O'Halloran, *Analytica Chimica Acta*, 1982, 140, 51.
441. S. Knox and D.R. Turner, *Estuarine and Marine Science*, 1980, 10, 317.
442. Department of the Environment and National Water Council, Standing Committee of Analysis, Manganese in Raw and Potable Waters by Spectrophotometry using Formaldoxime, 1977, Method for the Examination of Waters and Associated Materials HMSO, London, UK, 1978.
443. A. Nelson, *Analytica Chimica Acta*, 1985, 169, 287.
444. W. Frech and A. Ledergreu, *Analytica Chimica Acta*, 1977, 88, 57.
445. W.R. Hatch and W.L. Ott, *Analytical Chemistry*, 1968, 40, 2085.
446. Me. Hinckle and R.E. Learned, *US Geological Survey Professional Papers*, 1969, 605D, 251.
447. M.J. Fishman, *Analytical Chemistry*, 1970, 42, 1462.
448. G. Lindsteat, *Analyst*, 1970, 95, 264.
449. Y.K. Chau and H. Saitoh, *Environmental Science and Technology*, 1970, 4, 839.
450. S.H. Omang, *Analytica Chimica Acta*, 1971, 53, 415.
451. S.H. Omang and P.E. Paus, *Analytica Chimica Acta*, 1971, 56, 393.
452. J.H. Hwang, P.A. Ullucci, and A.L. Malenfant, *Canadian Spectroscopy*, 1971, 16, 1.
453. S.R. Aston and J.P. Riley, *Analytica Chimica Acta*, 1972, 59, 349.
454. V.I. Muscat, T.J. Vickers, and A. Andren, *Analytical Chemistry*, 1972, 44, 218.
455. G. Topping and J.M. Pirie, *Analytica Chimica Acta*, 1972, 62, 200.
456. R.A. Carr, J.B. Hoover, and P.E. Wilkniss, *Deep Sea Research*, 1972, 19, 747.
457. S.H. Omang, *Analytica Chimica Acta*, 1973, 63, 247.
458. D. Gardner and J.P. Riley, *Nature*, 1973, 241, 526.
459. J. Olafsson, *Analytica Chimica Acta*, 1974, 68, 207.
460. D. Voyce and H. Zeitlin, *Analytica Chimica Acta*, 1974, 69, 27.
461. W.F. Fitzgerald, W.B. Lyons, and C.D. Hunt, *Analytical Chemistry*, 1974, 46, 1882.
462. R.V. Coyne and J.A. Collins, *Analytical Chemistry*, 1972, 44, 1093.
463. R.A. Carr and P.E. Wilkniss, *Environmental Science and Technology*, 1973, 7, 62.
464. C. Feldman, *Analytical Chemistry*, 1974, 46, 99.

465. W.F. Fitzgerald, Distribution of mercury in natural waters, In *The Biochemistry of Mercury in the Environment* (ed J.O. Nriagu), Elsevier, Amsterdam, the Netherlands, 1979.

466. J. Olafsson, In *Proceedings of a NATO Advanced Research Institute on Trace Metals in Seawater* (eds. C.S. Wong et al.), Sicily, Italy, 30 March–30 April 1981, Plenum Press, New York, 1981, p. 476.

467. J. Olafsson, *Marine Chemistry*, 1982, 11, 129.

468. S. Blake, Water Research Centre, Medmenham UK, Technical Report TR229, Method for the Determination of Low Concentrations of Mercury in Fresh and Saline Waters, 1985, 28 pp (38994).

469. G.A. Gill and W.F. Fitzgerald, *Marine Chemistry*, 1987, 20, 277.

470. M. Filippelli, *Analyst*, 1984, 109, 515.

471. T. Hadeishi and R.D. McLaughlin, *Science*, 1971, 174, 404.

472. R.J. Watling, *Analytica Chimica Acta*, 1975, 75, 281.

473. M.J. Bloxam, A. Gachanijhu, S.J. Hill, and P.J. Worsfold, *Journal of Analytical Atomic Spectroscopy*, 1996, 11, 145.

474. E. Debrak and E.R. Denoyer, *Journal of Analytical Atomic Spectroscopy*, 1996, 11, 127.

475. I. Turyan and D. Mandler, *Nature (London)*, 1993, 362, 703.

476. H.Z. Wrembel, *Talanta*, 1983, 30, 481.

477. H.Z. Wrembel and W.A. Pajak, *Chemical Analysis, Warsaw*, 1986, 31, 409.

478. H.W. Nurnberg, P. Valenta, and M. Brancia, *Analytica Chimica Acta*, 1980, 115, 25.

479. S. Jaya, T.P. Rao, and P. Raog, *Analyst*, 1985, 110, 1361.

480. Y.L. Hiyai and Y. Murao, *Japan Analyst*, 1972, 21, 608.

481. H.V. Weiss, and T.E. Crozier, *Analytica Chimica Acta*, 1972, 58, 231.

482. M. Fujita and K. Iwashima, *Environmental Science and Technology*, 1981, 15, 929.

483. H. Agemian and J.A. Da Silva, *Analytica Chimica Acta*, 1979, 104, 285.

484. I.M. Davies and J.M. Pirie, *Marine Pollution Bulletin*, 1978, 9, 128.

485. G. Topping, J.M. Graham, and R.J. shepherd, An examination of the heavy metal levels, In *Muscle, Kidney and Liver of Saithe in Relation to Year, Class, Area of Sampling and Season*, ICE.5.CM, E37, 1975.

486. G. Topping and J.M. Pirie, *Analytica Chimica Acta*, 1972, 62, 200.

487. A. Quareshi, N.J. O'Driscoll, M. McLeod, Y.M. Neuhold, and K. Hingerbeuhler, *Environmental Science of Technology*, 2010, 44, 644.

488. H.M.A. Shriadah, M. Katoaka, and K. Ohzeki, *Analyst, (London)*, 1985, 110, 125.

489. T. Kiriyama and R. Kuroda, *Talanta*, 1984, 31, 472.

490. Y.S. Kim and H. Zeitlin, *Separation Science*, 1971, 6, 505.

491. R. Kuroda and T. Tarui, *Fresenius Zeitschrift für Analytische Chemie*, 1974, 269, 22.

492. H. Zeitlin, *Analytica Chimica Acta*, 1970, 51, 516.

493. J.P. Riley and D. Taylor, *Analytica Chimica Acta*, 1968, 40, 479.

494. K. Kawabuchi and R. Kuroda, *Analytica Chimica Acta*, 1969, 46, 23.

495. J. Korkisch, L. Godl, and H. Gross, *Talanta*, 1975, 22, 669.

496. I. Nucatsuka, A. Nishimura, and K. Ohzeki, *Analytica Chimica Acta*, 1995, 304, 243.

497. Y.K. Chau and K. Lum-Shui Chan, *Analytica Chimica Acta*, 1969, 48, 205.

498. H.A. Van der Sloot, G.D. Wats, and H.A. Das, *Progress in Water Technology*, 1977, 8, 193.

499. T. Nakahara and C.L. Chakrabarti, *Analytica Chimica Acta*, 1979, 104, 99. Private communication.

500. R.A.A. Muzzarelli and R. Rochetti, *Analytica Chimica Acta*, 1973, 64, 371.

501. R.J. Emerick, *Atomic Spectroscopy*, 1987, 8, 69.

502. R. Kuroda, N. Matsumoto, and K. Oguma, *Fresenius Zeitschrift für Analytische Chemie*, 1988, 330, 111.

503. Y. Nakagawa, M.L. Firdenus, K. Narisuye, Y. Sahrin, K. Irasawa, and T. Hirata, *Analytical Chemistry*, 2008, 80, 9213.

504. C.H.G. Van den Berg, *Analytical Chemistry*, 1985, 57, 1532.
505. S.N. Willie, S.S. Berman, J.A. Page, and G.W. Van Loon, *Canadian Journal of Chemistry*, 1987, 65, 957.
506. C. Hua, D. Jagner, and L. Renman, *Analytica Chimica Acta*, 1987, 192, 103.
507. J.L. Hidalgo, M.A. Gomez, M. Caballero, R. Cela, and J.A. Perez-Bustamante, *Talanta*, 1988, 35, 301.
508. H. Monien, R. Bovenker, K.P. Kringe, and D. Rath, *Fresenius Zeitschrift für Analytische Chemie*, 1980, 300, 363.
509. A. Kimura, A. Yoneda, Y. Maeda, and T. Azumi, *Nippon Kaisui Gakkaishi*, 1985, 40, 141.
510. A.I. Kulathilake and A. Chatt, *Analytical Chemistry*, 1980, 52, 828.
511. W.M. Mok and C.W. Wai, *Analytical Chemistry*, 1984, 56, 27.
512. Y. Nagaosa and T. Kabayasho, *International Journal of Environmental Analytical Chemistry*, 1994, 7, 231.
513. E.W. Bruland, R.P. Franks, G.A. Knover, and J.H. Martin, *Analytica Chimica Acta*, 1979, 105, 233.
514. E.A. Boyle and J.M. Edmond, *Analytica Chimica Acta*, 1977, 91, 189.
515. K. Bruland, *Science Letters*, 1980, 47, 176.
516. E.A. Boyle, S.S. Huested, and S.P. Jones, *Journal of Geographical Research*, 1981, 86, 8048.
517. E. Kentner, D.B. Armitage, and H. Zeitlin, *Analytica Chimica Acta*, 1969, 45, 343.
518. K.B. Yatsimirskii, E.M. Ewel'Yakov, V.K. Pavloca, and Y.S. Savichenko, Oekeanologiya IIII, *Zhurnal Analiticheskoi Khimii*, 1970, 203(11), 19GD Abstract No. 11G.
519. H. Rampon and R. Cavelier, *Analytica Chimica Acta*, 1972, 60, 226.
520. D.S. Lee, *Analytical Chemistry*, 1982, 54, 1182.
521. H. Nishoika, S. Assadamongkol, Y. Maeda, and T. Azumi, *Nippon Kaisui Gakkalshi*, 1987, 40, 286.
522. C.M.G. Van den Berg and M. Nimmo, *Science of Total Environment*, 1987, 60, 185.
523. M. Koide, E.D. Goldenberg, and R. Walker, *Deep Sea Research Part II*, 1996, 43, 53.
524. C, Chen and M. Sharman, *Analytical Chemistry*, 2009, 81, 5400.
525. C.M.G. Van den Berg and B.S. Jacinto, *Analytica Chimica Acta*, 1988, 211, 129.
526. Q. Chen, A. Aarkrog, S.P. Nielson, H. Dahgaard, S. Nies, Y. Yu, and K. Handrup, *Journal of Radioanalytical and Nuclear Chemistry*, 1993, 172, 281.
527. A. Delle Site, V. Marchionni, and C. Testa, *Analytica Chimica Acta*, 1980, 117, 217.
528. B. Skwarzek and R. Bojanoski, *Marine Biology*, 1988, 97, 30.
529. S. Tsunogai and Y. Nozaki, *Geochemical Journal*, 1971, 5, 165.
530. L.V. Shannon, R.D. Cherry, and M.J. Orrens, *Geochemica and Cosmochimica Acta*, 1970, 34, 701.
531. A. Flynn, *Analytical Abstracts*, 1970, 18, 1624.
532. J.P. Cowen, V.F. Hodge, and T.R. Falson, *Analytical Chemistry*, 1977, 49, 494.
533. A. Torbjoern and D. Jaguer, *Analytica Chimica Acta*, 1973, 66, 152.
534. A. Torbjoern and D. Jaguer, *Analytical Abstracts*, 1972, 23, 994.
535. G. Marquis and J. Lebel, *Analytical Letters (London)*, 1981, 14, 913.
536. G.K. Ward, Report 29966, National Technical Information Services, Springfield, VA (32696), 1979.
537. M. Marezak and E. Ziaga, *Chimica Analytica*, 1973, 18, 99.
538. R.W. Perkins, Report of the Atomic Energy Commission US BNWL105 (pt 2), pp. 23–27, 1969.
539. S. Colley and J. Thomson, *Science of the Total Environment*, 1994, 155, 273.
540. H. Koide, V. Hodge, J.S. Yang, and E.D. Goldberg, *Analytical Chemistry*, 1987, 59, 1802.
541. A.D. Matthews and J.P. Riley, *Analytica Chimica Acta*, 1970, 51, 455.

542. Z. Shen and P. Li, *Fenxi Huaxue*, 1986, 14, 55.
543. R.C. Smith, R.C. Pmai, T.J. Chow, and T.R. Folson, *Limnology and Oceanography*, 1965, 10, 226.
544. I. Shoenfeld and S. Held, *Israel Journal of Chemistry*, 1969, 7, 831.
545. Committee on Medical and Biologic Effects of Environmental Pollutants: Selenium. National Academy of Science, Washington DC, 1976.
546. T.D. Luckey and B. Venegopal, *Chemical Engineering News*, 1976, 54, 2.
547. J.P. Riley and G. Skirrow, *Chemical Oceanography*, 2nd edn., Vol. 1, Academic Press, New York, 1975, p. 418.
548. V.B. Stein, E. Canelli, and A.H. Richards, *Atomic Spectroscopy*, 1980, 1, 61.
549. J. Neve, M. Hanocq, and L. Molle, *International Journal of Environmental Analytical Chemistry*, 1980, 8, 177.
550. Y. Shimoishi, *Analytica Chimica Acta*, 1973, 64, 465.
551. C.I. Measures and J.D. Burton, *Analytica Chimica Acta*, 1980, 120, 177.
552. K.W.M. Sui and S.S. Berman, *Analytical Chemistry*, 1984, 56, 1806.
553. J.H. Tzeng and H. Zeitlin, *Analytica Chimica Acta*, 1978, 101, 71.
554. V.M. Goldschmidt and L.W. Strock, *Nachr Akad Wiss Goettingen Math-Physik KI IV NF*, 1993, 1, 123.
555. H. Wattenberg, *Zeitung Anorganic Chemie*, 1938, 236, 339.
556. M. Ishibashi, T. Shigematsu, and Y. Nakagawa, *Recent Oceonographic Works: Japan Special Number*, 1953, 1, 44.
557. Y.K. Chau and J.P. Riley, *Analytica Chimica Acta*, 1965, 33, 36.
558. D.F. Schutz and K.K. Turekian, Abstract of Papers XIII General Assembly IUGG, Berkeley, Paper 23, 1963.
559. D.F. Schtuz, PhD thesis, Department of Geology, Yale University, New Haven, CT, 1964.
560. R.E. Sturgeon, S.N. Willie, and S.S. Berman, *Analytical Chemistry*, 1985, 57, 6.
561. G.A. Cutter, *Analytica Chimica Acta*, 1978, 98, 59.
562. P. Bermejo-Barrera, J. Moreda-Pineiro, A. Moreda-Pineiro, and A. Bermejo-Barrera, *Talanta*, 1996, 43, 35.
563. L.A. Miller and K.W. Bruland, *Environmental Science and Technology*, 1995, 29, 2616.
564. K. Kawabuchi and J.P. Riley, *Analytica Chimica Acta*, 1973, 65, 271.
565. I.I. Kulev and U.D.C. Bakalov, *Acad Bulgarian Science*, 1973, 26, 787.
566. I.I. Kulev, D.S. Stanev, and U.D. Bakalov, *I Khimiya Industry*, 1974, 46, 65.
567. R.A. Carr, *Limnology and Oceanography*, 1970, 15, 318.
568. T.J. Chow and T.G. Thompson, *Analytical Chemistry*, 1955, 27, 18.
569. B. Xu, Z. Wong, D. Gao, Y. Dong, W. Li, and Y. Li, *International Journal of Environmental Analytical Chemistry*, 2011, 91, 291.
570. Q. Chen, H. Dahlgaard, and S.P. Nielson, *Analytica Chimica Acta*, 1994, 285, 177.
571. M.O. Andreae, *Analytical Chemistry*, 1984, 56, 2064.
572. L. Petit, *Revue Internationale d'Oceanographic Medicale*, 1985, 19, 79/80.
573. C.A. Huh and M.P. Bacon, *Analytical Chemistry*, 1985, 57, 2065.
574. Y. Kodama and H. Tsubota, *Japan Analyst*, 1971, 20, 1554.
575. D.T. Burns, M. Harriott, and F. Glockling, *Fresenius Zeitschrift für Analytische Chemie*, 1987, 327, 701.
576. S. Dogan and W. Haerdi, *International Journal of Environmental Analytical Chemistry*, 1980, 8, 249.
577. C. Bergerioux and W. Haerdi, *Analusis*, 1980, 8, 169.
578. T.M. Florence and Y.J. Farrer, *Journal of Electroanalytical Chemistry*, 1974, 51, 191.
579. T.M. Florence, *Analytical Abstracts*, 1971, 20, 4585.
580. T.M. Florence, *Analytical Abstracts*, 1972, 23, 5011.
581. C.Y. Yang, J.S. Shih, and Y.C. Yeh, *Analyst*, 1981, 106, 385.

582. P.L. Croot, *Analytical Chemistry*, 2011, 83, 6395.
583. H.A. Van der Sloot and H.A. Pas, *Progress in Water Technology*, 1977, 81, 193.
584. Y.K. Agrawal, D.B. Upadhyaya, and S.P. Chudasama, *Journal of Radioanalytical and Nuclear Chemistry*, 1993, 170, 79.
585. C. Jie, L. Zaijun, and L. Ming, *International Journal of Environmental Analytical Chemistry*, 2008, 88, 583.
586. C.M.G. Van den Berg and Z.Q. Zhang, *Analytica Chimica Acta*, 1984, 164, 209.
587. C. Hua, D. Jagner, and L. Renman, *Analytica Chimica Acta*, 1987, 197, 265.
588. C.M.G. Van den Berg and M. Nimmo, *Analytical Chemistry*, 1987, 59, 924.
589. A.T. Economon, P.R. Filden, and A. Packham, *Analyst*, 1994, 119, 279.
590. R. Djogic and M. Branica, *Analyst*, 1995, 120, 1989.
591. R. Spencer, *Talanta*, 1968, 15, 1307.
592. G.W.E. Milner, J.D. Wilson, and G.A. Barnett, *Journal of Electroanalytical Chemistry*, 1961, 2, 25.
593. G. Billon, B. Ouddane, N. Proix, J. Desieres, Y. Abdelnour, and A. Boughriet, *Internal Journal of Environmental Analytical Chemistry*, 2005, 85, 1013.
594. N. Nishimura, K. Matsunaga, T. Kudo, and F. Obara, *Analytica Chimica Acta*, 1973, 65, 466.
595. T. Kirkyama and R. Kuroda, *Analytica Chimica Acta*, 1972, 62, 464.
596. H.V. Weiss, H.A. Gutman, J. Korkisch, and I. Steffan, *Talanta*, 1977, 24, 509.
597. L. Machackova and M. Zemberyova, *International Journal of Environmental Analytical Chemistry*, 2012, 92, 405.
598. H. Monien and R. Stangel, *Fresenius Zeitschrift für Analytische Chemie*, 1982, 311, 209.
599. D.W. Hastings, S.R. Emerson, and B.K. Nelson, *Analytical Chemistry*, 1996, 68, 371.
600. C.B.G. Van den Berg and Q. Huang, *Analytical Chemistry*, 1984, 56, 2383.
601. K.W. Bruland, G.A. Knauer, and J.H. Martin, *Nature*, 1978, 271, 741.
602. J.R. Donat and K.W. Bruland, *Marine Chemistry*, 1990, 28, 301.
603. V.N. Sastry, T.M. Krishnamoorthy, and T.P. Sarma, *Current Science, Bombay*, 1969, 38, 279.
604. J.L. Nowicki, K.G. Johnson, K.H. Coale, N.A. Eiro, and S.H. Lieberman, *Analytical Chemistry*, 1994, 66, 2732.
605. S. Hirata and K. Honda, *Bunseki Kagaku*, 1987, 36, 213.
606. D.C. Burrell and G. Guener, *Analytica Chimica Acta*, 1969, 48, 45.
607. R. Guevremont, *Analytical Chemistry*, 1981, 53, 911.
608. Y.S. Kim and H. Zeitlin, *Separation Science*, 1972, 7, 1.
609. S.D. Huang and K.Y. Shih, *Spectrochimica Acta Part B*, 1995, 50B, 837.
610. K. Akatsuka, T. Kotoh, N. Nabuyama, T. Okanaka, M. Okumura, and S. Hoshi, *Analytical Science*, 1996, 12, 109.
611. S.H. Lieberman and A. Zirino, *Analytical Chemistry*, 1974, 46, 20.
612. R.A.A. Muzzarrelli and L. Sipos, *Talanta*, 1971, 18, 853.
613. R. Andruzzi and A. Trazza, *Talanta*, 1981, 28, 839.
614. S.H. Lieberman and A. Zirino, *Analytical Chemistry*, 1974, 46, 20.
615. C.M.G. Van der Berg, *Marine Chemistry*, 1985, 16, 121.
616. C.M.G. Van der Berg, *Talanta*, 1984, 31, 1069.
617. J. Zima and C.M.G. Van der Berg, *Analytica Chimica Acta*, 1994, 289, 291.
618. H. Armannsson, *Analytica Chimica Acta*, 1979, 110, 21.
619. S. Olsen, C.R. Luiz, J.R. Pessenda, and E.O. Hansen, *Analyst*, 1983, 108, 905.
620. Z. Fang, J. Ruzickaa, and E.H. Hansen, *Analytica Chimica Acta*, 1984, 164, 23.
621. L.M. Cabezon, M. Caballero, R. Cela, and J.A. Perez-Bustamante, *Talanta*, 1984, 31, 597.
622. M.J. Orren, *Journal of South Africa Chemical Institute*, 1971, 24, 96.

623. D.J. Hydes, *Analytical Chemistry*, 1980, 52, 959.
624. G. Capodaglio, C. Turette, G. Toscano, A. Gambaro, G. Scarponi, and P. Lescon, *International Journal of Environmental Analytical Chemistry*, 1998, 71, 195.
625. R. Frache, M.L. Abelmoschi, M. Grotti, C. Ranni, E. Riagi, F. Soggia, G. Capodaglio, C. Turetta, and C. Barlante, *International Journal of Environmental Analytical Chemistry*, 2001, 79, 301.
626. J. Yates, Technical Support TR181 Determination of Trace Metals (cadmium, copper, lead, nickel and zinc) in filtered saline water samples, August 1982, Water Research Centre, Medmenham, UK, 1982.
627. E.A. Boyle and J.M. Edmond, *Analytica Chimica Acta*, 1977, 91, 1801.
628. L. Brugmann, L.G. Danielsson, B. Magnusson, and S. Westerlund, *Marine Chemistry*, 1983, 13, 327.
629. K-W. Bruland, R.P. Franks, G.A. Knauer, and J.H. Martin, *Analytica Chimica Acta*, 1979, 105, 233.
630. J.R. Montgomery and G.N. Peterson, *Analytica Chimica Acta*, 1980, 117, 397.
631. V.B. Stein, G. Canelli, and A.H. Richards, *International Journal of Environmental Analytical Chemistry*, 1980, 8, 99.
632. M. Hoenig and R. Wollast, *Spectrochimica Acta*, 1982, 37B, 399.
633. G. Cachetti, N. Cvleic-Lazzari, and R. Rolic, *Metodi Analitica per le Aeque*, 1981, 2, 1.
634. P.G. Brewer, D.W. Spencer, and C.L. Smith, *American Society of Testing Materials*, 1969, 443, 70.
635. E.A.C. Cimadevilla, K. Wrobel, and A. Sanz-Medel, *Journal of Analytical Atomic Spectrometry*, 1995, 10, 149.
636. A.A. Tikhomirova, S.A. Patin, and N.P. Morozov, *Zhurnal Analiticheskoi Khimii*, 1976, 31, 282.
637. M. Tominaga, K. Bansho, and Y. Umezaki, *Analytica Chimica Acta*, 1985, 169, 171.
638. M. Tominaga, K. Banso, and Y. Umezaki, *Analytica Chimica Acta*, 1985, 169, 176.
639. K. Hayase, K. Shitashima, and H. Tsuibota, *Talanta*, 1986, 33, 754.
640. The Analytical Report No. 12, February 1988, Perkin-Elmer Ltd, Beaconsfield, UK, 1988, p. 6.
641. W. Slavin, G.R. Carnick, and D.C. Manning, *Analytical Chemistry*, 1982, 54, 621.
642. Techniques in Graphite Furnace Atomic Absorption Spectrometry, Perkin-Elmer, Part 0993-8150 April 1985.
643. V. Hodge, M. Stallard, M. Koide, and E.D. Goldberg, *Analytical Chemistry*, 1986, 58, 616.
644. T. Hadeishi, D.A. Church, R.D. McLaughlin, B.D. Zak, M. Nakamura, and B. Chang, *Science*, 1975, 187, 348.
645. H. Koizumi and K. Yasuda, *Analytical Chemistry*, 1975, 47, 1679.
646. R. Stephens and D.E. Ryan, *Talanta*, 1975, 22, 655.
647. R. Stephens and D.E. Ryan, *Talanta*, 1975, 22, 659.
648. D.E. Veinot and R. Stephens, *Talanta*, 1976, 23, 849.
649. H. Koizumi and K. Yasuda, *Spectrochimica Acta*, 1976, 31B, 237.
650. H. Koizumi and K. Yasuda, *Analytical Chemistry*, 1976, 48, 1178.
651. R. Stephens, *Talanta*, 1978, 25, 435.
652. R. Stephens, *Talanta*, 1979, 26, 57.
653. J.B. Lawson, E. Grassam, D.J. Ellis, and M.J. Keir, *Analyst*, 1976, 101, 315.
654. E. Grassam, J.B. Dawson, and D.J. Ellis, *Analyst*, 1977, 102, 804.
655. H. Koizumi and K. Yasuda, *Spectrochimica Acta*, 1976, 31B, 523.
656. H. Koizumi, K. Yasuda, and M. Katayama, *Analytical Chemistry*, 1977, 49, 1106.
657. F.J. Fernandez, W. Bohler, N.M. Beatty, and W.B. Barnett, *Atomic Spectroscopy*, 1981, 2, 73.
658. W. Slavin, D.C. Manning, and G.R. Carnrick, *Atomic Spectroscopy*, 1981, 2, 137.

659. W. Slavin, *Atomic Spectroscopy*, 1980, 1, 66.
660. Z. Grabenski, R. Lehmann, B. Radzuik, and U. Voellkopf, *Atomic Spectroscopy* 1984, 5, 87.
661. A.M. De Kersabiec, G. Blanc, and M. Pinta, *Fresenius Zeitschrift für Analytische Chemie*, 1985, 322, 731.
662. Z. Grobenski, R. Lehmann, B. Radzuck, and U. Voelkopf, *Atomic Spectroscopy Application Study No 686, Papers presented to Pittsberg Conference*, Atlantic City, NJ, 5–9 March 1984, The Determination of Trace Metals in Seawater using Zeeman Graphite Furnace AAS, March 1984.
663. F. Ferenandez et al., *Atomic Spectroscopy*, 1981, 2.3, 73.
664. Perkin-Elmer, Analytical methods for graphite furnace AAS, March 1984.
665. J.Y. Cabon and A. Le Bihan, *Spectrochemica Acta Part B*, 1995, 50B, 1703.
666. M.O.P. Andreae, *Proceedings of a NATO Advanced Research Institute on Trace Metals in Seawater* (eds. C.S. Wong et al.), Sicily, Italy, 30 March–30 April 1981, Plenum Press, New York, 1981, p. 500.
667. R.S. Braman, L.C. Justen, and C.C. Foreback, *Analytical Chemistry*, 1972, 44, 2195.
668. R.S. Braman and C.C. Foreback, *Science*, 1973, 182, 1247.
669. M.A. Tompkins, Environmental-analytical studies of antimony, germanium and tin, Thesis, University of South Florida, Tampa, FL, 1977.
670. M.O. Andreae, J.F. Asmode, P. Foster, and L. Van'tdack, *Analytical Chemistry*, 1981, 53, 1766.
671. H.H. Fleming and R.G. Ide, *Analytica Chimica Acta*, 1976, 75, 8367.
672. E.A. Crecelius, *Analytical Chemistry*, 1978, 50, 826.
673. S. Nakashima, R.E. Sturgeon, S.N. Willie, and S.R. Bermon, *Analytica Chimica Acta*, 1988, 207, 291.
674. R.K. Winge, V.A. Fassel, R.N. Kniseley, E. De Kalb, and W.J. Haas, *Spectrochimica Acta*, 1977, 32B, 327.
675. D.D. Nygaard, *Analytical Chemistry*, 1979, 51, 881.
676. C.W. McLeod, K. Otsuki, H. Okamoto, K. Fuwa, and H. Haraguchi, *Analyst*, 1981, 106, 419.
677. A. Mijazaki, A. Kimura, K. Bansho, and Y. Amezaki, *Analytica Chimica Acta*, 1981, 144, 213.
678. S.S. Berman, J.M. McLaren, and S.N. Willie, *Analytical Chemistry*, 1980, 52, 488.
679. J.W. McLaren and S.S. Braman, *Applied Spectroscopy*, 1981, 35, 403.
680. A. Sujimae, *Analytica Chimica Acta*, 1980, 121, 331.
681. H. Berndt, U. Harms, and M. Sonneborn. *Fresenius Zeitschrift für Analytische Chemie*, 1985, 322, 329.
682. J.A.C. Blackaert, F. Leis, and K. Hagua, *Talanta*, 1981, 28, 745.
683. N.S. Chang, H. L. Norton, and J.L. Anderson, *Analytical Chemistry*, 1990, 62, 1043.
684. N. Beauchemin and S.S. Berman, *Analytical Chemistry*, 1989, 61, 1857.
685. G. Chapple and J.P. Byrne, *Journal of Analytical Atomic Spectroscopy*, 1996, 11, 549.
686. L.C. Alves, L.A. Allen, and R.H. Honk, *Analytical Chemistry*, 1993, 65, 2468.
687. C. Vandecasteele, N.H. Vanhoe, and R. Dams, *Analytica Chimica Acta*, 1988, 211, 91.
688. A. Stroh and U. Voelikopf, *Journal of Analytical Atomic Spectroscopy*, 1993, 8, 35.
689. M. Bettinelli and S. Spezia, *Journal of Chromatography, A*, 1995, A709, 275.
690. M.J. Bloxam, S.J. Hill, and F.J. Worsfold, *Analytical Proceedings (London)*, 1993, 30, 159.
691. T.M. Florence, *Journal of Electroanalytical Chemistry*, 1972, 35, 237.
692. T. Rojahn, *Analytica Chimica Acta*, 1972, 62, 438.
693. W.R. Seitz, R. Jones, L.N. Klatt, and W.D. Mason, *Analytical Chemistry*, 1973, 45, 840.
694. G.C. Whitnack and R. Basselli, *Analytica Chimica Acta*, 1969, 47, 267.
695. M. Komatsu, T. Matsueda, and H. Kakiyami, *Japan Analyst*, 1971, 20, 987.

696. J. David and J.D. Redmond, *Journal of Electroanalytical Chemistry*, 1971, 33, 169.

697. R.G. Clem, *MPI Appalikcation N Otes*, 1973, 8, 1.

698. D.D. Nygaard and S.R. Hill, *Analytical Letters (London)*, 1979, 12, 491.

699. H.W. Nurnberg and B. Raspar, *Environmental Technology letters*, 1981, 2, 457.

700. R. Andruzzi, A. Trazza, and G. Marruso, *Talanta*, 1982, 29, 751.

701. R.A.A. Muzzareli and J. Sinos, *Talanta*, 1971, 18, 853.

702. R.G. Clem, G. Mitton, and L.D. Ornelas, *Analytical Chemistry*, 1973, 45, 1306.

703. G. Scarponi, G. Capoaglio, and P. Cescon, *Analytica Chimica Acta*, 1982, 135, 263.

704. R.G. Clem and A.T. Hodgson, *Analytical Chemistry*, 1978, 50, 102.

705. L. Brugmann, *Science of the Total Environment*, 1984, 37, 41.

706. K.W. Bruland, K.H. Coale, and L. Mart, *Marine Chemistry*, 1985, 17, 285.

707. E.P. Acterberg and C.M.G. Van den Berg, *Marine Pollution Bulletin*, 1996, 32, 471.

708. C. Brihaye, G. Gillain, and G. Duyckaerts, *Analytica Chimica Acta*, 1983, 148, 51.

709. J.C. Duinker and C.J. M. Kramer, *Marine Chemistry*, 1977, 5, 207.

710. L. Mart, W.H. Nurnberg, and D. Dryssen, In *Proceedings of a NATO Advanced Research Institute on Trace Metals in Seawater* (eds. C.S. Wong et al.), Sicily, Italy, 30 March–30 April 1981, Plenum Press, New York, 1981, p. 510.

711. P. Valenta, L. Mart, and H. Rutzel, *Journal of Electroanalytical Chemistry*, 1977, 82, 327.

712. B. Pihlar, P. Valenta, and H.W. Nurnberg, *Fresenius Zeitschrift für Analytische Chemie*, 1981, 307, 337.

713. S.R. Pietrowicz, M. Springer-Young, J.A. Puig, and M.J. Spencer, *Analytical Chemistry*, 1982, 54, 1367.

714. W. Lund and D. Aonshus, *Analytica Chimica Acta*, 1976, 86, 109.

715. C.M.G. Van den Berg, *Geochimica Coschimica Acta*, 1984, 48, 2613.

716. D. Krnaric, *Marine Chemistry*, 1984, 15, 117.

717. H.W. Nurnberg and B. Raspar, *Environmental Technology Letters*, 1981, 2, 457.

718. L. Mart, W.H. Nurnberg, and D. Dyrssen, In *Proceedings of a NATO Advanced Research Institute on Trace Metals in Seawater* (eds. C.S. Wong et al.), Sicily, Italy, 30 March–30 April 1981, p. 512, Plenum Press, New York, 1981.

719. V. Cuculic and M. Branica, *Analyst*, 1996, 121, 1127.

720. L. Sipos, J. Golimowski, P. Valenta, and H.W. Nurnberg, *Fresenius Zeitschrift für Analytische Chemie*, 1979, 298, 1.

721. G. Gillain, G. Duychaerts, and A. Diteche, *Analytica Chimica Acta*, 1979, 106, 23.

722. H. Eskilsson, H. Haroldsson, and D. Jagner, *Analytica Chimica Acta*, 1985, 175, 79.

723. M. Mooizumi, Isotope dilution mass spectrometry of copper, cadmium, thallium and lead in Marine Environments, *Presented at the American Chemical Society, Chemical Congress*, Honolulu, HI, 1964.

724. L. Prochazkova, *Analytical Chemistry*, 1964, 36, 865.

725. V.J. Stuckas and C.S. Wong, In *Proceedings of a NATO Advanced Research Institute on Trace Metals in Seawater* (eds. C.S. Wong et al.), Sicily, Italy, 30 March–30 April 1981, Plenum Press, New York, 1981, p. 513.

726. H. Elderfield and H.J. Greaves, In *Proceedings of a NATO Advanced Research Institute on Trace Metals in Seawater* (eds. C.S. Wong et al.), Sicily, Italy, 30 March–30 April 1981, Plenum Press, New York, 1981, p. 427, 520.

727. J.R. Donat, K.A. Lao, and K.W. Bruland, *Analytica Chimica Acta*, 1993, 284, 547.

728. J.R. Donat and K.W. Bruland, *Analytical Chemistry*, 1988, 60, 240.

729. Private communication.

730. O. Abollino, M. Aceto, G. Sacchero, E. Sarzanni, and E. Mentasti, *Analytica Chimica Acta*, 1995, 305, 200.

731. C.M.G. Van der Berg, *Science of the Total Environment*, 1986, 48, 89.

732. D. Jagner and A. Granelli, *Analytica Chimica Acta*, 1976, 83, 19.
733. D. Jagner, *Analytical Chemistry*, 1978, 50, 1924.
734. D. Jagner, M. Josefson, and S. Westerlund, *Analytica Chimica Acta*, 1981, 129, 153.
735. I. Drabekl, O. Pheiffer, P. Madsen, and J. Sorensen, *International Journal of Environmental Analytical Chemistry*, 1986, 15, 153.
736. B. Armitage and H. Zeitlin, *Analytica Chimica Acta*, 1971, 53, 47.
737. A.W. Morris, *Analytica Chimica Acta*, 1968, 42, 397.
738. M. Murate, M. Omatsu, and S. Muskimoto, *X-ray Spectrometry*, 1984, 13, 83.
739. A. Prange, A. Knockel and W. Michaelis, *Analytica Chimica Acta*, 1985, 172, 79.
740. M. Haarich, D. Schmidt, P. Freimann, and A. Jacobson, *Spectrochimica Acta, Part B*, 1993, 48B, 183.
741. J.M. Lo, J.C. Wei, and S.J. Yeh, *Analytical Chemistry*, 1977, 49, 1146.
742. R.R. Greenberg and H.M. Kingston, *Analytical Chemistry*, 1983, 55, 1160.
743. R.R. Greenberg and H.M. Kingston, *Journal of Radioanalytical Chemistry*, 1982, 71, 147.
744. D.S. Piper and G.G. Goles, *Analytica Chimica Acta*, 1969, 47, 560.
745. S. Ghoda, *Bulleting of the Chemical Society of Japan*, 1972, 45, 1704.
746. E. Brun, Report of Aktiebolaget Atomenerai, AE-466, 1972.
747. K.H. Lieser, W. Calmano, E. Heuss, and V. Neitzert, *Journal of Radioanalytical Chemistry*, 1977, 37, 717.
748. C. Lee, N.B. Kim, I.C. Lee, and K.S. Chung, *Talanta,* 1977, 24, 241.
749. J.M. Lo, J.C. Wei, and S.J. Yeh, *Analytical Chemistry*, 1977, 49, 1146.
750. R.R. Greenberg and H.M. Kingston, *Analytical Chemistry*, 1983, 55, 1160.
751. R.S.S. Murthey and D.E. Ryan, *Analytical Chemistry*, 1983, 55, 682.
752. M. Stiller, M. Mantel, and M.S. Rappaport, *Journal of Radioanalytical and Nuclear Chemistry*, 1984, 83, 345.
753. R.R. Greenberg and H.M. Kingston, *Journal of Radioanalytical Chemistry*, 1982, 71, 147.
754. J. Holzbecker and D.E. Ryan, *Journal of Radioanalytical Chemistry*, 1982, 74, 25.
755. R.E. Robertson, *Analytical Chemistry*, 1968, 40, 1067.
756. Y. Nagaosa, H. Kawabe, and A.M. Bond, *Analytical Chemistry*, 1991, 63, 28.
757. Y. Shijo, H. Sato, N. Uehara, and S. Aratake, *Analyst*, 1996, 121, 325.
758. A.A. Riccardo, G.R. Muzzarelli, and O. Tubertini, *Journal of Chromatography*, 1970, 47, 414.
759. R.M. Cassidy and S. Elchuk, *Journal of Chromatographic Science*, 1981, 19, 503.
760. R.M. Cassidy and S. Elchuk, *Journal of Chromatographic Science*, 1980, 18, 217.
761. D.J. Machey and H.W.T. Higgins, *Journal of Chromatography*, 1987, 436, 243.
762. J. Posta, A. Alimonti, F. Petrucci, and S. Caroli, *Analytica Chimica Acta*, 1996, 325, 185.
763. H. Tao, A. Migusaki, K. Bansho, and Y. Umezaki, *Analytical Chemistry*, 1984, 56, 181.
764. M. Soylak, S. Saracoglu, L. Elei, and M. Dogan, *International Journal of Environmental Analytical Chemistry*, 2002, 82, 225.
765. I. Dakova, P. Vasileva, I. Karadiova, M. Karadjov, and V. Slaveykova, *International Journal of Environmental Analytical Chemistry*, 2011, 91, 62.
766. L.A. Escaleira, R.E. Santelli, E.P. Oliveira, M. de Fatima, B. de Carvallio, and M.A. Bezeerra, *International Journal of Environmental Analytical Chemistry*, 2009, 89, 515.
767. B. Wen, X Q. Shan, and S-P Xu, *International Journal of Environmental Analytical Chemistry*, 2000, 77, 65.
768. Z. Wang, Y. Li, and Y. Guo, *Analytical Letters*, 1994, 77, 957.

769. C.C. Chen, H. Cheng, R.L. Edwards, S.B. Moron, H.R. Edmonds, J.A. Hoh, and R.B. Thomas, *Analytical Chemistry*, 2003, 75, 1075.
770. D.A. Pickett, M.T. Murrell, and R. Williams, *Analytical Chemistry*, 1994, 66, 1044.
771. C.D. Biggin, G.T. Cooke, A.B. Mackenzie, and J.M. Bates, *Analytical Chemistry*, 2007, 74, 671.
772. S. May, C. Engelmann, and G. Pinte, *Journal of Radioanalytical and Nuclear Chemistry*, 1987, 113, 343.

4 Metal Preconcentration Techniques in Seawater

The considerable difficulty of trace element analysis in a high-salt matrix such as seawater, estuarine water, or brine is clearly reflected in the literature. The extremely high concentrations of alkali metals, alkaline earth metals, and halogens, combined with the extremely low levels of transition metals and other elements of interest, make direct analysis by most analytical techniques difficult or impossible.

A variety of preconcentration procedures have been used, including solvent extraction of metal chelates, coprecipitation, chelation of ion-exchange resins, adsorption onto other solids such as silica-bonded organic complexing agents, and liquid–liquid extraction.

An ideal method for the preconcentration of trace metals from natural waters should have the following characteristics: it should simultaneously allow the isolation of the analyte from the matrix and yield an appropriate enrichment factor; it should be a simple process, requiring the introduction of few reagents in order to minimise contamination, hence producing a low sample blank and a correspondingly lower detection limit and it should produce a final solution that is readily matrix matched with solutions of the analytical calibration method.

Various preconcentration procedures have been discussed in the literature. These include the following:

1. Chelation by adsorption on solids followed by solvent extraction
2. Adsorption on ion-exchange resin
3. Adsorption on Chelex resins
4. Adsorption on solid phases
5. Coprecipitation methods
6. Flow injection analysis
7. Hydride generation procedure
8. Cloud point extraction
9. Hollow-fibre membrane extraction
10. Silica microsphere extraction

4.1 CHELATION: SEAWATER EXTRACTION

This extraction method can be generally classified into two major categories. The first one comprises the conversion of metals to metal dithiocarbamate chelates, the extraction of the metal-dithiocarbamate complexes from a larger volume of the aqueous phase into a smaller volume of oxygenated organic solvents such as methyl isobutyl ketone (thereby achieving the concentration of metals), and the analysis of the solvents directly. The second one includes the extraction of metal complexes or oxygenated chlorinated organic solvents such as chloroform and methyl isobutyl ketone followed by nitric acid back-extraction and then the analysis of trace elements in the acid solution. There are several drawbacks associated with the acid back-extraction of metal dithiocarbamates; the kinetics is generally slow and the efficiency of acid extraction is poor for certain metals.

Methods based on these principles for the preconcentration of metals in seawater are reviewed in Table 4.1. In many instances, detection limits are sufficiently low.

4.2   ADSORPTION ON ION-EXCHANGE RESINS

In this procedure, a solution of metals or their complexes in a large volume of samples are adsorbed in a column of ion-exchange resins, and then the metal is desorbed by a relatively small volume of dilute acid prior to analytical determination. Methods based on this principle are reviewed in Table 4.2.

4.3   ADSORPTION ON CHELEX ION-EXCHANGE RESINS

Studies of the use of ion-exchange resins for the preconcentration of metals from seawater have been mainly concerned with the use of Chelex-100.

The iminodiacetate-containing resin, Chelex-100, is the most commonly employed chelating resin for the removal and preconcentration of trace heavy metals from seawater. The work on the use of Chelex-100 resin for the preconcentration of metals from seawater is reviewed in Table 4.3. In each case, metals are desorbed from the resin with nitric acid (2–2.5 M) and then determined in the extract by graphite furnace atomic absorption spectrometry.

Preconcentration factors of up to 100–120 have been reported by this technique enabling metals to be determined at the μg/L or ng/L level.

4.4   PRECONCENTRATION ON OTHER SOLID PHASES

Sturgeon et al. [39] preconcentrated selenium and antimony from seawater on C_{18}-bonded silica gel prior to determination by graphite furnace atomic absorption spectrometry. The method was based on the complexation of selenium(IV), antimony(III), and antimony(V) with ammonium pyrrolidine dithiocarbamate. These complexes were adsorbed onto a column of C_{18}-bonded silica gel and then eluted with methanol, followed by evaporation to near dryness. The residue was taken up in 1% nitric acid. Concentration factors of 200 could be obtained. Detection limits for selenium(IV), antimony(III), and antimony(V) were 7, 50, and 50 ng^{-1}, respectively, based on a 300 mL sample volume.

Shabani et al. [40] preconcentrated trace earth elements in seawater by complexation with bis(2-ethylhexyl) hydrogen phosphate and 2-ethylhexyl dihydrogen phosphate adsorbed onto C_{18} cartridge, which were determined by indirectly coupled plasma mass spectrometry. Using 1 or 5 L, enrichments of 200- and 100-fold were achieved.

TABLE 4.1
Preconcentration of Metals in Seawater Chelation Solvent Extraction Techniques Followed by Direct Atomic Absorption Spectrometry and Graphite Furnace Atomic Absorption Spectrometry

Metals	Chelating Agent	Solvent	Detection Limit (μg/L)		References
Direct Atomic Absorption Spectrometry					
Mn, Fe, Co, Ni, Zn, Pb, Cu	Hexahydro-azepine-1-carbodithioate	Butyl acetate	Mn	0.2	[1]
			Fe	1.5	
			Co	0.6	
			Ni	0.6	
			Zn	0.4	
			Pb	2.6	
			Cu	0.5	
Fe, Pb, Cd, Co, Ni, Cr, Mn, Zn, Cu	Diethyldithiocarbamate	MIBK or xylene			[2]
Fe, Cu	Ammonium pyrrolidine dithiocarbamate	MIBK	Cu	<1	[3]
			Fe	<1	
Cd, Zn, Pb, Ca, Ni, Cu, Ag	Dithizone	Chloroform	Ag	0.05	[4]
			Cd	0.05	
			Zn	0.6	
			Pb	0.04	
			Cu	0.06	
			Ni	0.3	
			Co	0.04	
Cd, Cu, Pb, Ni, Zn	(a) Ammonium dipyrrolidine dithiocarbamate	MIBK	Cu	10	[5]
			Cd	2	
	(b) Ammonium dipyrrolidine dithiocarbamate plus diethyl dithiocarbamate		Pb	4	
			Ni	16	
			Zn	30	
Graphite Furnace Atomic Absorption Spectrometry					
Cu, Ni, Cd	Ammonium pyrrolidine dithiocarbamate				[6]
Ag, Cd, Cr, Cu, Fe, Ni, Pb, Zn	Ammonium dipyrrolidine dithiocarbamate MIBK	MIBK	Ag	0.02	[7]
			Cd	0.03	
			Cr	0.05	
			Cu	0.05	
			Fe	0.20	
			Ni	0.10	
			Pb	0.03	
			Zn	0.03	
			($2 \times$ SD used)		

(*Continued*)

TABLE 4.1 (*Continued*)
Preconcentration of Metals in Seawater Chelation Solvent Extraction Techniques Followed by Direct Atomic Absorption Spectrometry and Graphite Furnace Atomic Absorption Spectrometry

Metals	Chelating Agent	Solvent	Detection Limit (µg/L)		References
Cu, Cd, Zn, Ni	Diethyl dithiocarbamate plus ammonium pyrrolidine dithiocarbamate	Chloroform	Cu	1.0	[8]
			Cd	0.2	
			Zn	2	
			Ni	10	
			(2×SD used)		
Cd, Pb, Ni, Cu, Zn	Ammonium pyrrolidine dithiocarbamate plus diethyl dithiocarbamate	Freon	Not stated		[9]
Cu	Ammonium pyrrolidine dithiocarbamate	MIBK		<0.5	[10]
Cd, Cu, Ni, Zn	Dithizone	Chloroform	Cu	0.006	[11]
			Cd	0.0004	
			Ni	0.032	
			Zn	0.016	
			(2×SD used)		
Cd, Cu, Fe	Ammonium pyrrolidine dithiocarbamate plus diethyl dithiocarbamate	Freon	Not stated		[12]
Cd, Zn, Pb, Cu, Fe, Mn, Co, Cr, Ni	Ammonium pyrrolidine-*N*-carbodithioate plus 8-hydroxyquinoline	MIBK	Fe	0.08	[13]
			Cu	0.10	
			Pb	0.06	
			Cd	0.02	
			Zn	0.34	
Cd	Ammonium pyrrolidine dithiocarbamate	Carbon tetrachloride	Cd	0.006	[14]
Cd, Zn, Pb, Fe, Mn, Cu, Ni, Co, Cr	Dithiocarbamate	MIBK	Not stated		[15]
Cd, Co, Cu, Fe, Mn, NI, Pb, Zn	Ammonium pyrrolidine dithiocarbamate	Chloroform	Cd	0.0001	[16]
			Cu	<0.012	
			Fe	<0.02	
			Mn	<0.004	
			Ni	<0.012	
			Pb	<0.016	
			Zn	<0.08	
			(2×SD used)		

(*Continued*)

TABLE 4.1 (*Continued*)
Preconcentration of Metals in Seawater Chelation Solvent Extraction Techniques Followed by Direct Atomic Absorption Spectrometry and Graphite Furnace Atomic Absorption Spectrometry

Metals	Chelating Agent	Solvent	Detection Limit (µg/L)		References
Cd, Co, Cu, Fe, Mn, Ni, Pb, Zn	Ammonium pyrrolidine dithiocarbamate	Chloroform	Cd	0.02	[17]
			Cu	0.24	
			Fe	0.24	
			Mn	0.02	
			Ni	0.08	
			Pb	0.04	
			Zn	1.0	
Mn, Cd	Ammonium pyrrolidine dithiocarbamate and diethyl-ammonium diethyldithiocarbamate	Freon	Mn	0.07	[18]
			Cd	0.027	
			(2 × SD used)		
Cd, Cu, Fe, Pb, Ni, Zn	Ammonium pyrrolidine dithiocarbamate and diethyl-ammonium diethylammonium diethyldithiocarbamate	Freon	Not quoted		[19]

Wang et al. [41] used C_{18} column loaded with sodium diethyldithiocarbamate to preconcentrate cadmium and copper from seawater prior to determination by graphite furnace atomic absorption spectrometry. Detection limits with 0.5 mL samples are in the 4–24 ppt range.

Comber [42] used a C_2 column coated with an ammonium pyrrolidine-1-ylddithioformate-cetyltrimethyl ammonium bromide ion pair to preconcentrate copper, nickel, and cadmium from seawater. The metal-dithiocarbamate complexes were then separated on a C_{18} column by high-performance liquid chromatography using a UV detector. The detection limit with a 10 mL sample was $0.5 \, \mu L^{-1}$.

Murthy and Ryan [43] preconcentrated copper, cadmium, mercury, and lead from seawater on a column of dithiocarbamate cellulose derivative. Metal concentrations on the adsorbent material were determined by neutron activation analysis. The recoveries of added spikes are between 93% and 98% of Cu(II) and Pb(II).

Ogura and Oguma [44] preconcentrated molybdenum and vanadium in seawater on a cellulose phosphate column prior to determination by inductively coupled plasma atomic emission spectrometry.

Sturgeon et al. [45] preconcentrated cadmium, copper, zinc, lead, iron, manganese, nickel, and cobalt from seawater into silica-immobilised 8-hydroxylquinoline

TABLE 4.2

Preconcentration on Ion-Exchange Resins

Metals	Type of Resins	Extraction of Metals from Resins	References
Atomic Absorption Spectrometry			
Cu, Fe, Cd, Pb, Zn	Polyacrylamide oxime	1.1 hydrochloric acid	[21]
Cd, Mn, Cr, Cu	Amberlite	1% nitric acid	[22]
Ni, Fe, Co, Pb	XAD-7		
Ni, Fe, Co, Pb	XAD-1		[23]
Miscellaneous			[24]
Vi, Co, Cu, Zn, Cd	Anion exchange in resin thiocyanate form	2 M nitric acid	[25]
U	Strong base ion-exchange resin (chloride form)	1 M hydrochloric acid	[26]
Cd, Co, Ni	Formed sodium bis(2-hydroxyethyldi-thiocarbamate, complex adsorbed onto XAD4 resin)	–	[27]
Dimethyl arsenic acid	Strong cation-exchange resin		[28]

prior to determination by graphite furnace atomic absorption spectrometry. A good agreement was obtained with accepted values for seawater.

Resing and Measures [76] have described a highly sensitive fluorometric method for the determination of aluminium in seawater by flow injection analysis with in-line preconcentration. The method employs in-line preconcentration of aluminium on a column of resin-immobilised 8-hydroxyquinoline. The column is subsequently eluted into the flow injection system from the resin-acidified seawater. The eluted aluminium reacts with lumogallion to form chelate, which is detected by its fluorescence. The fluorescence is enhanced approximately fivefold by the addition of a micelle-forming detergent, Brij-35. The method has a detection limit of ~0.15 nM and a precision of 1.7% at 2.4 nM. The method has a time cycle of 3 min and can be readily automated. The ease of use and relative freedom from contamination artefacts make this method ideal for shipboard determination of aluminium in seawater.

The results obtained by using the immobilised 8-hydroxyquinoline concentration procedure were compared to 'accepted' values for these samples. A good agreement with accepted values is evident. The precision of analysis, expressed as a relative standard deviation, averages 6% (relative standard deviation (RSD) range: 2%–12%) over all elements, concentrations, and samples.

Nakashima et al. [46] determined trace metals in seawater by graphite furnace atomic absorption spectrometry with preconcentration on silica-immobilised 8-hydroxyquinoline in a flow system. Cadmium, lead, zinc, copper, iron, manganese, nickel, and cobalt were quantified by graphite furnace atomic absorption

TABLE 4.3

Application of Chelex-100 Resin to the Preconcentration of Metals in Seawater Prior to an Analysis by Graphite Furnace Atomic Absorption Spectrometry

Elements	Eluent	Detection Limit		References
Cd, Co, Cu. Fe, Mn, Ni, Pb, Zn	2.5 M nitric acid	Sub-nanogram µg/L		[29]
Cu, Cd, Zn, Ni,	2 M nitric acid	Cu	0.006	[30]
		Cd	0.006	
		Zn	0.015	
		Ni	0.015	
Cd, Zn, Pb, Cu, Fe, Mn, Co, Cr, Ni	2.5 M nitric acid	Not stated µg/L		[31]
Cd, Pb, Ni, Cu, Zn	2.5 M nitric acid	Cd	0.01	[20,32]
		Pb	0.16–0.28	
		Ni	0.24–0.68	
		Cu	0.6	
		Zn	1.8	
Co, Cr, Cu, Fe, Mn, Mu, Ni, Sc, Th, U, Pb, Cd, Th, V, Zn, Co, Zn, Vi, Cu, Cr, Mn, Cd, Cu, Ni		Cr	<1	[33,34]
		Mn	<1	
		Co	<1	
		Ni	<1	
		Cu	<1	
		Zn	<1	
		Cd	<1	
		Pb	<1	
		Th	0.6	
		V	2.5	
Mn, Te, Co, Zn, Ni, Cu, Cr				[35]
Mn, Ct, Cu, Ni		Ni	0.015 µg/L	[36]
		Cu	0.006 µg/L	
		Zn	0.015 µg/L	
		Cd	0.0006 µg/L	
Cr, Cu, Mn				[37]
Cd, Cu, Co, Mn, Ni, Pb, Zn				[38]
Pb		6 ppt		[59]
Pb, Cd, Cu, Zu		2 nitric acid		[69]

spectrometry following optimised preconcentration. The estimated detection limits ranged from 0.2 ng/L (cobalt) to 40 ng/L (iron) based on a 50 mL sample volume (100 mL for cobalt).

Lau and Yang [47] used 8-quinolinol immobilised on silica to determine 16–70 ppt of copper, nickel, and cadmium in seawater by inductively coupled plasma atomic emission spectrometry.

Volkan et al. [48] preconcentrated trace metals from seawater on a mercapto-modified silica gel. The procedure was developed for the preconcentration of cadmium, copper, lead, and zinc from natural waters by adsorption on silica gel modified by treatment with (3-mercaptopropyl) trimethoxysilane. Recoveries greater than 95% were common.

Johansson et al. [49] preconcentrated mercury and lead from seawater by pumping through a tubular membrane containing a resin-containing dithiocarbamate group.

Orians and Boyle [50] preconcentrated gallium, titanium, and indium in seawater into an 8-hydroxyquinoline-immobilised resin prior to determination in amounts down to 0.1–0.4 ppt by inductively coupled plasma mass spectrometry.

Elements in the preconcentrate were determined by individually coupled plasma atomic emission mass spectrometry. The optimum experimental conditions such as pH, sample flow rate, and volume of eluents were investigated. A concentration factor of at least 300 for analytes of interest in seawater and separation of matrix components can be achieved. The method was applied for the determination of trace elements in coastal seawater. The results indicated that the recovery ranged from 91% to 107%, and the relative standard deviation was found to be less than 5% for trace elements at the ng/L level.

Blair et al. [51] used a chelamine-chelating resin to preconcentrate the following from seawater with 90% recovery: heavy metals, cadmium, copper, manganese, nickel, lead, and zinc.

Isshiki et al. [52] preconcentrated trace metals from seawater with 7-dodecenyl-8-quinolinol-impregnated XA-4 macroporous resin. The extraction behaviour of this resin was compared with solvent extraction with DDQ (7-(vinyl-3,3,6,6,-tetramethylhexyl)-8-quinolinol) for silver, aluminium, bismuth, cadmium, copper, iron, gallium, manganese, nickel, lead, and thallium.

Greenberg and Kingston [53] studied the trace element analysis of seawater using neutron activation analysis with a chelating resin. Transition metals are thus separated from alkali metals, alkaline earth metals, and halogens.

Isshiki and Nakayama [54] have discussed the selective concentration of cobalt in seawater by complexation with various ligands or sorption on macroporous XAD resins. Complexed cobalt is collected after passage through a small XAD resin-packed column.

Fujita and Iwashima [55] preconcentrated mercury compounds in seawater by first forming diethyldithiocarbamate and then concentrating this on XAD-2 resin. The resin was eluted with methanol/3 M hydrochloric acid; organic mercury was extracted with benzene and then back-extracted with cysteine solutions. The organic mercury in the cysteine solution and the total mercury adsorbed on the resin were determined by flameless atomic absorption spectrometry.

The lower limit of detection in seawater is 0.1 ng/L for organic mercury, using 80 L of samples.

Fernandez Garcia et al. [56] studied the use of various chelating agents for mercury prior to preconcentration on silica C_{18} followed by continuous cold vapour atomic absorption spectrometry, down to 16 ppt mercury in seawater could be determined.

The work of Mackay [57] suggests that both metal-organic species and inorganic ions can be absorbed from seawater by Amberlite XAD-1. The low capacity of the resin for inorganic ions and the probable slow kinetics lead to competition for the limited sites, and the amounts of inorganic trace metal ions adsorbed by the resin, therefore, depend strongly on other parameters such as flow rate and the volume of seawater processed. The trace metals eluted from the resin by organic solvents probably consist of metal-organic compounds but it is clear that no combinations of common organic solvents can remove all the trace metals adsorbed by the resin. It is not known whether the adsorbed metals removed by methanolic hydrochloric acid are inorganic or a mixture of inorganic and strongly adsorbed metal-organic species.

Saylak et al. [58] have described a preconcentration/separation method for the atomic absorption spectrometric determinations of Cu(II), Ni(II), and Pb(II) ions in seawater by using a chromatographic column filled with Amberlite XAD-1180. The recoveries of Cu(II), Ni(II), and Pb(II) ions were quantitative (>95%) at pH 8–9. The influences of the various analytical parameters, including the amount of calmagite, amount of Amberlite XAD-1180, eluent type, and so on, were investigated. The influence of the seawater matrix was also examined. Simultaneous enrichment and determination of copper, nickel, and lead ions in seawater are possible with satisfactory results (recoveries >95%, RSD <9%).

Dakova et al. [78] determined the dissolved and labile concentrations of Cd(II), Ni(II), and Pb(II) in the Burgas Gulf of the Bulgarian Black Sea Coast. The solid-phase extraction procedure based on monodisperse, sub-micrometre silica spheres modified with 3-aminopropyltrimethoxysilane followed by the electrothermal atomic absorption spectrometry (ETAAS) was applied to quantify the total dissolved metal concentrations in seawater. Quantitative sorption of Cd, Cu, Ni, and Pb was achieved in the pH range of 7.6–8 for 30 min. Adsorbed elements were eluted with 2 mL of 2 mol/L nitric acid. Detection limits achieved for total dissolved metal quantification were as follows: Cd, 0.002; Cu, 0.005; Ni, 0.03; and Pb, 0.02 µg/L; relative standard deviations varied from 5% to 13% for all studied elements (for typical Cd, Cu, Ni, and Pb concentrations in Black Sea water).

4.5 COPRECIPITATION METHODS

Work in this field has been mainly concerned with coprecipitatives in seawater metals onto ferric hydroxide, cupric sulphide, and zirconium hydroxide. In a typical procedure, iron salt was added to seawater samples, the pH was adjusted, and the iron plus trace precipitate was separated by filtration on paper. The paper and precipitate were dissolved in nitric acid, and the acid solution was analysed directly by graphite furnace atomic absorption spectrometry. A 200-fold concentration was achieved with a recovery in excess of 90% of the metals in seawater. Manganese did not coprecipitate.

4.5.1 COPRECIPITATION ONTO IRON HYDROXIDE

Siu and Berman [60] determined selenium in seawater by gas chromatography after coprecipitation with hydrous iron(III) oxide. The method used rapid coprecipitation of selenium with hydrous iron(III) oxide. Following a brief stirring and settling period, the coprecipitate was filtered and dissolved in hydrochloric acid, derivatised to 5-nitropiazselenol, and extracted into toluene. Selenium was determined by gas chromatography–electron capture detection. The detection limit was 1 pg injected or 5 ng selenium/L of seawater using a 200 mL sample. The precision was 6% at 25 pg/L selenium.

Co-flotation with octadecylamine and ferric hydroxide as collectors has been used to separate copper, cadmium, and cobalt from seawater [61]. The method was based on the co-flotation or adsorbing colloid flotation technique. The substrates were dissolved in an acidified mixture of ethanol, water, and methyl isobutyl ketone to increase the sensitivity of the determination of these elements by flame atomic absorption spectrophotometry. Recoveries of the three metals were between 79% and 83%.

Coprecipitation with iron hydroxy has also been applied to the preconcentration of zirconium aluminium, lead, vanadium, and plutonium [62].

4.5.2 COPRECIPITATION ON ZIRCONIUM HYDROXIDE

Akagi et al. [63] used zirconium coprecipitation for simultaneous multi-element determinations of trace metals in seawater by inductively coupled plasma atomic emission spectrometry. The coprecipitation procedure, ageing and washing of coprecipitates, and optimal pH conditions are described, together with spectral interferences. Recoveries of most metals increased with an increase in pH.

4.5.3 OTHER COPRECIPITANTS

Other coprecipitants that have been used include 2(ethiazolyl)-2 naphthol (thorium, vanadium, uranium, and zinc) [64], Ni K_2 (Fe, Nb) (rubidium) [65], cobalt pyrolidine dithiocarbamate (copper, nickel, cadmium) [66], copper sulphide (mercury, lead, cadmium) [67], and gallium salts (aluminium, thallium, chromium, manganese, iron, cobalt, nickel, lead, copper, zinc, yttrium) [68].

4.6 PRECONCENTRATION BY FLOW INJECTION ANALYSIS

Flow injection analysis has been applied as an automatic sample injection technique for atomic absorption spectrometry [70–73], atomic absorption spectrometry.

Fang et al. [74] have described a flow injection system with online ion-exchange preconcentration on dual columns for the determination of trace amounts of heavy metals at μg/L and sub-μg/L levels by flame atomic absorption spectrometry. The degree of preconcentration ranges from 50- to 105-fold for different elements at a sampling frequency of 60 samples/h.

Fang et al. [74] concluded that the flow injection atomic absorption spectrometry system with online preconcentration will challenge the position of the graphite furnace technique because it yields comparable sensitivity for much lower cost by using simpler apparatus and separation mode. The method offers unusual advantages when matrices with high-salt contents such as seawater are analysed because the matrix components do not reach the nebuliser.

Marshall and Mettola [75] have used silica-immobilised 8-quinolol as a means of preconcentrating metals for analysis by flow injection atomic absorption spectrometry. This has proved to be a particularly useful material for sample preparation, matrix isolation, and preconcentration of trace metal ions.

The selectivity of silica-immobilised 8-quinolol for transition metal ions over the alkali metals and alkaline earth metal ions makes it useful for samples containing large quantities of the latter such as seawater.

These researchers evaluated breakthrough capacities under different flow, temperature, and geometric characteristics of the preconcentration columns. The columns have relatively high capacities for metals and do not suffer from complications due to swelling. Excellent agreement was obtained in the determination of copper and standard environmental samples in the concentration range of 9–250 μg/L.

4.7 CLOUD POINT EXTRACTION

The method is based on cloud point extraction of these metals as complexes of diethyldithiocarbamate in micellar media of nonionic surfactant octylphenoxypolyethoxyethanol (Triton X-114). The effect of interference from residual salinity in surfactant-rich phase was also investigated. The developed procedure allows to achieve enhancement factors of 20.0, 20.4, 19.5, and 20.6, along with the limits of detection of 0.030, 2.1, 0.62, and 0.27 μg/L, and precision was expressed as relative standard deviation (%RSD, $n = 10$) of 3.7% (40.0 μg/L), 5.7% (20.0 μg/L), 6.6% (20.0 μg/L), and 3.1% (10.0 μg/L) for Cd, Pb, NI, and CU, respectively. The accuracy was evaluated by spike tests on the seawater (salinity of 35%). Recoveries were between 79% and 105%.

REFERENCES

1. D.L. Tsalev, J.P. Almarin, and S.I. Neuman, *Zhur Analytical Khim*, 1972, 27, 1223.
2. F.F. El Enamy, K.F. Mamoud, and M.M. Varma, *Journal of Water Pollution Control Federation*, 1979, 51, 2545.
3. R.E. Pellenberg and T.M. Church, *Analytica Chimica Acta*, 1978, 97, 81.
4. H. Armansson, *Analytica Chimica Acta*, 1979, 110, 21.
5. L. Brugmans, L.G. Danielsson-Bagnusson, and S. Westerlund, *Marine Chemistry*, 1983, 13, 327.
6. E.A. Boyle and J.M. Edmond, *Analytica Chimica Acta*, 1977, 91, 789.
7. T.K. Jan and D.R. Young, *Analytical Chemistry*, 1978, 50, 1250.
8. K.W. Bruland, R.P. Franks, G.A. Knauer, and J.H. Martin, *Analytica Chimica Acta*, 1979, 105, 233.
9. L. Rasmussen, *Analytica Chimica Acta*, 1981, 125, 117.
10. R.D. Ediger, G.E. Persson, and J.D. Kerber, *Atomic Absorption Newsletter*, 1974, 13, 61.
11. R.G. Smith and H.L. Windom, *Analytica Chimica Acta*, 1980, 113, 39.
12. L.G. Danielsson, B. Magnusson, and S. Westerlund, *Analytica Chimica Acta*, 1978, 98, 48.
13. R.F. Sturgeon and S.S. Berman, A. Desauliniers, and D.S. Russell, *Talanta*, 1980, 27, 85.

14. K.R. Sperling, *Fresenius Zeitschrift für Analytische Chemie*, 1980, 301, 254.
15. R.E. Sturgeon, S.S. Berman, J.H.H. Desauliniers, A.P. Mykytink, J.M. McLaren, and D.M. Russell, *Analytical Chemistry*, 1980, 52, 1585.
16. J.M. Lo, J.C. Yu, E.I. Hutchinson, and C.M. Wal, *Analytica Chimica Acta*, 1982, 54, 2536.
17. J.M. Lo, J.C. Yu, E.I. Hutchinson, and C.M. Wal, *Analytica Chimica Acta*, 1982, 144, 183.
18. P.T. Statham, *Analytica Chimica Acta*, 1985, 169, 149.
19. L.G. Danielsson, B. Magnusson, and S. Westerlund, *Analytica Chimica Acta*, 1982, 144, 183.
20. E.A. Boyle and J.M. Edmond, *Analytica Chimica Acta*, 1977, 91, 189.
21. M.B. Collela, S. Siggia, and R.M. Barnes, *Analytical Chemistry*, 1980, 52, 2347.
22. C.C. Wan, S. Chaing, and A. Corsini, *Analytical Chemistry*, 1985, 57, 719.
23. D.J. Mackey, *Marine Chemistry*, 1982, 11, 164.
24. A. Kirivama and R. Kuroda, *Mikrochimica Acta*, 1985, 1, 405.
25. M. Koide, D.S. Lee, and M.O. Stallard, *Analytical Chemistry*, 1984, 56, 1956.
26. R. Kyrodo, K. Oguma, N. Mukai, and M. Iwamoto, *Talanta*, 1987, 34, 443.
27. A. Van Green and E. Boyle, *Analytical Chemistry*, 1990, 62, 1705.
28. J.A. Persson and K. Irgum, *Analytica Chimica Acta*, 1982, 138, 111.
29. R. Benesch and P. Mangelsdorf, *Helgolander Wiss Meersunters*, 1972, 23, 365.
30. R.W. Bruland, R.R. Franks, G.A. Knauer, and J.H. Martin, *Analytica Chimica Acta*, 1979, 105, 233.
31. R.E. Sturgeon, S.S. Berman, A.A. Desauliniers, and J.H. Martin, *Analytica Chimica Acta*, 1980, 27, 85.
32. L. Rasmussen, *Analytica Chimica Acta*, 1981, 125, 117.
33. B.R. Greenberg and H.M. Kingston, *Journal of Radioanalytical Chemistry*, 1982, 71, 147.
34. H. Kingston and P.A. Pella, *Analytical Chemistry*, 1981, 53, 223.
35. R. Boniforti, R. Ferraroli, P. Frigier, D. Heltav, and C. Queriazza, *Analytica Chimica Acta*, 1984, 162, 33.
36. A.J. Paulson, *Analytica Chemistry*, 1986, 58, 183.
37. F. Bafti, A.M. Cardinale, and R. Bruzzone, *Analytica Chimica Acta*, 1992, 270, 79.
38. S.E. Pai, *Analytica Chimica Acta*, 1988, 211, 271.
39. R.E. Sturgeon, S.N. Willie, and S.S. Berman, *Analytical Chemistry*, 1985, 57, 6.
40. M.B. Shabani, T. Tkagi, and A. Masuda, *Analytical Chemistry*, 1992, 64, 737.
41. M. Wang, A.I. Yusetousky, and R.G. Michael, *Microchemical Journal*, 1993, 48, 326.
42. S. Comber, *Analyst*, 1993, 118, 505.
43. R.S.S. Murthy and D.E. Ryan, *Analytica Chimica Acta*, 1982, 140, 163.
44. H. Ogura and K. Oguma, *Microchemical Journal*, 1994, 49, 220.
45. R.E. Sturgeon, S.S. Berman, S.N. Willie, and J.A. Desauliniers, *Analytical Chemistry*, 1981, 53, 2337.
46. S. Nakashima, R.E. Sturgeon, S.N. Willie, and S.S. Berman, *Fresenius Zeitschrift für Analytische Chemie*, 1988, 330, 592.
47. C.R. Lau and M. Yang, *Analytica Chimica Acta*, 1994, 287, 111.
48. M. Volkan, O.Y. Ataman, and A.G. Howard, *Analyst (London)*, 1987, 112, 1409.
49. M. Johansson, H. Emteborg, B. Glad, F. Reinholdsson, and D.C. Baxter, *Fresenius Journal of Analytical Chemistry*, 1995, 351, 461.
50. K.J. Orians and E.A. Boyle, *Analytica Chimica Acta*, 1993, 282, 63.
51. S. Blair, P. Appriou, and H. Handel, *Analytica Chimica Acta*, 1993, 272, 91.
52. K. Isshiki, F. Tsuji, T. Kuwamoto, and E. Nakayama, *Analytical Chemistry*, 1987, 59, 2491.
53. R.E. Greenberg and H.M. Kingston, *Analytical Chemistry*, 1983, 55, 1160.

54. K. Isshiki and E. Nakayama, *Analytical Chemistry*, 1987, 59, 291.
55. M. Fujita and K. Iwashima, *Environmental Science and Technology*, 1981, 15, 929.
56. M. Fernandez Garcia, R. Peleiro-Garcia, N. Bandel Garcia, and A. Sanz-Medel, *Talanta*, 1994, 41, 1833.
57. D.J. Mackay, *Marine Chemistry*, 1982, 11, 169.
58. M. Saylak, S. Saracoglu, L. Elei, and M. Dogon, *International Journal of Environmental Analytical Chemistry*, 2002, 82, 225.
59. P.A. Reimer and A. Miyazaki, *Journal of Analytical Atomic Spectroscopy*, 1992, 7, 1238.
60. K.W.H. Siu and S.S. Berman, *Analytical Chemistry*, 1984, 56, 1806.
61. L.M. Cabezon, M. Caballero, R. Cela, and J.A. Bustamante, *Talanta*, 1984, 37, 597.
62. Q. Chen Aarkog, S.F. Neilson, H. Dahlgaard, S. Nies, Y. Yu, and K. Handrup, *Journal of Radioanalytical and Nuclear Chemistry*, 1993, 172, 281.
63. T. Akagi, Y. Nojiri, M. Matsui, and H. Hargaduchi, *Applied Spectroscopy*, 1985, 39, 662.
64. R.R. Rao and A. Chat, *Journal of Radioanalytical Nuclear Chemistry*, 1993, 168, 439.
65. V.A. Lebedev, A.V. Alvarez, A.V. Krainyanskii, T.I. Shurypuva, and I.V. Golubstov, *Vesin Mosk University Serz Khim*, 1987, 28, 174.
66. E.A. Boyle and J.M. Edmond, *Analytica Chimica Acta*, 1977, 91, 189.
67. A.A. Patin and M.P. Morozov, *Zhurnal Analitcheskoi Khimi*, 1976, 31, 282.
68. T. Akaji and H. Haraguchi, *Analytical Chemistry*, 1990, 62, 81.
69. S. Olsen, C.R. Latz, J.B. Pessenda, and E.O. Hansen, *Analyst*, 1983, 108, 905.
70. J.F. Tyson, J.M.H. Appleton, and A.B. Idris, *Analyst*, 1983, 108, 153.
71. H. Kimura, K. Oguma, and R. Kyroda, *Bensek Kagacku*, 1983, 32, 179.
72. N. Zhou, W. Frech, and E. Lundberg, *Analytica Chimica Acta*, 1983, 33, 153.
73. J.F. Tyson and A.B. Idris, *Analyst*, 1984, 109, 26.
74. Z. Fang, J. Ruzicka, and E.H. Hansen, *Analytica Chimica Acta*, 1984, 164, 23.
75. H. Marshall and H.A. Mettola, *Analytical Chemistry*, 1985, 57, 729.
76. J.A. Resing and O.J. Measures *Analytical Chemistry*, 1994, 6, 4105.
77. I. Dakova, P. Vasileva, I. Karadjova, M. Karadjova, and V. Slaveykova, *International Journal of Environmental Analytical Chemistry*, 2011, 91, 62.

5 Determination of Organic Compounds in Seawater

5.1 ALIPHATIC AND AROMATIC HYDROCARBONS

The variation between crude oils and their resistance to the sea-weathering process is not enough in many cases for providing the unequivocal identification of the pollutant. The *n*-paraffins can, apparently, be removed by biodegradation, as well as the lower acyclic isoprenoids at, respectively, slower rates [1]. The flame photometric detector chromatogram is less sensitive to modification by bacterial metabolism but can be affected by evaporation, in spite of its higher retention range. However, the last part of the flame ionisation detector chromatogram appears to be highly promising in overcoming these limitations. In fact, this part corresponds to a hydrocarbon fraction that boils at more than 400°C, so it cannot easily be evaporated under environmental marine conditions. Moreover, it contains a wealth of compounds of geochemical significance, namely isoparaffins and polycyclic alkanes of isoprenoids, sterane, and triterpane, structure [2–4] as a result of a complete reduction of precursor isoprenyl alcohols, sterols, and triterpanes, respectively. Therefore, their occurrence and final distribution in crude oils will be related to their particular genetic history, that is, to the original sedimentary organic matter and to the processes undergone during its geochemical cycle. These factors provide unique hydrocarbon compositions for each crude oil by which the unambiguous identification of the samples can be brought about. Besides their geochemical stability, these compounds do also remain unaltered after biodegradation [5], being, in this respect, valuable passive tags for characterising marine pollutants.

The objective of a study by Venosa et al. [6] was to measure the biodegradability of the 19-year lingering crude oil in laboratory microcosms. Samples of beach substrate were collected from representative sites in Prince William Sound contaminated with oil residues of varying weathering states according to the mass weathering index model. Enough sacrificial microcosms were set up to accommodate two treatments for each site (natural attenuation and biosimulation). Results indicated that lingering oil is biodegradable. Nutrient addition stimulated biodegradation compared to natural attenuation in all treatments regardless of the degree of weathering. The most weathered oil according to the mass weathering index was the most biodegradable.

Substantial biodegradation occurred in the natural attenuation microcosms due to the high sediment content. Total Kjeldahl nitrogen served as a nitrogen source for biodegradation. Most of the observed biodegradation was due to the presence of dissolved oxygen. Nitrogen was a limiting factor, but oxygen was the predominant one.

A spectrofluorometer capable of dispersed-spectrum, simultaneous, multiwavelength UV excitation and collection of luminescence has been used to qualitatively and quantitatively determine aromatic hydrocarbon pollutants dissolved in ocean water. Hydrocarbon fluorescence data produced by this instrument are in the form of excitation emission matrices, which provide more spectral information about these complex mixtures than is available from conventional excitation or emission fluorescence profiles.

This method is capable of determining low parts-per-billion concentrations of two primary fluorescent compounds, naphthalene and styrene, found in ocean water exposed to gasoline despite the presence of uncalibrated interference from similar aromatic compounds, the ocean water matrix, and the instrumental background.

Further development of this method is an excitation/emission matrix imaging spectrofluorometer employed for the quantitation of the two fluorescent compounds, naphthalene and styrene, contained in ocean water exposed to gasoline. Multidimensional parallel factor analysis models were used to resolve the naphthalene and styrene fluorescence spectra from a complex background signal and overlapping spectral interference not included in the calibration set. Linearity was demonstrated more than two orders of magnitude for the determination of naphthalene with a detection limit of 8 ppb. Similarly, nearly two orders of magnitude of linearity was demonstrated in the determination of styrene with an 11 ppb limit of detection. Furthermore, the synthesis of the execution/emission matrix spectrofluorometer and the multidimensional parallel factor analysis could be used for the unbiased prediction of naphthalene and styrene concentration in mixture samples containing uncalibrated spectral interferents.

5.1.1 GAS CHROMATOGRAPHY

Fam et al. [7] determined hydrocarbons in runoff water from catchments in San Francisco Bay using liquid chromatography and high-resolution gas chromatography.

Wade and Quinn [8] measured the hydrocarbon content of sea surface and subsurface samples. Hydrocarbons were extracted from the samples and analysed by thin-layer and gas–liquid chromatography. The hydrocarbon content of the surface microlayer samples ranged from 14 to 599 µg/L with an average of 155 µg/L, and the concentration in the subsurface samples ranged from 13 to 239 µg/L and averaged 73 µg/L. Several isolated hydrocarbon fractions were analysed by infrared spectrometry, and each fraction was found to contain a minimum of 95% hydrocarbon material, including both alkanes and aromatics.

Although the weathering of marine oil pollutants can be such that the oil is rendered unrecognisable, the time required to achieve this was found to be so long as to be insignificant in regard to pollutant identification.

Occasionally, the mound of unresolved components on the chromatogram supporting the superimposed n-alkane peaks confuses the true n-alkane profile. This has been overcome by separating the n-alkanes using molecular sieves prior to gas chromatography. However, separation of n-alkanes in this way, or by urea complex formation [9], is reported as being more applicable to distillates rather than residual materials, and also the separation is not entirely specific. In the case of marine pollutants, it has been found advantageous to chromatograph a distilled residue (bp > 343°C) [10] or fraction (pb 254–370°C), which avoids problems caused by the evaporation of the lower ends by weathering.

Zafiron and Oliver [11] have developed a method for characterising environmental hydrocarbons using gas chromatography. Solutions of samples containing oil were separated on an open tubular column coated with OV–1010 and temperature programmed from 75°C to 275°C at 6°C per minute; helium was used as the carrier gas and detection was done by flame ionisation.

Rasmussen [12] has described gas chromatography methods for the identification of hydrocarbon oil spills. The spill samples are analysed on a 30.5 m Dexsil-300 support-coated open tube (SCOT) column to obtain maximum resolution.

5.1.2 MASS SPECTROMETRY

Walker et al. [13] examined several methods and solvents for use in the extraction of petroleum hydrocarbons from estuarine water and sediments during an *in situ* study of petroleum degradation in seawater. The use of hexane, benzene, and chloroform as solvents was discussed and compared, and quantitative and qualitative differences were determined by analysing using low-resolution computerised mass spectrometry. Using these data and data obtained following the total recovery of petroleum hydrocarbons, it is concluded that benzene–methanol azeotropes are the most effective solvents.

Smith [14] classified large sets of hydrocarbon oil spectra data by computer into 'correlation sets' for individual classes of compounds. The correlation sets were then used for determining the class to which an unknown compound belonged from its mass spectral parameters. A correlation set was constructed by the use of ion series summation, in which a low-resolution mass spectrum was expressed as a set of numbers representing the contribution to the total ionisation of each of the 14 ion series. The technique is particularly valuable in the examination of results from coupled gas chromatography–mass spectrometry of complex organic mixtures.

5.1.3 ELECTROLYTIC STRIPPING

Wasik [15] used an electrolytic stripping cell to determine hydrocarbons in seawater. Dissolved hydrocarbons in a known quantity of seawater were equilibrated with hydrogen bubbles, evolved electrolytically from a gold electrode, rising through a cylindrical cell. In an upper headspace compartment of the cell, the hydrocarbon concentration was determined by gas chromatography. The major advantages of this cell are that the hydrocarbons in the upper compartment are in equilibrium with the hydrocarbons in solution, and that the hydrogen used as an extracting solvent does not introduce impurities into samples.

Wasik [15] used this method to successfully determine the ppb of gasoline in seawater. He found that a convenient method for the concentration of hydrocarbons was to recycle the hydrogen steam containing the hydrocarbons many times over a small amount of charcoal (2.3 mg). The charcoal, while still in the filter tube, was extracted three times with 5 μL of carbon disulphide. A 2–3 μL aliquot of this solution was then injected into a SCOT capillary column.

5.1.4 Purge and Trap Analysis

Dewulf and van Langenhove [16] described a purge and trap-gc-MS technique for the analysis of volatile organic compounds. The method was to measure simultaneously concentrations of $13C_1$- and C_2-chlorinated hydrocarbons and monocyclic aromatic hydrocarbons down to the 1 ng/L level. Low reproducibility was caused by the water vapour present in the gas stream after purging samples. An offline purge system, combined with thermal desorption by means of the online apparatus, gave reproducible measurements (standard deviations from 2.6% to 15.7%). Contamination was lower than 0.76 ng/L for all volatile organic compounds except for benzene, toluene, and chloroform. Limits of detection of 0.48 (tetrachloroethylene) to 1.25 ng/L (o-xylene) were obtained, except for chloroform, benzene, and toluene. Their limits of detection were 4.93, 4.79, and 2.68 ng/L because of the contamination. The results of the analysis of water samples taken in the North Sea region and the Scheldt estuary illustrate the applicability of the method. Target compounds were detected in concentrations from 0.5 to 100 ng/L.

5.1.5 Gel Permeation Chromatography

Done and Reid [17] applied gel permeation chromatography in the identification of crude oils and products isolated from estuary and seawater. The technique, which appears more suited to the analysis of crude oils, is based on the separation of oil components in order of their molecules.

5.1.6 Spectrofluorometry

Muroski et al. [18,19] carried out single-measurement activation/emission matrix fluorometry to determine hydrocarbon in ocean water.

5.1.7 Photochemical Oxidation

Rontani and Giral [20] studied the significance of photosensitised oxidation of alkanes during the photochemical degradation of petroleum hydrocarbon fractions in seawater.

In this study, a crude oil hydrocarbon fraction as a surface film on seawater was photooxidised under simulated environmental conditions. After irradiation, gas chromatography and gas chromatography/mass spectrometry analyses demonstrated the presence of relatively high quantities of compounds deriving from the photosensitised oxidation of n-alkanes, pristane and phytane, being among the non-acidic photoproducts of this fraction. The results suggest that the photochemical degradation of alkanes should be considered in studies of the fate petroleum crudes in the marine environment.

El Anba-Iurot [21] studied the photochemical oxidation of 3.3^1 and 4.4^1 dimethylbiphenyl in seawater.

Degradation studies have been conducted in natural seawater, without photosensitiser and under artificial UV irradiation. The photooxidation process takes place mainly on benzene ring side chains. Photoproduct analyses carried out by GC-MS, FTIR, and GC-FTIR confirm three successive oxidation states of methyl substituents: alcohol, aldehyde, and carboxylic acids.

5.2 AROMATIC HYDROCARBONS

5.2.1 Fluorescence Spectrometry

Hiltabrand [22] has investigated the fluorimetric determination of PAHs in seawater.

Payne [23] carried out a field investigation of benzopyrene hydrolysate induction as a monitor for marine petroleum pollution. Isaaq et al. [24] isolated stable mutagenic ultraviolet photodecomposition products of benzo(a)pyrene by thin-layer chromatography.

Other aspects that have been covered in the literature are water column processes of aliphatic and polycyclic aromatic hydrocarbons in a water column [25] and the atmospheric and marine sedimentation fluxes of polycyclic aromatic hydrocarbons in seawater [26].

Ballesteros-Gomez and Rubio [59] used hemimicelles of alkyl [C10–C18] carboxylates chemisorbed into magnetic nanoparticles to extract polycyclic aromatic hydrocarbons from surface and seawater. Tetradecanoate hemimicelles were used to extract carcinogenic polycyclic aromatic hydrocarbons prior to analysis by liquid chromatography using a C_{18} Supelcosil LC-PAH column and a gradient elution program with water and acetonitrile and fluorescence detection. The procedure involved stirring filtered aqueous samples (350 mL) with 200 mg of tetradecanoate-coated magnetic nanoparticles for 15 min, isolating the sorbent with an Nd–Fe–B magnet and eluting the carcinogenic polycyclic aromatic hydrocarbons with a mixture of acetonitrile and tetrahydrofuran. The preconcentration factor was 116. No cleanup of the extracts was needed, and the method proved matrix independent. The limits of quantitation thus obtained, 0.2–0.5 ng/L, meet the stringent water quality requirements established by the recently amended European Water Framework Directive 2000/60/EC and also the U.S. EPA for the determination of carcinogenic PAHs in surface water and groundwater.

5.3 DISSOLVED GASES

The rise and dissolution of carbon dioxide and methane particles (bubbles or drops) from the deep ocean is important for understanding the rate of the release of these components from marine sediments.

Wernecke et al. [27] have described a device for the continuous *in situ* measurement of methane in seawater using laser absorption spectrometry. The detection limit of this method was 30 ppt.

Bozzano and Dente [28] studied the dissolution of carbon dioxide and methane bubbles rising from the oceans.

Bell et al. [29] used membrane inlet mass spectrometry for the quantitative measurements of dissolved gases and volatile organics over a wide range of ocean depths. This requires characterisation of the influence of hydrostatic pressure on the permeability of membrane inlet mass spectrometry systems. To simulate measurement conditions in the field, a laboratory apparatus was constructed for the control of sample flow rate, temperature, pressure, and the concentrations of a variety of dissolved gases and volatile organic compounds. Membrane inlet mass spectrometry data generated with this apparatus demonstrated that the permeability of polydimethylsiloxane membranes was strongly dependent on hydrostatic pressure. For the range of pressure encountered between the surface and 2000 m ocean depths, the pressure-dependent behaviour of polydimethylsiloxane membranes could not be satisfactorily described using previously published theoretical models of membrane behaviour. The observed influence of hydrostatic pressure on signal intensity could, nonetheless, be quantitatively modelled using a relatively simple semi-empirical relationship between permeability and hydrostatic pressure. The semi-empirical membrane inlet mass spectrometry calibration developed in this study was applied to *in situ* underwater mass spectrometer data to generate high-resolution, vertical profiles of dissolved gases in the Gulf of Mexico. These measurements constitute the first quantitative observations of dissolved gas profiles in the oceans obtained by *in situ* membrane inlet mass spectrometry. Alternative techniques used to produce dissolved gas profiles were in good accord with underwater measurements.

5.4 PHENOLS

Boyd [30] used aqueous acetylation followed by gas chromatography–mass spectrometry to determine ppt concentrations of phenols in seawater.

Dimou et al. [31] discussed the determination of phenol compounds in the marine environment of Thermaikos Gulf in northern Greece.

5.5 CARBOXYLIC ACIDS

5.5.1 Gas Chromatography

Quinn and Meyers [33] discussed a gas–liquid chromatographic method for the determination of dissolved organic acids in seawater.

5.5.2 Atomic Absorption Spectrometry

Treguer and Le Corre [34] determined the total dissolved free fatty acids in seawater. The sample (1 L) was shaken with chloroform to remove the free fatty acids, and the extract was evaporated to dryness under reduced pressure at 50°C. Chloroform–heptane (29:21) (2 mL) and fresh (M triethanolamine-M acetic acetate – 6.8% $CuSO_4 \cdot 5H_2O$) solution (9:1:10) (0.5 mL) was added to the residue. The solution was shaken vigorously for 3 min and centrifuged at 3000 rpm for 5 min. A portion of the organic phase was evaporated to dryness, and 1% ammonium diethyldithiocarbamate solution in isobutyl methyl ketone was added to the residue to form a yellow copper complex. The copper in the solution was determined by atomic absorption spectrometry at 324.8 nm (air–acetylene flame). Palmitic acid was used to prepare a calibration graph. The standard deviation for samples containing 30 µg/L of free fatty acids (as palmitic acid) was ± 1 µg/L.

5.5.3   DIFFUSION METHOD

Xiao et al. [36] determined nanomolar quantities of individual low-molecular-weight carboxylic acids (and amines) in seawater. This method is based on the diffusion of acids across a hydrophobic membrane to concentrate them and separate them from inorganic salts and most other dissolved organic compounds. Acetic, propionic, butyric, valeric, pyruvic, acrylic, and benzoic acids were all found in measureable amounts of seawater.

Kieber et al. [37] determined formate in natural waters, including seawater, by a procedure based on coupled enzymatic high-performance liquid chromatography. The procedure is capable of determining down to 4–10 μm carboxylic acids.

In this procedure, formate is oxidised with formate dehydrogenase, and this is accompanied by a corresponding reduction of β-nicotinamideadenine dinucleotide (β(NAD)$^+$) to reduced βNADH. βNADH is quantified by high-performance liquid chromatography.

5.5.4   HIGH-PERFORMANCE LIQUID CHROMATOGRAPHY

Low-molecular-weight C_1–C_8 aliphatic and aromatic amines and carboxylic acids are important metabolic intermediates in decomposition processes and are widely distributed in the marine environment [76–79]. However, the analysis of individual low-molecular-weight amines and organic acids in seawater is difficult because of their extremely low concentrations (nanomolar) and the presence of high concentrations of inorganic salts. The high water solubilities and adsorption tendencies of amines and organic acids make analysis difficult. Short-chain organic acids in sediment pore water and in coastal water sample have been determined by high-performance liquid chromatography [80] and gas chromatography [81,82]. After derivatisation or vacuum distillation of acidified samples, atmospheric short-chain organic acids in sample and in glacial meltwater have been determined by ion chromatography [83,84]. However, these techniques cannot generally be applied to water column samples due to their low concentrations in seawater and high blank volume.

Goncharova and Khomemkô [35] described a column chromatographic method for the determination of acetic, propionic, and butyric acids in seawater and thin-layer chromatographic methods for determining lactic, aconitic, malonic, oxalic, tartaric, citric, and malic acids. The pH of the sample is adjusted to 8–9 with sodium hydroxide solution. It is then evaporated almost to dryness at 50°C–60°C and the residue was washed on a filter paper with water acidified with hydrochloric acid. The pH of the resulting solution is adjusted to 2–3 with hydrochloric acid (1:1), the organic acids are extracted into butanol, and then back-extracted into sodium hydroxide solution; this solution is concentrated to 0.5–0.7 mL and acidified, and the acids are separated on a chromatographic column.

5.5.5   ION-EXCHANGE CHROMATOGRAPHY

Some typical results obtained by this procedure are given in Table 5.1.

TABLE 5.1

Chromatography for the Determination of Free and Combined Fatty Acids in Seawater

Determined	Range	Details	Method of Analysis	References
18 FFA	CFA 20–85 µg/L		Ion-exchange chromatography	[65]
FFA tank	20–310 µg/L	Pre-extracting with ethyl acetate	Ion-exchange chromatography	[66]
North Sea water	1.8–8.5 µg/L	Cu-Chelex method	Ion-exchange chromatography	[67,68]
15 FFA		Ligand-exchange Cu-Chelex and trace recovery technique		
Up to 30 FFA	FFA 20–180 µg/L	Desalting with Dowex 500WX8 at pH 4; elution with NH4; lyophilisation with added glycerine	Ion-exchange chromatography, ninhydrin detection 3.5 h analyses, 100 pmol sensitivity	[69]
Up to 29 FFA and CFA identified, Baltic Sea water	CFA ca. 500 µg/L	Desalting with Dowex as earlier; sample size 40 mL seawater; glycerine added before evaporation	Ion-exchange chromatography fluorimetric detection – 3 h analyses, sensitivity 20 pmol	[70]
FA, FFA, and CFA, oceanic	0–5 µg/L FFA; 10–120 µg/L CFA (totals)	Ligand-exchange Cu-Chelex (no wash water)	Ion-exchange chromatography	[71]
18 FFA and CFA	5–92 µg/L FFA; 28–200 µg/L CFA	Semiautomated Cu-Chelex method	Ion-exchange chromatography	[72]

Notes: FFA, free amino acids; CFA, combined amino acids.

5.5.6 Gas Chromatography

Yang et al. [75] developed a concentrating technique using diffusion of carboxylic acids (and amines) across a hydrophobic membrane for the gas chromatographic determination of these compounds in seawater. Concentration factors of up to 1000 were achieved.

5.6 CARBOHYDRATES

5.6.1 Gas Chromatography

Eklund et al. [38] developed a method for the sensitive chromatographic analysis of monosaccharides in seawater, using trifluoroacetyl derivatisation and electron capture detection. It is difficult to determine accurately the monosaccharide concentrations by this method because a number of chromatographic peaks result from each monosaccharide.

5.6.2 Partition Chromatography

Josefsson et al. [39] determined soluble carbohydrates in seawater by partition chromatography after desalting by ion-exchange membranes. The electrodialysis cell used had a sample volume of 430 mL and an effective membrane surface area of 52 cm^2. Perinaplex A-20 and C-20 ion-exchange membranes were used. The water-cooled carbon electrodes were operated at up to 250 mA at 500 V. The desalting procedure normally took less than 30 h. After the desalting, the samples were evaporated nearly to dryness at 40°C *in vacuo*, then taken up in 2 mL of 85% ethanol, and the solution was subjected to chromatography on anion-exchange resins (sulphate form) with 85% ethanol as the mobile phase. With this procedure, it was possible to determine eight monosaccharides in the range of 0.15–46.5 μg/L with errors of less than 10% and detect traces of sorbose, fucose, sucrose, diethylene glycol, and glycerol in seawater.

5.6.3 Miscellaneous

Williams [40] studied the rate of oxidation of ^{14}C-labelled glucose in seawater by persulphate. After oxidation, carbon dioxide was blown off and residual activity was measured. For glucose concentrations of 2000, 200, and 20 μg/L, residual radioactivities (as a percentage of total original radioactivity) were 0.04, 0.05, and 0.025, respectively, showing that biochemical compounds are extensively oxidised by persulphate. With the exception of change in temperature, modifications of conditions had little or no effect. Oxidation for 2.5 h at 100°C was the most efficient.

5.7 CHLORO COMPOUNDS

5.7.1 Aliphatic Chloro Compounds

Lovelock [41,42] determined methyl fluoride, methyl chloride, methyl bromide, methyl iodide, and carbon tetrachloride in the Atlantic Ocean. This shows a global distribution of these compounds. Murray and Riley [43,44] confirmed the presence of carbon tetrachloride and also found low concentrations of chloroform and tri- and tetrachloroethylene in Atlantic surface waters.

The Bellar purge and trap method [45] has been applied for determining vinyl chloride, acetone, methylene chloride, chloroform, carbon tetrachloride, and trichloroethene in seawater. Using the Hall electrolytic conductivity detector, no response was obtained for the acetone used to prepare the vinyl chloride standard solution.

Dawson et al. [46] described samplers for large volume collection of seawater samples for chlorinated hydrocarbon analyses. The samplers use the macroreticular absorbent Amberlite XAD-2. Operation of the towed 'fish' type sampler causes minimal interruption to a ship's program and allows a large area to be surveyed. The second type is a self-powered *in situ* pump, which can be left unattended to extract large volumes of water at a fixed station.

Zocoolillo et al. [49] analysed aqueous matrices from Antarctica for three volatile chlorinated hydrocarbons: tetrachloromethane, trichloroethylene, and tetrachloroethylene. The matrices analysed were snow from Rennick Nèvè and Rennick Glacier sampled during the Italian Expeditions of 1995/1996 and 1996/1997, respectively, and seawater, pack ice, sea-microlayer, sub-superficial water, and freshwater collected during the Italian Expedition of 1997/1998. An extraction procedure was developed, suitable for large volumes of water (10 L), in order to combine the extraction of other classes of organic compounds (polychlorinated biphenyls, polycyclic aromatic hydrocarbons, and chlorinated pesticides) with those of direct interest. The volatile chlorinated hydrocarbon organic extracts were analysed in Italy by GC-ECD and GC-MS. The analyses confirmed the presence of the three halocarbons in Antarctica in quantities up to 10–15 ng/kg. The results were evaluated with respect to the local distribution of these compounds and their diffusion on a global scale.

Sobek et al. [57] pointed out that the sampling step was likely to be the longest sampler of concentration in the analytical chain other than studying hydrophobic contamination in open sea surface waters.

5.7.2   HEXACHLOROCYCLOHEXANES

Rice and Shigaev [47] studied the spacial distribution of hexachlorocylcohexane in the Bering and Chukchi Sea Shelf ecosystem. The concentration of alpha and gamma hexachlorocylcohexanes in sea ice in the Arctic has been measured using a sump-hole technique [48]. Wiberg et al. [60] examined concentrations and fluxes of hexachlorocyclohexane and chlorohexadiene in environmental samples from the southern Pacific Sea.

5.7.3   POLYCHLOROBIPHENYLS

Both gas chromatography [50] and high-performance liquid chromatography [51] have been used to determine polychlorobiphenyls in seawater. Polychlorobiphenyls have persistent compounds which cause toxic effects in various marine organisms.

Sobek et al. [58] pointed out that the sampling step was likely to be the longest source of uncertainty in the analytical chain when studying hydrophobic contaminants in open sea surface waters. Their four different water-sampling systems for the collection of particle-bound and dissolved polychlorinated biphenyls were evaluated in the surface water of the Baltic Sea. Individual concentrations of 16 polychlorobiphenyl congeners sampled with stainless steel tubing (system A) spanned between 0.04 and 3 fM (0.02 and 1 pg/L) in the particulate fraction and between 0.04 and 8 fM (0.01 and 2 pg/L) in the dissolved fraction. The sampling variance introduced to the data was expressed as relative standard deviation (RSD%) for single polychlorobiphenyl congeners in the system. System A varied between 14% and 77% (average 35%) in the dissolved fraction and between 8% and 20% (average 13%) in the particulate fraction. The implication of constrained sampling and total variances is that variations in environmental data that are larger than about 50% RSDD may be interpreted as reflecting natural processes as opposed to merely methodological variance.

Schneider et al. [52] described a method of measuring dissolved polychlorinated biphenyls in natural waters on 30 min timeframes using negligible depletion nonequilibrium solid-phase microextraction. The detection limits range from 0.6 to 5.6 ng/L. Solid-phase microextraction fibres made from optical cable were inserted into a glass tube and attached to the shaft of a motor that revolved at 130 rpm to move the solid-phase microextraction fibres through the sampled water at a constant rate. To test for matrix interferences, measurements were made in three solutions with the same known dissolved polychlorobiphenyl concentration but different matrices. Dissolved polychlorobiphenyl measurements made in the presence of 8 mg/L of dissolved organic carbon and 200 mg/L of suspended solids were not significantly different from the measurements made in deionised water, demonstrating that neither matrix interfered with solid-phase microextraction measurements, suggesting that XAD measurements included DOC-associated PCBs.

Fuoco et al. [53] analysed polychlorobiphenyl in seawater samples gathered during the 1990/1991 Italian Expedition in Terra Nova Bay–Gerlache Inlet. These showed a typical total concentration of 550 pg/L. A slight increase in polychlorobiphenyl content was observed in the surface layer of the water column before pack-ice melting, while an increase by a factor of about 2 was found in the same water layer after pack-ice melting. Polychlorobiphenyl congener distributions were centred on 3- to 6-substituted and on 4- to 7-substituted congener classes for seawater and sediment/soil sampled, respectively, and were typical for each matrix considered.

Tenax-Celite [54,55] and XAD-2 resin [50] have been used to preconcentrate polychlorobiphenyls from seawater.

The Ob and Yenisei Rivers contribute 37% of riverine freshwater inputs to the Arctic Basin and thus represent an important pathway for the land–Arctic Ocean exchange of contaminants.

The Yenisei (2003) and Ob (2005) River estuaries and Kara Sea contribute 37% of riverine inputs to the Arctic Basin [56], and these represent an important pathway for the land–Arctic Ocean exchange of contaminants. Contaminant analyses were performed by high-resolution mass spectrometry on sample extracts taken from filtered large volume water samples (50–100 L) and concentrated *in situ* onto XAD-2 resin columns. Hexachlorocyclohexanes, polychlorinated biphenyl mixtures Sovol and trichlorodiphenyl, dichlorodiphenyltrichloroethane, as well as 'penta' brominated technical mixtures of polybrominated diphenyl ethers are important contributors to persistent organohalogen contamination for these waterways. Dissolved fluxes to the Kara Sea were estimated at hexachlorocyclohexane biphenyl 63 kg/year, dichlorobiphenyl trichloroether 16 kg/year, hexachlorobenzene 8 kg/year, α-endosulfan 8 kg/year, Dieldrin 5 kg/year, polybrominated bisphenylether 4 kg/year, and chlordanes 4 kg/year. Contaminant fluxes from these rivers are similar to those reported for major Canadian rivers, confirming expectations that the Ob and Yenisei are also major point sources for the Arctic Basin.

5.7.4 CHLORINATED BISPHENOL A PRODUCTS

Liu et al. [32] investigated the formation of chlorinated intermediates from bisphenol A in surface waters under simulated solar light irradiation.

To illustrate the possibility of photochemical formation of organochlorine compounds in natural water, Liu et al. [32] studied the phototransformation of bisphenol A in aqueous saline solution containing Fe(III) and fulvic acid, and in coastal seawater under simulated solar light irradiation. 2-(3-Chloro-4-hydroxyphenyl)-2-(4-hydroxyphenol) propane (3-CIBPA) and 2,2-bis(3-chloro-4-hydroxyphenol) propane (3,3-diCIBPA) were the main chlorinated derivatives during the processes. Laser flash photolysis and electron spin resonance results indicated that the chlorination of bisphenol A was most likely due to the formation of $Cl_2^{\bullet-}$ radical as a consequence of Fe(III) irradiation, yielding $Cl^{\bullet}$ and $OH^{\bullet}$ radical species and finally forming $Cl_2^{\bullet-}$ radical upon further reaction with chloride. The formation of Fe(III)–fulvic acid complex, which is a normal coexistence configuration of Fe(III) and fulvic acid in natural water, promoted the bisphenol A chlorination through producing more $Cl_2^{\bullet-}$ radicals. Moreover, fulvic acid had two opposite effects: forming Fe(III)–fulvic acid complex to enhance $Cl_2^{\bullet-}$ formation and competing radical with bisphenol A which resulted in different overall effects at different concentrations. Bisphenol A chlorination was enhanced by the increase in fulvic acid concentration when the salty acid concentration was less than 3.2 ng/L, and when the concentration of fulvic acid was as high as 10 mg/L, it slowed down, obviously. The described bisphenol A photochlorination process took place from pH 6.3 to 8.5 and increased with the increase in chloride concentration, indicating it could occur universally in natural saline surface water. These results propose a natural photochemical source for organochlorine compounds.

5.8 BROMO COMPOUNDS

Illna and Hunziker [61] examined the spatial and temporal distribution of the flame-retardant hexabromocyclododecane in the North Sea during 1995 to 2005 using a pollutant transport model FANTOM. Model calculations allow conclusions on relevant sinks and fluxes in and out of the North Sea and on the time needed to establish a steady state. Calculations were performed for two additional scenarios with different rates of primary degradation ranging from fast degrading to absolute persistency. Concentrations calculated in the scenarios with degradation are in line with the monitoring data available for hexabromododecane. Concentrations calculated in the 'persistent' scenario disagree with measured data. According to model calculations, a steady state is established within months for the water and the top layer sediment with no evidence for a temporal trend, except for the 'persistent' scenario, in which concentrations increase continuously in the southeastern part of the North Sea, where hydrographic and circulation characteristics produce areas of converging currents. This study enables a better understanding of the fate of hexabromododecane in the North Sea, its potential for transport, and overall elimination.

5.9 FLUORO COMPOUNDS

Atmospheric deposition is a major pathway of polychlorododecyl fluorides to the Baltic Sea. Sobek et al. [57,58] studied the aerosol–water distribution for aerosols collected close to the Baltic Sea in order to investigate the availability of pollutants sorbed to aerosols deposited in water. Aerosols were analysed for both total concentration (Soxhlet extraction) and the freely dissolved water concentration (extraction with 17 µm polyoxymethylene equilibrium passive samplers). Concentrations of polychlorododecyl fluoride and sum polychlorobiphenyl-7 in aerosols were 65–1300 pg/g dw TEQ and 22–100 ng/g dw, respectively. Organic carbon–normalised aerosol–water distribution ratios ($K_{aer-water,OC}$) were consistently lower (factor 2–60) than previously determined sediment organic carbon–water ratios ($K_{sed,OC}$). Hence, polychlorododecyl fluorides and polychlorobiphenyls entering the Baltic Sea through aerosols disposition seem to be more available for desorption to the water phase than polychlorododecyl fluoride and polychlorobiphenyls sorbed sediment. Further, whether aerosol–water distribution may be predicted from the air–aerosol partitioning constant multiplied by Henry's law constant was investigated. This proposed model for aerosol–water distribution underestimated the measured value for polychlorododecyl and polychlorobiphenyl by more than a factor 10. These findings can be used to improve the future fate modelling of polychlorobiphenyls and polychlorododecyl in marine environments and specifically in the Baltic Sea.

Ahrens et al. [63] determined the longitudinal and latitudinal distribution of perfluoroalkyl compounds in the surface water of the Atlantic Ocean. In this study, perfluoroalkyl compounds were determined in 2 L surface water samples collected in the Atlantic Ocean onboard the research vessels *Maria S. Merian* along the longitudinal gradient from Las Palmas (Spain) to St John's (Canada) (15° W–52° W) and *Polarstern* along the latitudinal gradient from the Bay of Biscay to the South Atlantic Ocean (46° N–26° S) in the spring and fall of 2007, respectively. After filtration, the dissolved and particulate phases were extracted separately, and perfluoroalkyl compounds concentrations were determined using high-performance liquid chromatography interfaced to tandem mass spectrometry. No perfluoroalkyl compounds were detected in the particulate phase. This study provides the first concentration data of perfluorooctanesulfonamide, perfluoroheaxanoic acid, and perfluoroheptanoic acid. Results indicate that trans-Atlantic Ocean currents caused the decreasing concentration gradient from the Bay of Biscay to the South Atlantic Ocean and the concentration drop-off close to the Labrador Sea. Maximum concentrations were found for perfluro octane sulphonate and perfluorooctanoic acid to 302, 291, and 229 pg/L, respectively. However, the concentration of each single compound was usually in the tens of picograms per litre range. South of the equator, only perfluoroalkyl compounds could be detected.

Bullister and Weiss [62–64] have reported on the spatial distribution of C_4, C_6, and C_8 perfluoroalkyl sulphonates, C_6–C_{14} perfluoroalkyl carboxylates, and perfluorooctanesulfonamide in the Atlantic and Arctic Oceans, including previously unstudied coastal waters of North and South America, and the Canadian Arctic Archipelago. Perfluorooctanoate and perfluro octane sulphonate were typically the dominant perfluoroalkyl acids in Atlantic water. In the mid-northwest Atlantic/Gulf Stream, some perfluoroalkyl acids, concentrations (ΣPFAAs) were low (77–190 pg/L) but increased rapidly upon crossing into U.S. coastal waters (up to 5800 pg/L near Rhode Island). Perfluoroalkyl acids in the northeast Atlantic were highest north of the Canary Islands (280–980 pg/L) and decreased with latitude. In the South Atlantic, concentrations increased near Rio de la Plata (Argentina/Uruguay; 350–540 perfluoroalkyl acids) possibly attributed to insecticides containing N-ethyl perfluorooctanesulfonamide, or proximity to Montevideo and Buenos Aries. In all other southern hemisphere locations, perfluoroalkyl acids were less than 210 pg/L. Perfluorooctanoate/perfluorooctane sulphonate ratios were typically ≤1 in the northern hemisphere, ~1 near the equator, and ≤ in the southern hemisphere. In the Canadian Arctic, perfluoroalkyl acid ranged from 40 to 250 pg/L, with perfluoroheptanoate, perfluorooctanoate, and perfluorooctane sulphonate among the perfluoroalkyl acids detected at the highest concentrations. Perfluorooctanoate/perfluorooctane sulphonate ratios (typically 1) decreased from Baffin Bay to the Amundsen Gulf, possibly attributable to increased atmospheric inputs. These data help validate global emissions models and contribute to the understanding of long-range transport pathways and sources of perfluoroalkyl acids to remote regions.

5.10   NITROGEN COMPOUNDS

5.10.1   ALIPHATIC AND AROMATIC AMINES

Varney and Preston [73] discussed the measurement of trace aromatic amines in estuary and seawater using high-performance liquid chromatography. Aniline, methyl aniline, 1-naphthylamine, and diphenylamine at trace levels were determined using this technique with electrochemical detection. Two electrochemical detectors (a thin-layer, dual glassy carbon electrode cell and a dual porous electrode system) were compared. The electrochemical behaviour of the compounds was investigated using hydrodynamic and cyclic voltammetry. Detection limits of 15 and 1.5 nM were achieved using coulometric and amperometric cells, respectively, when using an in-line preconcentration step.

Petty et al. [74] used flow injection sample processing with fluorescence detection for the determination of total primary amines in seawater. The effects of carrier stream flow rate and dispersion tube length on sensitivity and sampling rates were studied. Relative selective responses of several amino acids and other primary amines were determined using two dispersion tube lengths. Linear calibration curves were obtained over the ranges of 0–10^{-6} M and 0–10^{-8} M glycine. The precisions of better than 2% at 10^{-6} M and a detection limit of 10^{-8} M glycine were obtained.

Yang et al. [75] developed a concentration technique using diffusion of amines and organic acids across a hydrophobic membrane for the measurement of these compounds in seawater. Using this approach, the amines and organic acids are concentrated from seawater and are also separated from inorganic salts and most other dissolved organic compounds. Amines and organic acids in the concentrated sample are then quantified by gas chromatography. Concentration factors of up to 1000 can be achieved.

Natural concentrations of amines and most organic acids in the 10 nM can be measured in 500 mL of seawater for acetic acid.

Measurable levels are higher, about 700 nM, because of the higher blanks. However, detection limits can be achieved using a circulation diffusion method with larger volumes of seawater concentration factors of various amines, and organic acids obtained using this method are reported for coastal and estuarine seawaters and sediment pore waters.

5.10.2 AMINO ACIDS

The classical work of Dawson and Pritchard [85] on the determination of α-amino acids used a standard amino acid analyser modified to incorporate a fluorometric detection system. In this method, the samples were desalinated on cation-exchange resins and concentrated prior to analysis. The output of the fluorimeter was fed through a potential divider and low-pass filter to a compensation recorder.

Dawson and Pritchard [85] pointed out that all of the procedures used for concentrating organic components from the estuary of seawater, however mild and uncontaminated, were open to criticism, simply because of the ignorance as to the nature of these components in the sample. It is, for instance, feasible that during the process of desalting on ion-exchange resins under weakly acidic conditions, labile peptide linkages are disrupted or metal chelates are dissociated and thereby larger quantities of 'free' components are released and analysed.

Mopper [86] has discussed developments in the reversed-phase high-performance liquid chromatographic determination of amino acids in estuarine and seawaters. He describes the development of a simple, highly sensitive procedure based on the conversation of dissolved free amino acids to highly fluorescent, moderately hydrophobic isoindoles by a derivatisation reaction with excess o-phthalaldehyde and a thiol, directly in seawater. Reacted samples were injected without further treatment into a reversed-phase high-performance liquid chromatography column, followed by a gradient elution. The eluted amino acid derivatives were detected fluorometrically. Detection limit for most amino acids was 0.1–0.2 nmol/500 μL injection.

Crawford et al. [66] have described a method employing fluorescence on a fluorophore for the determination of μg/L levels of amino acids in seawater.

Hammers and Luck [87] described a method for the determination of total dissolved amino acids in seawater. This method is based on the fluorometric detection of primary amines with phthalicdianhydride.

5.10.3 AMINO SUGARS

Kaiser and Benner [88] have described a method for determining amino sugars in seawater by high-performance anion-exchange chromatography with pulsed amperometric detection and on online sample cleanup procedure.

Samples were hydrolysed with 3 M hydrochloric acid for 5 h (100°C) and neutralised with an ion-retardation resin. Before injection, salts and organic contaminants were removed with a strong cation exchanger in the Na^+ form. Detection limits for amino sugars were between 1 and 4 nM (signal-to-noise ratio 3), allowing for the first time quantification of amino sugars in seawater without preconcentration. The precision was 2%–11% at the 20 nM level. The relatively simple and rapid sample preparation makes it suitable for routine analyses.

5.10.4 ETHYLENEDIAMINE TETRAACETIC ACID AND NITRILOACETIC ACID

Differential pulse polarography has also been used to determine EDTA and nitriloacetic acid in synthetic seawater and phytoplankton media [89]. Cadmium is used to convert a large fraction of either ligand to the reducible cadmium complex. The presence of competing metal cations, including copper, is not detrimental if the method of standard additions is used. The method was used in correlating the concentration of complexed and uncomplexed species of copper with phytoplankton productivity and the production of extracellular metal-binding organic compounds.

Nowack et al. [90] have described a method for determining ethylenediamine tetraacetic acid species in various environmental samples at low molar concentrations by high-performance liquid chromatography. Distinction between Fe(III) EDTA and all the other species can be made. NiEDTA can be detected semiquantitatively. The fraction of ethylenediamine tetraacetic acid adsorbed to suspended particles or to sediments can be determined after desorption with phosphate. After complexation with Fe(III), the ethylenediamine tetraacetic acid is detected by reversed-phase ion-pair liquid chromatography as the FE(III) EDTA complex at a wavelength of 258 nm. The behaviour of a variety of metal-EDTA complexes during analysis was checked. Fe(III) EDTA was found to be the main species at 30%–70%: NiEDTA was less than 10% in most of the samples.

5.10.5 UREA

Mitauure [91] determined urea in the low nM concentrations in seawater using spectrofluorimetry with the diacetyl–urea reaction product.

5.10.6 MISCELLANEOUS

Acrylonitrile monomer [92,94,95] and acid amides [93] have both been determined in seawater by methods based on high-performance liquid chromatography.

Adamski [96] has reported a simplified Kjeldahl nitrogen determination of organic nitrogen in seawater using semiautomated persulphate digestion followed by the determination of ammonia by the indophenol blue procedure [97]. This method gave an accuracy of ±8.1% and a precision of ±8.2% for seawater samples spiked with 3.35 μg/L of organic nitrogen.

Various other researchers have described automated procedures for the determination of low levels of organic nitrogen in seawater [98–100].

Vahatalo et al. [101] have pointed out that solar radiation photochemistry can be considered a new source of nutrients when photochemical reactions release bioavailable nitrogen from biologically nonreactive dissolved organic nitrogen. Pretreatments of Baltic Sea waters in the dark indicated that more than 72% of absorbed organic nitrogen was recalcitrant to biological mineralisation. When this absorbed organic nitrogen (16–21.5 µM) was exposed to simulated solar radiation, the concentration of ammonium ion increased to 0.5–2.5 µM more in irradiated waters than in the dark controls. The photochemical production of ammonium ion and the dose of absorbed photons were used to calculate the apparent quantum yield spectrum for photoammonification [mol NH_4^+ (mol photons)$^{-1}$ nm^{-1}] at wavelengths (λ) of 290–700 nm (µmol NH_4 n−2d−1). The modelled mean rates of photoammonification based on µmol NH_4 β were 143 and 53 µmol ammonium ion/m^2/day at the surface and in the whole water column, respectively, at Baltic Sea stations during the summer. The results of this study indicate that the rate of photoammonification approximately equals and periodically exceeds the rate of atmospheric deposition of reactive inorganic nitrogen to the northern Baltic Sea. For these stratified surface waters beyond the riverine input of labile nitrogen, photoammonification can periodically be the largest source of new bioavailable nitrogen.

5.11 SULPHUR-CONTAINING COMPOUNDS

Various sulphur-containing compounds have been determined in seawater, including thiols [102,103], thiabendazole [104], mercaptans, and 1.4 thioxane [105].

Simo et al. [106] adopted a reported borohydride reduction method for the trace determination of aqueous dimethyl sulphoxide for use with different sample preparation and analytical systems. The adaption and optimisation steps they followed gave them further insight into the method. Increasing the proportion of reducing agent was critical. A number of compounds with potential for analytical interference were tested, but all proved negative.

Water blanks were problematic, with substantial dimethyl sulphoxide contamination observed in all but very recently purified water. Preliminary comparison with the highly specific and precise enzyme-linked method gave very good agreement. When dimethyl sulphoxide analysis was done sequentially after the analysis of dimethyl sulphide and alkali hydrolysis for dimethylsulfoniopropionate, it was found that the dimethyl sulphoxide concentration was not affected by increasing the length of the hydrolysis step. This allows storage and/or transport of hydrolysed samples in gaslight containers. The adapted method was applied to the determination of dimethyl sulphoxide on glass fibre filters, and this revealed a significant pool of particulate dimethyl sulphoxide in marine particles.

5.12 PHOSPHOROUS-CONTAINING COMPOUNDS

Ahern et al. [107] have observed that organophosphorus compounds are coprecipitated with cobalt and pyrrolidine dithiocarbamate and suggested that this might be a suitable means of preconcentrating the phosphorus fraction of harbour water prior to analysis by X-ray fluorescence spectrometry.

To determine isopropylmethylphosphonofluoridate (GB) and S-2-(diisopropyl-amino)ethyl-O-ethylmethylphosphonothioate at the parts per trillion (U.S.) level at seawater, Michael et al. [108] mixed the sample (0.1 mL) with 0.1 mL of eel cho-linesterase solution in a buffer solution of pH 7.2 (0.1 M in morpholinopropane sul-phonic acid, 0.01 M in EDTA, and containing 0.1% of gelatine). The mixture was incubated at 25°C for up to 30 h. To determine residual enzyme activity, 0.1 mL of substrate mixture (2 nM in 5,5′ dithiol-bis-(2-nitrobenzoic acid) was added and after an hour, the extinction of the mixture was measured at 412 nm. To determine S-2-disopropylethyl-O-ethyl phosphonothioate alone, seawater is extracted with dichlo-romethane and combined extracts are mixed with 0.1 mL of 01 M hydrochloric acid and the mixture is evaporated under nitrogen at room temperature. The residue is dissolved in water, and the aliquot is analysed as described earlier.

5.13 SURFACE ACTIVE AGENTS

5.13.1 Ionic Detergents

5.13.1.1 Spectrophotometric Methods

The picrate spectrophotometric method described by Favretto et al. [109] has been applied in the determination of nonionic detergents in seawater.

A mean value of 93% ± 1% was obtained in recovery experiments on $C_{12}E_9$ (at an aqueous concentration of 0.10 mg/L) extracted from synthetic seawater. Therefore, a multiplication factor of 1.07 was adopted in correcting for the extraction losses.

Boyd-Bowland and Eckert [112] have described a method for the determination of nonionic detergents in seawater in the concentration range of 0.1–1 mg/L. The surfactant is extracted into trichloroethane as a neutral adduct with potassium iodide and the analysis performance by measuring the absorbance at 380 nm.

5.13.1.2 Atomic Absorption Spectrometry

Courtot Coupez and Le Bihan [111] determined nonionic detergents in seawater and freshwater samples at concentrations down to 0.002 ppm by benzene extraction of the tetrathiocyanatocobaltate (II) $(NH_4)_2[Co(SCN)_4]$)-detergent ion pair followed by atomic absorption spectrophotometric determination of cobalt.

Crisp [110] has described a method for the determination of nonionic detergent concentrations between 0.05 and 2 mg/L in freshwater, estuarine water, and sea-water based on the extraction of the detergent-potassium tetrathiocyanatozincate (II) complex followed by the determination of extracted zinc by atomic absorption spectrometry.

5.13.1.3 Gas Chromatography

Favretto et al. [109] applied gas–liquid chromatography in the evaluation of the polydispersity of polyoxyethylene nonionic surfactants as $C_{16}E_3$–$C_{16}E_{11}$ and $C_{12}E_1$–$C_{12}E_{15}$. Per 500 mL volume of water was extracted three times with 1,2-dichloro-ethane. The combined organic layers were extracted with 0.10 M sulphuric acid and then 1.10 M sodium hydroxide solution. The purified extract was concentrated under vacuum, transferred into a conical glass mini-vial, and evaporated to a small volume (0.1 mL) by means of a stream of nitrogen. An aliquot of this solution was injected into the gas chromatographic column.

Chromatographic peaks attributable to the polydisperse surfactant $C_m \pm E_n$ were first recognised by using as an internal standard a similar $C_m \pm E_n$ surfactant (the comparison of the elution temperature of the peaks is a doubtful criterion as elution temperatures vary with the age of the column) mixed (approximately 1:1 by mass) with a portion of residue solution. The values n of the peaks pertaining to the distribution were then determined by introducing a monodisperse $C_m \pm E_n$ as an internal standard in another aliquot of the residue solution (1:10).

The gas–liquid chromatographic evaluation of $\bar{n}$ can obviously be performed on a single distribution $C_m \pm E_n$ but not on mixtures of distributions. Commercial surfactants always consist of mixed distribution. Only when the procedure is applicable, $n = \Sigma x_n n$ is calculated from the observed distribution of the molecular fraction (x_n) for various values of n. It should be checked that $\Sigma x_n = 1$.

5.13.1.4   Solid-Phase Microextraction

Troge et al. [113] analysed freely dissolved alcohol ethoxylate homology in various seawater matrices using solid-phase microextraction. Partitioning a wide range of alcohol ethoxylate homologues into 35 μm polyacrylate fibre coating was linearly related to aqueous concentrations as low as sub-micrograms per litre, with high reproducibility. Specific attention was given to the influence of various matrices on the analysis via solid-phase microextraction. The presence of sediment increases the uptake of alcohol ethoxylate homologues for which diffusion in the aqueous phase is rate limiting. The K_{fw} in equilibrated systems was not affected by the presence of other homologues, micelles, or varying amounts of sediment phase. Solid-phase microextraction is, therefore, a suitable tool for the analysis of alcohol ethoxylate in sorption studies and sediment toxicity tests. A strong linear relation was observed between K_{fw} and the hydrophobicity of the alcohol ethoxylate homologue using estimated octanol–water partition coefficients. This relation can be used to predict the partitioning coefficient of any anionic detergents homologue to the alcohol ethoxylate solid-phase microextraction fibre, which facilitates the analysis of complex mixtures.

5.13.2   Anionic Detergents

5.13.2.1   Spectrophotometric Methods

Bhat et al. [114] used complexation with the bis(ethylenediamine) copper II cation as the basis of a method for estimating anionic surfactants in fresh estuarine water and seawater samples. The complex is extracted into chloroform and copper measured spectrophotometrically in the extract using 1,2(pyridylazo)-2-naphthol.

Bhat et al. [114], using the same extraction system, were able to improve the detection limit of the method to 5 μg/L (as linear alkylsulphonic acid) in fresh estuarine water and seawater samples.

Kazarac [116] have discussed methylene-blue-based methods for determining anionic detergents in seawater samples. This technique is based on solvent extraction and preconcentration of detergents using Wickbold's apparatus [117]. A method using azure A instead of methylene blue has been proposed by Den Tonkelaar and Bergshoeff [118,119]. Researchers at the Water Research Centre [122] have described a methylene-blue-based autoanalysis method for determining 0–1 µg/L anionic detergents in water and sewage effluents. This method was based on the work of Longwell and Maniece [120] and subsequently modified by Södergren et al. [121].

Xu et al. [123] described a method involving extraction photometry with ethyl violet to determine nonionic detergents in seawater and estuarine water.

5.13.2.2 Atomic Absorption Spectrometry

In a method described by Le Bihan and Courtot Coupez [124–126], samples of copper 1.10 phenanthroline sulphate are added to 1 L of seawater. Atomic absorption spectrometry was carried out on a methyl ethyl ketone extract of the couplex-1. Anionic detergents do not interfere in this procedure.

Crisp et al. [127] were the first to use bis(ethylenediamine) Cu(II) ion for the determination of anionic detergents. They determined the concentration of detergents by flame atomic adsorption spectroscopy or by colorimetric method. The colorimetric method was more sensitive with a limit of detection of 0.03 µg/L (as linear alkyl sulphonic acid) compared to 0.06 µg/L for atomic absorption spectroscopy. Their method is applicable to freshwater and seawater. Crisp et al. [127, 128] determined anionic detergents in fresh estuarine water and seawater at the ppb level. The detergent anions in a 750 mL water sample were extracted with chloroform as an ion association compound with the bis(ethylenediamine) Cu(II) cation and determinate by atomic adsorption spectrometry using a graphite furnace atomiser. The limit of detection (as linear alkylsulphonic acids) is 2 µg/L at 1 µg/L but reaches 90% at 10 µg/L. The recovery is 97% or better at higher concentrations. At detergent concentrations of milligrams per litre, environmental copper concentrations produce no problem.

Various other techniques applied in the determination of anionic detergent in seawater include polarography [129] and combined gas chromatography–mass spectrometry [115].

Gagnon [131] used a method based on the bis/ethylene amine/Cu(II) complex to determine anionic detergents in seawater with a detection limit of 0.3 µg/L.

5.13.2.3 Cationic Detergents

Wang and Pek [130] have described a simple titration method in the analysis of cationic surfactants in seawater. Methyl orange and azure A were used as primary dye and secondary dye, respectively. The method is free from interference by high levels of inorganic salts in seawater.

Yamamoto and Motomizu [132] have described a method spectrophotometric method based on ethyl violet to determine sodium lauryl sulphate in seawater.

Ion-pair formation has been used as the basis for the determination of down to 1.4 ppt of sodium dodecyl sulphate in seawater [133].

5.14 CHLOROPHYLLS

Abayachi and Riley [134] used performance liquid chromatography to determine chlorophylls and their degradation products, and carotenoids in phytoplankton and marine particulate matter.

Pigment extraction was carried out with acetone and methanol. After evaporation of the combined extracts under reduced pressure, the pigments were separated on a Partisil 10 stationary phase with a mobile phase consisting of light petroleum (bp 60°C–80°C), acetone, dimethyl sulphoxide, and diethylamine. When chlorophyll c is present, a further development is performed with a similar, but more polar, solvent mixture. Detection is carried out spectrophotometrically at 440 nm. The method has sensitivity for the chlorophylls of ca. 80 ng and for the carotenoids of ca. 5 ng. The coefficient of variation of the chromatographic stage of the procedure lies in the range of 0.6%–1.8%.

Abayachi and Riley [134] and others [108,136–141] compared the results obtained by the high-performance liquid chromatographic method with those obtained by a reflectometric thin-layer chromatographic method and the SCOR/UNESCO polychromatic procedure. The results obtained from the latter were evaluated by the SCOR/UNESCOP equations. The carotenoids were determined collectively from the absorption of a 90% acetone extract at 480 nm by means of equations of Strickland and Parsons [135]. The results of these comparative studies show that there is satisfactory agreement for all pigments between the two chromatographic methods.

5.15 INSECTICIDES

5.15.1 CHLORINATED PESTICIDES

Girenko et al. [142] noted that it was difficult to determine chlorinated insecticides in samples of seawater because they are severely polluted by various co-extractive substances, chiefly chlorinated biphenyls. To determine organochlorine insecticide reduces by gas chromatography with an electron capture detector, the chlorinated biphenyls were eluted from the column together with the insecticides. They produce inseparable peaks with equal retention times, hence interfering with the identification and quantitative determination of the organochlorine insecticides. The presence of chlorinated biphenyls is indicated by additional peaks on the chromatographs of the water samples and aquatic organic organisms. Some of the peaks coincide with the peaks of the o,p' and p,p' isomers DDE, DDD, and DDT and some of the constituents are eluted after p,p' DDT.

Liquid scintillation counting of [^{14}C] DDT has been used to study the pick-up and metabolism of DDT by freshwater algae and to determine DDT in seawater [143]. The analysis of samples several days later gave only partial recovery owing to the adsorption of DDT onto suspended matter.

Wilson and Forester [144] discussed the determination of Aldrin, Chlordane, Dieldrin, Endrin, Lindane, o,p and p,p' isomers of DDT and its metabolites, Mirex, and Toxaphene in seawater. The concentrated solvent extracts were analysed by electron capture gas chromatography using columns of different liquid phases. The following columns were used: DX–200, QF–1, EGS, OV–101, mixed DC–200-Q–1, and mixed OV–101–OV–17. Just prior to extraction, all samples were fortified with o,p'-DDE to evaluate the integrity of the analysis. The recovery rates of o,p'-DDE in all tests were greater than 89%, indicating no significant loss during analyses.

Wilson and Forester [145] discovered that organochlorine pesticides were not stable in water. Since petroleum ether was the solvent used in these earlier studies for extracting DDT from seawater, Wilson and Forester [144] initiated further studies to evaluate the extraction efficiencies of the other solvent systems, viz petroleum ether, 15% ethyl ether, in hexane followed by hexane or methylene chloride.

Acer and Picer [146] described a method for measuring the radioactivity of labelled DDT-contaminated sediments in relatively large volumes of water using a liquid scintillation spectrometer. External standard ratios and counting efficiencies of the systems investigated were obtained as was the relation of efficiency factor to external standard ratio for each system studied.

Organochlorine insecticides have been preconcentrated on Amberlite XAD-2 resin [147] and Tenax-Celite [148,149].

Samuelson et al. [150] studied the degradation of Trichlorophon and Dichlorvos in seawater in which the seawater sample is directly injected into a high-performance liquid chromatographic column with a methyl nitrile-phosphate buffer at pH 5.8. Separated compounds were located by UV detection at 205 nm.

Carroll et al. [56] carried out a study of the amounts of organochlorine pesticides and polychlorobiphenyls released into the Arctic Ocean by various Russian rivers.

Zhang et al. [151] collected surface seawater and marine boundary layer air samples on the ice-breaker R/V *Xuelong (Snow Dragon)* from the East China Sea to the high Arctic (33.23° N–84.5° N) in July to September 2010. These samples were analysed for six currently used pesticides: trifluralin, endosulfan, chlorothalonil, chlorpyrifos, dacthal, and dicofol. In all oceanic samples, six of these were detected, showing highest level (>100 pg/m^3) in the Sea of Japan. Gaseous pesticides basically decreased from East Asia (between 36.6° N and 45.1° N) towards the Bering and Chukchi Seas. The dissolved pesticides in ocean water ranged widely from < MDL to 111 pg/L.

Latitudinal trends of α-endosulfan, chlorpyrifos, and dicofol in seawater were roughly consistent with their latitudinal trends in air. Trifluralin in seawater was relatively high in the Sea of Japan (35.2° N) and evenly distributed between 36.9° N and 72.5° N, but it remained below the detection limit at the highest northern latitudes in Chukchi Sea. In contract with other pesticides, concentrations of chlorothalonil and dacthal were more abundant in Chukchi Sea and in East Asia. The air–sea gas exchange of pesticides was generally dominated by net deposition. Latitudinal trends of fugacity ratios of α-endosulfan, chlorothalonil, and dacthal showed stronger deposition of these compounds in East Asia than in Chukchi Sea, while trifluralin showed stronger deposition in Chukchi Sea (−455 ± 245 pg/m$_2$/day) than in the North Pacific (−241 ± 158 pg/m^2/day). Air–sea gas exchange of chlorpyrifos varied from net volatilisation in East Asia ($<40°$ N) to equilibrium or net deposition in the North Pacific and the Arctic.

5.15.2 PHOSPHORUS-CONTAINING PESTICIDES

Weber [152] has described a kinetic method for studying the degradation of Parathion in seawater. Weber observed two pathways whereby Parathion is hydrolysed. The first reaction proceeds via dearylation with loss of p-nitrophenol:

$$\left(C_2H_5O\right)_2 - PS - OC_6H_4 - NO_2 + H_2O$$

$$= \left(C_2H_5O\right)_2 - PS - OH + HOC_6H_4NO_2$$

Additionally, they observed a second pathway, hydrolysis through dealkylation leading to a secondary ester of phosphoric acid, which still contains the p-nitrophenyl, moiety, that is, de-ethyl Parathion (O-ethyl-O-p nitrophenyl-monothiophosphoric acid):

$$\left(C_2H_5O\right)_2 - PS - OC_6H_4NO_2 + H_2O$$

$$= NO_2 - C_6H_4O - PS\left(OC_2H_5\right)OH + C_2H_5OH$$

Aliquots of ethanolic solution of Parathion were separated into undecomposed insecticides and decomposed insecticide products. Among the products, free p-nitrophenol, chemically bound in acidic phosphorus compounds and in non-hydrolysed neutral phosphorus compounds, was detected in the same way after saponification. Saponification after removal of ether without separation of neutral and acidic compounds yielded total p-nitrophenyl equivalents.

Jain and Ali [153] determined pesticides and their degradation products by gas chromatography. The review covers the determination of organochlorine, nitrogen-containing pesticides in various waterbodies, sediments, and soils. The sources of pollution, sampling procedures, extraction purification, and preconcentration techniques and other gas chromatographic conditions for pesticides determination have been discussed.

5.15.3 AZINE-TYPE PESTICIDES

Wu et al. [154] carried out measurements of the enrichment of Atrazine on the microsurface water of an estuary. These authors used a microsurface water-sampling technique with a 16-mesh stainless steel screen, collecting bulk samples from the top 100–150 μm of the surface [155,156,157]. The enrichment of Atrazine in the microsurface varied from none to 110 times with the highest enrichment between mid-September and late October. Atrazine concentration in the actual microsurface was estimated to vary in the range of 150–8850 μg/L.

5.16   ANTIBIOTICS

Only one reference has been found to the determination of 0.01 µg/L oxytetracycline in open seawater.

In this method, the antibiotics for each method were eluted and separated using an LC gradient with mobile phases A and B. Mobile phase A was 5 mM ammonium acetate for the β-Lactones and macrolides class and 0.3% formic acid was used for the quinolines and oxytetracycline classes of antibiotics. Acetonitrile was used as mobile phase B for the β-lactam, macrolide, and quinoline classes, and methanol was used for the oxytetracycline class of antibiotics. The initial flow rates of mobile phases A and B were decreased and contained a higher proportion of mobile phase B to elute the Prospekt SPE. Compounds discussed include sulphadimethoxine, ormetoprim, oxytetracycline, and tetracycline.

This investigation covered the changes in concentration of these compounds with time. Generally speaking, a severe decrease in oxytetracycline and tetracycline occurred after about 10 days.

5.17   PHARMACEUTICALS

Concalves et al. [158] have developed a method for acidic, basic, and neutral pharmaceuticals in river water. This method with little or no modification might have an application in seawater analyses. This method involves application of solid-phase microextraction. J T Baker H_2O-philic provided the best recoveries (average 86%). 1°, 2° amino absorbents provide the most suitable cleanup media for water samples prior to enrichment by solid-phase microextraction. The overall method provided limits of detection generally below 5 ng/L and precision below or around 10% RSD in three non-consecutive days at 500 ng/L and around 12% at 50 ng/L concentration levels.

The first known results regarding the occurrence of pharmaceuticals in Leça River confirmed its expected high contamination load. Concentrations up to 770 ng/L of benzafibrate, 925 ng/L of paracetamol, 389 ng/L of hydrochlorothiazide, and 283 ng/L of furosemide were measured. The most ubiquitous in both seasons, February and June 2009, were bisoprolol, furosemide, bezafibrate, and gemfibrozil. In Douro River, the abundance and contamination level of pharmaceuticals were much lower, which gives a clear indication that the hydraulic features of the river provide enough attenuation of the several contamination sources, whereas Leça River is highly impacted by insufficiently treated anthropogenic effluents.

Buser and Muller [165] investigated the occurrence of the pharmaceutical drug clofebric and the herbicide Mecoprop in seawater.

5.18   OESTROGENS

Various researchers have discussed the determination of oestrogens in seawater [159–162].

Pojana et al. [160] developed an analytical method for the simultaneous determination of estrogenic compounds of natural (estradiol, estriol, estrone) and synthetic origin, both steroidal (ethinyl estradiol, mestranol) and non-steroidal (benzophenone, bisphenol A, diethylstilbestrol, octylphenol, nonylphenol, nonylphenol monoethoxylate, carboxylate), in environmental aqueous samples by high-performance liquid chromatography coupled with ion trap–mass spectrometry *via* electrospray interference. Quantitative mass spectrometry detection was performed in the negative mode for all compounds except mestranol and benzophenone, which were detected under positive ion conditions. Very low method detection limits, between 0.1 and 2.6 ng/L, were achieved in coastal lagoon water samples, while the solid-phase extraction procedure permitted simultaneous recovery of all analytes from spiked water samples with yields greater than 70% (7–11 RSD%), except estriol and benzophenone, which were recovered with 60% (9 RSD%) and 50% (11 RSD%) yields, respectively. This method was applied to the analysis of Venice (Italy) lagoon waters, where average concentrations of selected compounds in the 2.8–33 ng/L concentration range were found.

Roche et al. [161] have discussed the development and optimisation of gas chromatography–mass spectrometry methods for the evaluation of endocrine disruptive compounds, namely endocrine and persistent pollutants in estuary and seawater. The analysed endocrine disrupting compounds are oestrone (E_1), 17β-estradiol (E_2), 17α-ethynylestradiol (EE_2), 4-tert-octylphenol, 4-*n*-octylphenol, 4-nonylphenol, bisphenol A, and finally mono- and diethoxylates of 4-nonylphenol and 4-octylphenol. The method includes the preconcentration of water samples, 1000-fold factor, in OASIS HLB cartridges by solid-phase extraction, the derivatisation of all endocrine disrupting compounds by N,O-bis(trimethylsilyl)trifluoroacetamide added with 1% trimethylchlorosilane and pyridine (at 65°C for 30 min) and finally the stabilisation of the endocrine disrupting compounds–silylated derivatives, in hexane, for 72 h. The validation parameters revealed that this method was highly specific for all target compounds using real samples. The linearity of the calibration curves (r^2) showed correlation factors higher than 0.990. The detection limits from 0.10 to 1.45 ng/L, depending on each analysed compound, and recoveries were satisfactory for most of the assayed endocrine disrupting compounds (>60%). Analysis of samples from four polluted areas of the Douro River estuary and from two points of the Atlantic Ocean (Portugal) showed high amounts of oestrone (up to 1.96 ng/L), E_2 (up to 14.36 ng/L), and ethylene extrdiol (up to 2.76 ng/L).

5.19 PHOSPHOLIPIDS

Erka et al. [163] used electrochemical impedance spectrometry to characterise the physical properties of *ex situ* reconstruction of sea surface microlayer. Samples were extracted with *n*-hexane and dichloromethane, respectively, and transferred to mercury electrodes. Results obtained by electrochemical impedance spectroscopy were compared with those of a model phospholipid dioleoyl phosphatidyl choline, which forms a near defect-free monolayers on a mercury surface. The subsurface microlayer extracts formed in homogenous monolayers on the mercury surface and the dichloromethane sea surface microextract monolayers showed greater surface roughness than the hexane sea surface microextract. The hexane sea surface microlayer extract introduced defects and a greater surface roughness in the mixed dioleoyl phosphatidyl choline matter sea surface microlayer extraction. This was due to the lower compatibility of the non-polar extract with the dodecylphosphatidylcholines than that of the polar sea surface microlayer extracts.

In addition, the dichloromethane sea surface microlayer showed an association with pyrene added to the solutions.

5.20   POLYCHLORINATED DIBENZO-*p*-DIOXINS AND POLYCHLORINATED DIBENZOFURANS

Castro-Jiminez et al. [164] investigated the atmospheric occurrences and deposition of polychlorinated dibenzo-*p*-dioxins and polychlorinated dibenzo furans in open Mediterranean Seawater samples.

The $\Sigma 2,3,7,8$ polychlorinated dibenzo-*p*-dioxins and dibenzo furans/air (gas + aerosol) concentrations over the Mediterranean Sea ranged from 60 to 1040 fg/m^3. The highest value (1555 fg/m^3) was measured in a reference sample taken in the SW Black Sea. No consistent trend regarding the diel cycle of polychlorinated dibenzo-*p*-dioxins and dibenzo furans was observed. Polychlorinated dibenzo-*p*-dioxins and dibenzo furans transported to the open sea waters from continental areas and across the Atlantic as well as ship emissions may be significant sources to the open Mediterranean. Seawater concentrations in the Mediterranean Sea ranged from 42 to 64 fg/L. The $\Sigma 2,3,7,8$ polychlorinated dibenzo-*p*-dioxins and dibenzo furans dry deposition fluxes in the Marmara and Black Seas (210 kg/year) are from 2 to 55 times higher than dry fluxes in the Mediterranean Sea (4–156 kg/year). Analysis of estimated diffusive air-wave fluxes and/or water fugacity ratios show that a net volatilisation of some polychlorinated dibenzo-*p*-dioxins and dibenzo furans congeners is feasible. However, evidence of a net absorption flux for the rest of polychlorinated dibenzo-*p*-dioxins and dibenzo furans is found. When both atmospheric deposition processes are considered together, the open Mediterranean Sea is a net sink of polychlorinated dibenzo-*p*-dioxins and dibenzo furans, due to the importance of dry deposition fluxes of aerosol-bound polychlorinated dibenzo-*p*-dioxins and dibenzo furans.

5.21   LIPIDS

Delmas et al. [166] determined lipid class concentrations of these in seawater by thin-layer chromatography with flame ionisation detection.

A three-step separation on silica-coated chromarod was used to resolve dissolved and particulate seawater lipids into five neutral and two polar lipid classes.

The chromarods are conditioned before development by equilibration with an atmosphere of constant humidity and by equilibration with the developing solvent. The response obtained from flame ionisation detection of lipids on chromarods is usually linear in the range of 0.2–20 µg. Individual chromarods give 5%–15% precision for a range of lipid classes at low loads (3 µg). Land-based and shipboard analyses of seawater samples indicate considerable special variability in lipid class consumption in the top 100 m of water column over the Scotian Shelf.

5.22 VOLATILE ORGANIC COMPOUNDS

Wanatabe et al. [167] determined volatile organic compounds in seawater in amounts down to 0.1 pg/L.

5.23 FLAVINS

Vastano et al. [169] have reported a method based on solid-phase extraction with ion-pair high-performance liquid chromatography using fluorescence detection for the determination of flavins in seawater. Concentrations in the pm range could be determined.

5.24 PECTENOTOXINS

Suzuki et al. [168] developed a solid-phase extraction liquid chromatography–mass spectrometry method for the determination of pectenotoxin-2 and pectenotoxin-6 produced by toxic phytoplankton in water.

5.25 BIOGENIC AND ANTHROPOGENIC COMPOUNDS

Desideri et al. [Private Communication] analysed pack-ice and seawater samples collected at different depths from Terra Nova Bay and Ross Sea, during 1990/1991 Italian Antarctic Expedition, using HRGG and GC-MS. Several classes of biogenic and anthropogenic organic compounds were identified and measured in both matrices. The results showed the changes in the organic composition at varying depths of pack ice and seawater and the enrichment of organic compounds in the pack.

5.26 OXYGEN DEMAND MEASUREMENT

5.26.1 BIOLOGICAL OXYGEN DEMAND

It has been reported that while the dilution bottle method for BOD yields satisfactory results, on fresh and low saline waters, a discrepancy exists when the test is performed on waters containing elevated levels of sodium chloride and other salts.

Seymour et al. [170] found that an increase in salinity resulted in a decrease in BOD values. The degradation rate decreased as the salinity level increased, yet only the particulate fraction seemed to be affected. The degradation of the soluble portion was not affected as adversely by salt content. Lysis of bacteria will occur in a growth medium when sodium chloride is added. It will also occur when salt-acclimatised bacteria are placed in freshwater. If the change in salt concentration is less than 10,000 mg/L, the degree of lysis is negligible, but if greater, the lysis can be extensive. Cellular constituents released following lysis are metabolised by the remaining microbial population in preference to the substrate present, thus yielding erratic BOD results [171,172].

Davies et al. [173] set out to quantify the effects of fresh and saline dilution waters on the BOD of saltwater and to quantitatively establish the role of bacterial seed numbers and species associated with changes in BOD results. Significant BOD differences were found when saline samples were diluted with standard (nonsaline) BOD dilution water. Bacterial populations to general were monitored, and it was shown that equivalent numbers of bacteria did not have the same capability to degrade a given amount of organic matter with increases in salt concentrations to the 3% level. Seeding of hypersaline samples with known salt-tolerant species is recommended for constant BOD results. The BOD values obtained in sewage seed organic standards showed a significant trend BOD values decreased as salt concentrations increased.

The dilution of sewage-seeded saline organic standards with standard dilution water resulted in BOD values higher than those of corresponding nonsaline organic standard due to increases in bacteria populations and increased organic removal (BOD values) in the presence of low levels of salt. The bacteria populations of sewage seeds correlated with corresponding BOD values and concentrations of salt in organic standard solutions showing that the addition of 1% and 3% salt resulted in a decreased initial bacteria population, decreased growth rates, and a decreased ability of an equivalent number of bacteria to degrade an equivalent amount of organic matter. Each salt-tolerant bacteria species yielded comparable BOD values for each organic standard conducted at three salt concentrations. The population increased and the rates of biological oxidation were similar at the three salt concentrations when using either the standard dilution water or the saline dilution water.

Sewage seed does not compete favourably with salt-tolerant bacteria in determining the BOD of saline waters; a salt-tolerant bacteria seed should be used to ensure an accurate and reproducible BOD value when saline waters are being used.

5.26.2 DISSOLVED ORGANIC CARBON

Maestro et al. [94] developed a new system for the determination of total organic and inorganic carbon or total inorganic carbon in waters. Only nonvolatile organic compounds can be detected by this method. The system is based on the measurement of the carbon atomic emission intensity in inductively coupled plasma atomic emission spectrometry. In this way, the organic matter does not undergo any preoxidation step. A semiautomatic accessory connected to the spectrometer separates the different carbon fractions (i.e., organic and inorganic). Because most of the solutions used by Maestro et al. [94] did not contain suspended solid particles, the actual parameter that was determined was the dissolved organic carbon. The new system exhibits good sensitivities compared to those provided by conventional total organic carbon and inorganic carbon determination methods. The limits of detection obtained have been 0.07 and 0.0007 mg/L carbon in terms of total organic carbon and inorganic carbon, respectively. Furthermore, the system is able to handle high-salt-content solutions. This fact suggests that it would be possible to analyse seawater samples, avoiding some of the problems encountered with conventional methods, such as system blocking or interference. The total organic carbon and inorganic carbon values found for natural samples are very close to those measured using conventional methods. The inductively coupled plasma atomic emission spectrometry method was successfully used in two applications: (1) monitoring the efficiency of a water treatment plant and

(2) determining the contents of dissolved carbon dioxide, on one hand, and that of carbonate and bicarbonate, on the other, in the same sample.

5.26.3 TOTAL ORGANIC CARBON

Coulomb et al. [95] have described a procedure for on-site qualitative, quantitative analysis of organic matter from the discharge of municipal wastewater in seawater. This method is based on the knowledge of the UV signal of both seawater and anthropogenic absorbing matter and on the mathematical de-convolution of the sample spectrum using reference spectra. The main application is the estimation of total organic carbon at the direct outlet of the discharge. This quantitative application is obviously limited by the nature of organic compounds, but the UV estimation allowed one to have an overview of the composition and evolution of organic matter into the polluted area. The application of this procedure has been carried out to study the wastewater dilution into an area receiving urban discharges. Experiments showed satisfactory analytical features with a range of total organic carbon values from 75 to 1500 µM C, and the comparison of the results with those obtained by reference method presented a reasonable correlation ($r^2 = 0.9636$) in the marine discharge. The results facilitated the estimation of the plume evolution at the sea surface and in depth. This alternative method could be integrated in a portable device for on-site analysis and multiplication of measurements for relevant results, or in a continuous flow analyser.

REFERENCES

1. N.J.L. Bailey, A.M. Jobson, and M.A. Rogers, *Chemical Geology*, 1973, 11, 203.
2. W. Henderson, V. Wollrab, and G. Egilnton, In *Advances in Organic Geochemistry, 1968* (eds. P.A. Schenk and I. Havenaar), Pergamon, Oxford, UK, 1969, p. 181.
3. E.J. Gallegos, *Analytical Chemistry*, 1971, 43, 1151.
4. J. Albaigés and P. Albrecht, *International Analytical Chemistry*, 1979, 6, 171.
5. I. Rubinstein, P.O. Strausz, C. Spychkerell, R.J. Crawford, and D.W.S. Westlake, *Geochimica et Cosmochimica Acta*, 1977, 41, 1341.
6. A.D. Venosa, P. Compo, and M.T. Suidan, *Environmental Science and Technology*, 2010, 47B, 7613.
7. S. Fam, M.K. Stenstran, and G. Silverman, *Journal of Environmental Engineering*, 1987, 113, 1032.
8. T.L. Wade and J.G. Quinn, *Marine Pollution Bulletin*, 1975, 6, 54.
9. D.J. Duckworth, Aspects of petroleum pollutants analysis, In *Water Pollution in Oil* (ed. C.P. Heppler), Institute of Petroleum, London, UK, 1971, p. 115.
10. J.V. Brunnock, D.F. Duckworth, and G.G. Stephens, *Journal of the Institute of Petroleum*, 1966, 54, 310.
11. O.C. Zafiron and C. Oliver, *Analytical Chemistry*, 1973, 45, 952.
12. D.V. Rasmussen, *Analytical Chemistry*, 1976, 48, 1562.
13. J.D. Walker, R.P. Colwell, M.C. Hamming, and H.I. Ford, *Bulletin of Environmental Contamination and Toxicology*, 1975, 13, 245.
14. D.H. Smith, *Analytical Chemistry*, 1972, 44, 536.
15. S.L. Wasik, *Journal of Chromatography*, 1974, 12, 845.
16. J. Dewulf and H. van Langenhove, *International Journal of Environmental Analytical Chemistry*, 1995, 61, 35.
17. J.M. Done and W.K. Reid, *Separation Science*, 1970, 58, 25.

18. A.R. Muroski, K.S. Booksh, and M.I. Myrick, *Analytical Chemistry*, 1996, 68, 3534.

19. K.S. Booksh, A.R. Muroski, and M.L. Myrick, *International Chemistry*, 1996, 68, 3539.

20. J.F. Rontani and J.P. Giral, *International Journal of Environmental Analytical Chemistry*, 1990, 42, 61.

21. F. El Anba-Iurot, M. Euilion, P. Doumenq, and G. Mille, *International Journal of Environmental Analytical Chemistry*, 1995, 61, 27.

22. R.T. Hiltabrand, *Marine Pollution Bulletin*, 1978, 9, 19.

23. J.F. Payne, *Science*, 1976, 191, 945.

24. H.J. Isaaq, A.W. Andrews, G.N. Janini, and E.W. Barr, *Journal of Chromatography*, 1979, 2, 319.

25. L Maldozado, M. Bayona, and L. Bodineau, *Environmental Science and Technology*, 1999, 33, 2693.

26. M. Tspakis, M. Aposlalaki, S. Elsenreich, and E.G. Stepanon, *Environmental Science and Technology*, 2006, 40, 4922.

27. G. Wernecke, G. Floeser, S. Kern, C. Wennekamp, and M. Michaelis, *Bulletin of the Geological Society of Denmark*, 1994, 51, 5.

28. G. Bozzano and M. Dente, *Industrial and Engineering Chemistry Research*, Articles ASAP. Publication date (web) February 7, 2014 Article DOI: 10.1021/i.e. 4032900.

29. R.J. Bell, T. Short, H. Frisco, M.W. Van Amerom, and R.H. Byrne, *Environmental Science and Technology*, 2007, 41, 8123.

30. T.O. Boyd, *Journal of Chromatography*, 1994, 662, 281.

31. A.D. Dimou, T.M. Sakellarides, F.K. Vasniakos, N. Glanneubis, E. Lenet, and T. Abanis, *International Journal of Environmental Analytical Chemistry*, 2006, 86, 119.

32. H. Liu, A. Zhao, Q. Quian, V. Zhang, and S. Chen, *Environmental Science and Technology*, 2009, 43, 7712.

33. J.G. Quinn and P.A. Meyers, *Limnology and Oceanography*, 1971, 16, 129.

34. P. Treguer and P. Le Corre, *Journal of Marine Biological Association*, 1972, 52, 1045.

35. I.A. Goncharova and A.N. Khomenkô, *Gidrokhimicheskie Materialy*, 1970, 53, 36.

36. Hua Yang Xiao, C. Lee, and M.I. Scranton, *Analytical Chemistry*, 1993, 65, 857.

37. D.I. Kieber, G.M. Vaughan, and K. Mopper, *Analytical Chemistry*, 1988, 60, 1654.

38. G. Eklund, B.O. Josefsson, and C. Ross, *Journal of Chromatography*, 1977, 142, 595.

39. B.O. Josefsson, *Analytica Chimica Acta*, 1970, 52, 65.

40. LeB. Williams, *Limnology and Oceanography*, 1964, 9, 138.

41. J.E. Lovelock, *Nature (London)*, 1975, 256, 193.

42. J.E. Lovelock, R.J. Maggo, and R.J. Wade, *Nature (London)*, 1973, 241, 194.

43. A.J. Murray and J.P. Riley, *Analytica Chimica Acta*, 1973, 65, 261.

44. A.J. Murray and J.P. Riley, *Nature (London)*, 1973, 242, 37.

45. T.A. Bellar, J.J. Lichtenberg, and J.W. Eichelberger, *Environmental Science and Technology*, 1976, 10, 926.

46. R. Dawson, J.P. Diley, and R.H. Tennant, *Marine Chemistry*, 1976, 4, 83.

47. C.P. Rice and V.A. Shigaev, *Environmental Science and Technology*, 1997, 31, 2092.

48. M. Pucko, A. Stern, R.W. Macdonald, and D.G. Barber, *Environmental Science and Technology*, 2010, 44, 9258.

49. L. Zoccolillo, C. Abete, L. Amendola, R. Rucco, A. Shrilli, and M. Termine, *International Journal of Environmental Analytical Chemistry*, 2004, 84, 513.

50. D. Elder, *Marine Pollution Bulletin*, 1976, 7, 63.

51. G. Petrick, D.E. Schulz, and J.C. Duinker, *Journal of Chromatography*, 1988, 435, 241.

52. A.R. Schneider, A. Padiechi, and J.E. Barker, *International Journal of Environmental Analytical Chemistry*, 2006, 86, 789.

53. F. Fuoco, M.P. Colombini, and C. Albete, *International Journal of Environmental Analytical Chemistry*, 1994, 55, 15.
54. V. Leoni, G. Pucetti, R.J. Calombo, and O. Ovidio, *Journal of Chromatography*, 1976, 125, 399.
55. V. Leoni, G. Pucetti, and A. Grella, *Journal of Chromatography*, 1975, 106, 119.
56. J. Carroll, V. Savinov, T. Savinova, S. Dahle, R. McCrea, and D.C.G. Muir *Environmental Science and Technology*, 2008, 42, 69.
57. A Sobek, O. Gustaffson, and J. Alexman, *International Journal of Environmental Analytical Chemistry*, 2003, 83, 177.
58. A. Sobek, H.P.H. Arp, K. Wiberg, S.J. Hedman, and G. Carnelissen, *Environmental Science and Technology*, 2013, 47, 781.
59. A.R. Ballesteros and A. Rubio, *Analytical Chemistry*, 2009, 81, 9012.
60. K. Wiberg, E. Brostrom-Lunder, I. Wangberg, T.F. Bidleman, and D. Haglund, *Environmental Science and Technology*, 2001, 35, 4739.
61. T. Illna and R.W. Hunziker, *Environmental Science and Technology*, 2010, 44, 4622.
62. J.L. Bullister and R.F. Weiss, *Deep Sea Research Part A*, 1988, 35, 839.
63. L. Ahrens, J.L. Barher, Z. Yie, and R. Ebinghaus, *Environmental Science and Technology*, 2009, 43, 3122.
64. J.L. Bullister and R.F. Weiss, *Deep Sea Research Part A*, 1988, 35, 841.
65. B.B. North, *Limnology and Oceanography*, 1975, 20, 20.
66. C.C. Crawford, T.E. Hobbie, and K.L. Webb, *Ecology*, 1974, 55, 551.
67. U.H. Brockman, K. Eberlein, H.D. Junge, E. Majer-Reimer, D. Siebers, and H. Trageser, Entwhichling naturlicher Plantonpolulationene in einem outdoor-tank mit nahrstof-farmem Meerwasser. II. Kinzentrationsveranderungen von gelosten neutralen kohlen-hydraten und freien gelosten Aminosauren. Berichte aus dem Sonder-forschungsbereich Meeresforshung, SFB94, Universtat Hamburg, Hamburg, Germany, 6, 1974, 166–184.
68. P.J. Williams, T. Berman, and O. Holm-Hansen, *Marine Biology*, 1976, 35, 41.
69. C. Garrasi and E.T. Degens, Analytische Methoden zur saulenchoromatographischen Bestimming von Arminosauren und Zuckern im Meerwasser und Sdemiment. Berichte aus dem Projeekt DFG-DE74/3: Litorolforschung-Abwasser in Kustennahe, DEG-Abshlusskolloquium, Bremerhaven, 1976.
70. R. Dawson and K. Mopper, *Biochemistry*, 1973, 83, 100.
71. C. Lee and J.L. Bada, *Limnology and Oceanography*, 1977, 22, 502.
72. R.A. Daumas, *Marine Chemistry*, 1976, 4, 225.
73. M.S. Varney and M.R. Preston, *Journal of Chromatography*, 1980, 348, 265.
74. R.L. Petty, W.R. Michel, P. Snow, and K.J. Johnson, *Analytica Chimica Acta*, 1982, 142, 299.
75. X.H. Yang, C. Lee, and M.I. Scranton, *Analytical Chemistry*, 1993, 65, 572.
76. C.M. King and M. King, *Applied and Environmental Microbiology*, 1983, 45, 1848.
77. T.A. Smith, *Photochemistry*, 1975, 14, 865.
78. X.C. Want and C. Lee, *Geochemistry and Cosmochimica Acta*, 1990, 54, 2743.
79. F.J. Sonsone and C.S. Martinez, *Geochimica Cosmochimica Acta*, 1982, 46, 1575.
80. R. Kondo, H. Kitado, A. Kiwada, Suisan Gakkaishi, and Y. Hata, *Nippon*, 1990, 56, 519.
81. D. Chritienson and T.H. Blackburn, *Marine Biology*, 1982, 71, 193.
82. A. Michelson and J.E. Mackin, *Limnology and Oceanography*, 1989, 34, 747.
83. C.A. Bardwell, J.R. Mabem, J.A. Hurt, W.E. Keene, J.N. Galloway, J.F. Boatman, and D.L. Wellman, *Global Biogeochemical Cycles*, 1990, 4, 156.
84. C. Saigne, S. Mirchner, and M. Legrand, *Analytica Chimica Acta*, 1987, 203, 11.
85. R. Dawson and R.G. Pritchard, *Marine Chemistry*, 1978, 6, 27.
86. K. Mopper, *Science of the Total Environment*, 1986, 49, 115.

87. K.D. Hammers and S. Luck, *Fresenius Zeitschrift für Analytische Chemie*, 1987, 327, 518.
88. K. Kaiser and R. Benner, *Analytical Chemistry*, 200, 71, 2566.
89. P.J. Stolzberg, *Analytical Chemistry*, 1977, 92, 139.
90. B. Nowack, F.G. Karl, S.U. Hilger, and L. Sigg, *Analytical Chemistry*, 1996, 68, 561.
91. O. Mitauure, *Osaka Kyoku Kago Dai-Kuma*, 1987, 31, 159.
92. L. Brown, M. Rhead, *Analyst*, 1979, 104, 391.
93. L. Brown and M. Rhead, *Analyst (London)*, 1979, 104, 393.
94. S.E. Maestra, J. Mora, V. Hernandez, and J.L. Todoli, *Analytical Chemistry*, 2003, 75, 111.
95. M. Coulomb, M. Richards, C. Brack-Papa, J.L. Bourdenne, and F. Therulaz, *International Journal of Environmental Analytical Chemistry*, 2006, 86, 1079.
96. J.H. Adamski, *Analytical Chemistry*, 1976, 48, 1194.
97. B. O'Connor and P. Miloski, Unpublished work, Suffolk County Department of Environment Control, Suffolk, UK, October 1974.
98. F.A.J. Armstrong, P.M. Williams, and J.D.H. Strictland, *Nature (London)*, 1966, 211, 481.
99. R.J. Stevens, *Water Research*, 1975, 10, 171.
100. A.M. Tenny, *Automated Analytical Chemistry*, 1966, 38, 580.
101. A.V. Vahatalo and R.G. Zepp, *Environmental Science and Technology*, 2005, 39, 6985.
102. D. Shae, W.A. McCrehan, *Analytical Chemistry*, 1998, 691, 1449.
103. I.L. Duane and J.T. Stoch, *Analytical Chemistry*, 1978, 50, 1891.
104. T. Capitan, E. Alonso, R. Avidad, L.F. Capitan-Valley, and J.L. Vilchez, *Analytical Chemistry*, 1993, 65, 1336.
105. X. Zhang, K.C. Hester, O. Mancillas, E.T.Peltzer, and P.M. Walz, *Environmental Science and Technology*, 1997, 31, 2190.
106. R. Simo, G. Malin, and P.S. List, *Analytical Chemistry*, 1998, 70, 4864.
107. E. Ahern, J.M. Exhael, S.F. Hain, K.E.A. Leggett, N. Payne, and K.L. Williams, *Analytica Chimica Acta*, 1987, 199, 259.
108. H.O. Michael, E. Gordon, and J. Epstein, *Journal of Environmental Science and Technology*, 1973, 7, 1045.
109. L. Favretto, B. Stanchee, and R. Tunis, *Analyst*, 1978, 103, 955.
110. P.T. Crisp, *Analytica Chimica Acta*, 1979, 104, 711.
111. J. Courtot Coupez and A. Le Bihan, *Rev Ital Soztanse Grasse*, 1971, 20, 672.
112. A.A. Boyd-Bowland and J.M. Eckert, *Analytica Chimica Acta*, 1993, 27, 311.
113. S.T.J. Troge, T.L. Sinnige, L. Joop, and M. Hermans, *Analytical Chemistry*, 2007, 79, 2885.
114. S.R. Bhat, J.M. Exkert, and N.A. Gibson, *Analytica Chimica Acta*, 1980, 116, 192.
115. H. Hom Nami and T. Hanya, *Water Research*, 1980, 14, 1251.
116. Z. Kazarac, *Marine Science Communications*, 1975, 1, 147.
117. R. Wickbold, *Tenside Detergents*, 1971, 8, 61.
118. W.A.M. Den Tonkelaar and G. Bergshoeff, *Water Research*, 1969, 3, 31.
119. W.A.M. Den Tonkelaar and G. Bergshoeff, *Water Research*, 1969, 3, 135.
120. J. Longwell and W.D. Maniece, *Analyst*, 1995, 80, 167.
121. A. Södergren, *Analyst*, 1996, 91, 113.
122. K.W. Petts and D. Parkes, Determination of anionic detergents within a concentration range 0–1 mgL^{-1} in sewage effluents and water by autoanalysis, Water Research Centre Report No. RT75, 1970.
123. T. Xu, B. Yu, W. Tong, and Y. Fang, *Huanjing Kexue*, 1987, 8, 76.
124. A. Le Bihan and J. Courtot Coupez, *Bulletin de la Societe Chimique de France*, 1970, 1, 406.
125. J. Courtot Coupez and A. Le Bihan, *Analyst Letters*, 1969, 2, 211.

126. A. Le Bihan and J. Courtot Coupez, *Analyst Letters*, 1977, 10, 759.
127. P.T. Crisp, J.M. Eckert, and N.A. Gibson, *Analytica Chimica Acta*, 1975, 78, 391.
128. P.T. Crisp, J.M. Eckert, N.A. Gibson, G.I. Kirkbright, and T.S. West, *Analytica Chimica Acta*, 1976, 87, 97.
129. T. Zvenaris, V. Zulic, and M. Branica, *Thallasia Javoslavica*, 1973, 9, 65.
130. L.K. Wang and S.L. Pek, *Industrial and Engineering Chemistry*, 1975, 14, 308.
131. M.J. Gagnon, *Water Research*, 1979, 13, 53.
132. K. Yamamoto and S. Motomizu, *Bunseki Kagaku*, 1987, 36, 335.
133. W.H. Ho, S.M. Hsiao, J.H. Chen, and J. Chin, *Chemical Society (Taipai)*, 1993, 46, 251.
134. J.K. Abayachi and J.P. Riley, *Analytica Chimica Acta*, 1979, 107, 1.
135. J.D.H. Strickland and T.F. Parsons, *A Practical Handbook of Seawater Analysis*, Ottawa, Fisheries Research Board of Canada, Ottawa, Quebec, Canada, 1968.
136. C. Garside and J.P. Riley, *Analytica Chimica Acta*, 1969, 46, 179.
137. W.T. Shoaf and B.I.W. Lium, *Journal of Research of US Geological Survey*, 1977, 5, 263.
138. K.G. Boto and J.S. Bunt, *Analytical Chemistry*, 1978, 50, 392.
139. N. Ton and Y. Takahasi, International Laboratory, September 1949.
140. P.D. Sullivan and B. Brookkshank, *Analytical Chemistry*, 1975, 47, 1943.
141. R.P. Schwarzenbach, R.H. Bromund, P.M. Gschwend, and O.C. Zafiron, *Organic Geo Chemistry*, 1978, 1, 93.
142. P.B. Girenko, M.A. Klisneko, and Y.K. Discholke, *Hydrobiological Journal*, 1975, 11, 60.
143. A.J. Wilson, *Bulletin of Environmental Contamination and Toxicology*, 1976, 15, 515.
144. A.J. Wilson and J. Forester, *Journal of the Association of Official Analytical Chemists*, 1971, 54, 525.
145. A.J. Wilson, J. Forester, and J. Knight, U.S. Fish Wildlife Circular, 355, 18–20, Centre for Estuaries and Research Gulf Breeze, FL, 1969.
146. N. Acer and M. Picer, *Bulletin of Environmental Contamination and Toxicology*, 1975, 14, 565.
147. G.R. Harvey, *Adsorption of Chlorinated Insecticides from Sea Water by a Cross-linked Polymer*, Woods Hole Massachusetts, Unpublished Manuscript.
148. V. Leoni, G. Pincetti, R.J. Columbo, and O. Ovidio, *Journal of Chromatography*, 1976, 125, 399.
149. V. Leoni, *Journal of Chromatography*, 1975, 106, 119.
150. S.B. Samuelson, *Aquaculture*, 1987, 60, 161.
151. G. Zhang, Z. Xie, M. Cai, A. Moller, K. Sturm, J. Tang, G. Zhong, J. He, and R. Ebinghaus, *Environmental Science and Technology*, 2012, 46, 259.
152. K. Weber, *Water Research*, 1976, 10, 237.
153. C.K. Jain and J. Ali, *International Journal of Environmental Analytical Chemistry*, 1997, 68, 83.
154. T.L. Wu, L. Lambert, D. Hastings, and D. Banning, *Bulletin of Environmental Contamination and Toxicology*, 1980, 24, 411.
155. W.D. Garrett, *Limnology and Oceanography*, 1965, 10, 602.
156. J.E. Dietz, E.A. Scribner, M.T. Meyer, and D.A. Koplin, *International Journal of Environmental Analytical Chemistry*, 2005, 85, 1141.
157. H. Poulignen, L. Pinault, and A. Falsif, *Export Chim. Toxicol.* 1992, 85, 111.
158. E.M.O. Concalves, M.A.D. Sousa, and M. de F. PSM Aepencluraddi, *International Journal of Environmental Analytical Chemistry*, 2013, 93, 1.
159. H. Del Bubba, A. Cincenelli, L. Checchini, L. Lepri, and P. Desideri, *International Journal of Environmental Analytical Chemistry*, 2004, 83, 1033.
160. G. Pojana, A. Bonfa, F. Busetti, A. Callarin, and M. Marimini, *International Journal of Environmental Analytical Chemistry*, 2004, 84, 717.

161. M.J. Roche, C. Ribeiro, and M. Ribeiro, *International Journal of Environmental Analytical Chemistry*, 2011, 91, 1191.

162. R.L. Gomes, J.W. Birkett, M.D. Gerimshaw, and J.N. Lester, *International Journal of Environmental Analytical Chemistry*, 2005, 85, 1.

163. S. Erka, A. Nelson, and Z. Kozarac, *International Journal of Environmental Analytical Chemistry*, 2006, 86, 325.

164. J. Castro-Jiminez, S.J. Eisenreich, M. Ghainai, G. Mariani, H. Skejo, G. Umiaf, J. Wollgast et al., *Environmental Science and Technology*, 2010, 44, 5456.

165. H.R. Buser and M. Muller, *Environmental Science and Technology*, 1998, 32, 188.

166. R.P. Delmas, C.C. Parrish, and R.G. Ackman, *Analytical Chemistry*, 1984, 56, 1272.

167. S. Wanatabe, K. Demura, and S. Tsunogai, *Hakkaido Daigaky, Suisangakubo:* Kenkyni Iho, 1987, 38, 288.

168. T. Suzuki, T. Mitsuya, H. Matsubara, M. Yamazaki, *Journal of Chromatography*, 1998, 815, 155.

169. R. Vastano, P.J. Miln, and K. Mopper, *Analytica Chimica Acta*, 1987, 201, 127.

170. M.P. Seymour, The Biodegradation of particulate organic matter in seawater media, Thesis, Rice University, Houston, TX, 1977.

171. D.F. Kincannon and A.F. Gaudy, *Biotechnology and Bioengineering*, 1966, 8, 371.

172. D.F. Kincannon and A. Gaudy, *Journal of Water Pollution Control Federation*, 1966, 38, 1148.

173. B.M. Davies, J.R. Bishop, R.K. Gurhrie, and R. Forthafers, *Water Research*, 1978, 12, 917.

6 Determination of Organometallic Compounds in Seawater

6.1 ORGANOTIN COMPOUNDS

Organotin compounds have been extensively used as an antifouling agent incorporated in paints used in ships and in the protection of harbour works. They are toxic to marine animals. Despite the fact that many countries no longer use organotin compounds in their paint formulations, due to their stability, the concentration and fate of these compounds in seawater is still a matter of great interest. Concentrations of organotin compounds found in seawater are in the range of 0.000002–20 µg/L [1–17]. Hence, very sensitive methods of analysis are required when analysing such samples. Various methods of analysis are now discussed.

6.1.1 ATOMIC ABSORPTION SPECTROMETRY

Valkirs et al. [8,9] conducted an inter-laboratory comparison on determinations of di- and tributyltin species in marine and estuarine waters using two methods, namely hydride generation with atomic absorption detection and gas chromatography with flame photometric detection. A good agreement was obtained between the results of the two methods. Studies on the effect of storing frozen samples prior to analysis showed that samples could be stored in polycarbonate containers at −20°C for 2–3 months without significant loss of tributyltin.

Dowling and Uden [10] have shown that a combination of online hydride generation and gas chromatography with microwave-induced atomic emission detection is capable of achieving a detection limit of 0.5 pg/L of organotin in seawater.

6.1.2 HYDRIDE GENERATION GAS CHROMATOGRAPHY–ATOMIC ABSORPTION SPECTROMETRY

Studies by Braman and Tompkins [11] have shown that non-volatile methyltin species $Me_nSn_{(4-n)}$ ($n = 1$–3) are ubiquitous at ng/L concentrations in natural waters including both marine and freshwater sources. Their work, however, failed to establish whether tetramethyltin was present in natural waters because of the inability of the methods used to effectively trap this compound during the combined preconcentration purge and reductive derivatisation steps employed to generate volatile organotin hydride necessary for tin-specific detection.

Tin compounds are converted to the corresponding volatile hydrides (SnH_4, CH_3, SnH_3, $(CH_3)_2$, SnH_2, and $(CH_3)_3SnH$) by reaction with sodium borohydride at pH 6.5 followed by separation of the hydrides by gas chromatography and then detection by atomic absorption spectroscopy using a hydrogen-rich, hydrogen–air flame emission type detector (Sn–H band).

This technique has a detection limit of 0.01 ng as tin and hence parts per trillion of organotin species can be determined in water samples.

Braman and Tompkins [11] found that stannane (SnH_4) and methylstannanes (CH_3SnH_3, $(CH_3)_2SnH_2$, and $(CH_3)_3SnH$) could be separated well on a column comprising silicone oil OV-3 (20% w/w) supported on Chromosorb W. Average tin recoveries from seawater are in the range of 96%–109%.

A number of estuarial waters, from in and around Tampa Bay, Florida, area were analysed by this method for tin content.

The average total tin content of estuarine waters was 12 ng/L. Approximately 17%–60% of the total tin present was found to be in methylated forms. This procedure, although valuable in itself, is incomplete in that any monobutyltin present escapes detection. Excellent recoveries of monobutyltin species are achieved with tropolone.

Pinel and Madiec [12] pointed out that tin speciation in aquatic environment is very complex. To the natural Sn^{IV} and methylated compounds, human activities add mainly butylated, octylated, phenylated, or even methylated derivatives. The most environmentally significant, due to their high toxicity and direct introduction in water through biocidal use, are the trisubstituted ones. Several sophisticated speciation procedures have been proposed, and they are not susceptible of common use.

Pinel and Madiec [12] proposed a simple and fast procedure allowing for the routine global distinction of 'heavy' tin species that are most susceptible of exerting harmful effects. This atomic absorption spectrometric method uses the differences in volatility of stannanes generated by reduction with sodium borohydride Sn^{IV}; methylated species have very close response coefficients, whereas 'heavy' compounds respond very slightly at room temperature and are eliminated in a −40°C cold trap.

'Heavy' tin determination in water is thus obviously obtained by the difference between two hydride A experiments: one performed on the untreated sample (light tin) and the other on a UV mineralised subsample (total tin). The mineralisations of organotins are realised by UV irradiation for 2 h in a quartz container with a yield of 95%.

6.1.3 Gas Chromatography

Brinkmann [13] used a gas chromatographic method with or without hydride derivatisation for determining volatile organotin compounds (e.g. tetramethyltin) in seawater. For non-volatile organotin compounds, a direct liquid chromatographic method was used. This system employs a Tenax GC polymeric sorbent in the automatic purge and trap sampler coupled to a conventional glass column gas chromatograph equipped with a flame photometric detector. Flame conditions in the flame photometric detector were tuned to permit a maximum response to SnH emission in a hydrogen-rich plasma as detected through narrow bandpass interference filters (610 ± 5 nm) [14]. Two modes of analysis were used: (1) volatile stannanes were trapped directly from sparged 10–15 mL water samples with no pre-treatment and (2) volatilised tin species were trapped from the same or replicate water samples following rapid injection of aqueous excess sodium borohydride solution directed into the purge and trap vessel immediately prior to beginning the purge and trap cycle [15].

Tolosa et al. [16] have reported a method for the determination of down to 0.5 ng/L of tributyltin in seawater based on capillary gas chromatography using a flame photometric detector.

Ali Sheikh et al. [17] used the gas chromatographic method to study the special seasonal behaviour of organotin compounds in protected estuarine waters in Okinawa, Japan.

The spatial and temporal behaviours of the organotin compounds (butyl- and phenyltin) were investigated in the Manko and Okukubi protected estuarine ecosystems on Okinawa Island, Japan, from February to October 2006. Butyltin compounds were frequently detected in all seasons, while phenyltin compounds were found in the winter and early spring. In the Manko Estuary, the total mean concentrations of butyltins and phenyltins were 22.78 ± 30.85 (mean ± SD, $n = 53$) and 0.08 ± 0.27 ng(Sn)/L, respectively. In the Okukubi Estuary, the concentrations of butyltins and phenyltins were 12.58 ± 23.96 and 0.47 ± 1.67 ($n = 55$) ng(Sn)/L, respectively.

Tao et al. [18] performed tin speciation experiments in the femtogram range in open seawater using gas chromatography/inductively coupled plasma mass spectrometry using a shield torch at normal plasma conditions. The sensitivity of the gas chromatography/inductively coupled plasma mass spectrometry was improved by more than 100-fold by operating the shield torch at normal plasma conditions, compared with that obtained without using it. The absolute detection limit of tin was in the sub-fentogram (fg) level. Furthermore, the detection limit in terms of relative concentration was improved 100-fold by using programmed temperature vaporisation method, which enabled the injection of a large sample volume of as much as 100 μL without loss of analyte. When the organotin species in seawater were extracted into hexane with a preconcentration factor of 1000 after ethylation with $NaBEt_4$ and a 100 μL aliquot of the extract was injected into the gas chromatography, the instrumental detection limit in relative concentration reached 0.01 pg/L in seawater. Sources of contamination of organotin species during the sample preparation were examined, and a purification method of sodium tetraethylborate was developed. The method was successfully applied to open ocean seawater samples containing organotin species at the level of 1–100 pg L.

6.1.4 HIGH-PERFORMANCE LIQUID CHROMATOGRAPHY

Various researchers have used high-performance liquid chromatography to determine organotin compounds in seawater [19–22].

Ebdon and Alonso [22] have determined tributyltin ions in estuarine waters by high-performance liquid chromatography and fluorimetric detection using morin as a micellar solution. Tributyltin ions were quantitatively retained from 100 to 500 mL of sample on a 4 cm long silane column. After washing off the salts with 20 mL of distilled water, the ODS column was back-flushed with methanol–water (80 + 20) containing 0.15 M ammonium acetate on a 25 cm long Partisil SCX analytical column. The eluent from the column (1 mL/min) was mixed with fluorimetric reagent (acetic acid, 0.01 M; morin, 0.0025% m/v; Triton X-100, 0.7% m/v 2.5 mL/min) for detection at 524 nm with excitation at 408 nm. The detection limit (2σ) is 16 ng of tributyltin (as Sn).

6.1.5   Nuclear Magnetic Resonance Spectroscopy

Laughlin et al. [23] analysed the chloroform extracts of tributyltin dissolved in seawater using nuclear magnetic resonance spectroscopy. It was shown that an equilibrium mixture occurs which contains tributyltin chloride, tributyltin hydroxide, the aquo complex, and a tributyltin carbonate species. Down to 0.4×10^{-8}–2×10^{-8} µg/L of organotin compounds could be determined.

6.1.6   Anodic Stripping Voltammetry

Kenis and Zirino [24] determined tri-*n*-butyltin oxide directly in seawater at microgram per litre levels at a hanging drop mercury electrode and at a mercury film rotating disc glassy electrode by differential pulse anodic stripping voltammetry. The hanging drop mercury electrode responded to tri-*n*-butyltin oxide additions in seawater purged either with nitrogen (pH 8.2) or with carbon dioxide (pH 4.8) with two distinct stripping peaks.

6.1.7   Miscellaneous

Smith [25] discussed the determination of tin in water. In the determination of low concentrations of the order of 40 µg of trialkyltin chlorides in seawater, it has been observed that these compounds are very volatile and are easily lost upon evaporation with acid. Quantitative recovery of tin is, however, obtained in the absence of chloride ion during evaporation with acid. Preliminary removal of chlorides from seawater by passage down a column of IRA resin before digestion with acid completely overcame the loss of tin in subsequent evaporation with acid giving a tin recovery of 90%.

Duhamel et al. [26] investigated the behaviour of (tributyltin) oxide and tributyltin chloride in saline water. The effects of salinity, pH, light, and oxygen were investigated. Debutylation due to the formation of insoluble compounds occurred under saline conditions.

Tremolada et al. [27] have described a simple model to predict the losses of organotin compounds in aquatic ecotoxicological tests. In these experiments, three nominal concentrations (100, 225, and 500 ng/L) of triphenyltin chloride were employed for an exposure period of 28 days in 50 L aquaria with the echinoderm *Antedon mediterranea* as test species. Extracts from water and biota samples collected during the experiments were analysed by GC-MS/MS, after the extraction/derivatisation step. An indicative mean BCF(V/V) on a fresh weight base of $3.5 \times 10^4 \pm 0.8 \times 10^4$ (standard deviation) could be calculated for *Antedon mediterranea*. Three different compartments (air, water, and biota) and main advection/reaction processes are taken into account in the model design, and the comparison between predicted and measured concentrations in both water and biota for the three concentrations tested confirmed that the assumptions given in this model application were valid and useful for further applications.

Aguliar-Martinez et al. [28] carried out field trials to assess the performance of the passive sampler Chemcatcher as aquatic mentoring technology for inorganic mercury and the organotin pollutants monobutyltin, dibutyltin, and tributyltin in different types of water across 10 locations in Europe. Two versions of the sampler were used. One for mercury that consists of 47 mm Empore™ disks of iminodi-acetic chelating groups as the receiving phase overlaid by a diffusion membrane of polyethersulphone and the other for organotin compounds comprising a C_{18} disk and a cellulose acetate membrane. Both membranes were held in a disposable poly-carbonate body. The two samplers were calibrated in the laboratory to estimate the pollutant concentration. For field sampling, the samplers were deployed for 14 days. In parallel, spot samples were periodically collected during the deployment period for comparison purposes. No significant biofouling on the samplers was observed for the locations monitored. In general, water concentrations estimated by Chemcatcher were lower than those found in spot water samples as the device only collected the soluble bioavailable fraction of target pollutants. However, the preconcentra-tion capability of Chemcatcher allowed the determination of the tested pollutants at levels where spot sampling fails, even in difficult water bodies such as sewage treatment plants. These advantages lead to consider this emerging methodology as a complementary tool to traditional spot sampling.

6.2 ORGANOMERCURY COMPOUNDS

The range of concentration of organomercury compounds found in the ocean range between 0.005 and 0.06 µg/L [29,30].

Fish frequently have 80%–100% of the total mercury in their bodies in the form of methylmercury, regardless of whether the sites at which they were caught were pol-luted with mercury or not. Methylmercury in the marine environment may originate from industrial discharges or be synthesised by natural methylation processes. Fish do not themselves methylate inorganic mercury but can accumulate methylmer-cury from both seawater and food. Methylmercury has been detected in seawater only from Minamata Bay, Japan, an area with a history of gross mercury pollu-tion from industrial discharge. It has been found in some sediments but at very low concentrations, mainly from the areas of known mercury pollution [3]. It represents usually less than 1% of the total mercury in the sediment and frequently less than 0.1% [31]. Microorganisms within the sediments are considered to be responsible for the methylation [51,62], and it has been suggested that methylmercury may be released by the sediments in the seawater, either in dissolved form or attached to particulate material and thereafter rapidly taken up by organisms [46–54,56,61].

Campeau and Bartha [32] studied the effects of different sea salt anions on the formation and stability of methylmercury. The extent of methylation was reduced in the presence of sulphide under anaerobic conditions and of bicarbonate under both aerobic and anaerobic conditions; other anions had no significant effect. In the dark, monomethylmercury chloride was chemically stable in the presence of all the anions tested.

Davies et al. [33,55] set out to determine the concentrations of methylmercury in seawater samples much less polluted than Minimata Bay, viz the Firth of Forth, Scotland. They described a tentative bioassay method for determining methylmercury at the 0.06 µg/g level. Mussels from a clean environment were suspended in cages. A small number of them were removed periodically, homogenised, and analysed for methylmercury. The rate of accumulation of methylmercury was determined, and by dividing this with mussel filtration rate the total concentration of methylmercury in the seawater was calculated. The mussels were extracted with benzene, and the extract was analysed by gas chromatography by Westhoo [58].

The methylmercury concentration in caged mussels increased from low levels (less than 0.01 µg/g) to 0.06–0.08 µg/g in 150 days giving a mean uptake organometal rate of 0.4 ng/g/day. The average percentage of total mercury in the form of methylmercury increased from less than 10% after 20 days to 33% after 150 days.

Davies et al. [33] calculated the total methylmercury concentration in the seawater as 60 pg/L (i.e. 0.1%–0.3% of the total mercury concentration) opposed to 32 ng/L methylmercury found in Minamata Bay, Japan.

Olsen et al. [34] have reported that losses of up to 20% of mercury occur during the storage of seawater samples for up to 96 h prior to analysis. They attributed losses to either a salting out effect which increased the volatilisation of mercury or biochemical reactions which either increase the volatility of or precipitate mercury.

Various methods for the determination of organic mercury in seawater are now discussed.

6.2.1 ATOMIC ABSORPTION SPECTROMETRY

A method developed by Dean and Rues [35] is suitable for the determination of 10–100 ng/L dissolved inorganic mercury and those organomercury compounds which form dithizonates in saline water, seawater, and estuary water. In this method, inorganic mercury is extracted from acidified saline water as its dithizonate into carbon tetrachloride. Organomercury compounds were also extracted by carbon tetrachloride, but not all these compounds form dithizonates and those which do not (e.g. dialkylmercury compounds) form dithizonates may not be determined by this method. Mercury is recovered from organic extracts by back-extraction with acid and determined by stannous chloride reduction followed by cold vapour atomic absorption spectrometry.

Agemian and Chau [36] showed that organomercurials could be quantitatively decomposed by ultraviolet radiation and that the rate of decomposition of organomercurials increased rapidly in the presence of sulphuric acid and with increased surface area of the ultraviolet irradiation. They developed a flow-through ultraviolet digester which had a delay time of 3 min to carry out the photooxidation in the automated system. The ultraviolet radiation had no effect on chloride. The method, therefore, can be applied to both freshwater and saline water without the chloride interference.

With an atomic absorption spectrometric finish, this method was capable of determining down to 0.02 µg/L mercury.

Yamagami et al. [37] evolved a technique involving amalgamation of methylmercury in seawater onto gold followed by atomic absorption spectrophotometry for the determination of picogram quantities of the organomercury compound.

Samples of seawater, groundwater, and river water were analysed for methylmercury and total mercury. Methylmercury is extracted with benzene and concentrated by a succession of three partitions between benzene and cysteine solution. Total mercury is extracted by wet combustion of the sample with sulphuric acid and potassium permanganate. The proportion of methylmercury to total mercury in the coastal seawater sampled around 1%.

Graphite furnace atomic absorption spectrophotometry has also been applied to the determination of trace levels of divalent mercury in inorganic and organomercury compounds in seawater. Filippelli [38] has described a technique in which mercury is first preconcentrated using the ammonium tetramethylenedithiocarbamate (ammonium pyrrolidine dithiocarbamate)–chloroform system and then determined by graphite furnace atomic absorption spectrometry. The technique is capable of detecting mercury(II) in the range of 5–1500 ng in 2.5 mL of chloroform extract and can be adapted to detect subnanogram levels. Atmospheric pressure helium microwave–induced plasma emission spectrometry has been used as an element-selective detector for gas chromatography of organomercury compounds in seawater.

Researchers at the Department of the Environment (UK) have described a method for the determination of organic mercury in seawater and estuarine water [39]. This method is suitable for determining dissolved inorganic mercury and those organomercury compounds which form dithizonates. In this method, inorganic mercury is extracted from the acidified saline water as its dithizonates into carbon tetrachloride. In general, organomercury compounds of the type R–Hg–X, in which X is a simple anion, form dithizonates, whereas the type $R–H_gR_2$ does not. Monomethylmercury ion is extracted though it only appears to have a transient existence in aerobic saline water. The dithizonates are decomposed by the addition of hydrochloric acid and sodium nitrite and the mercury or organomercury compound returned to the aqueous phase. Some organomercury compounds may not be completely re-extracted into the aqueous phase. The mercury in this aqueous phase is determined by the stannous chloride reduction–atomic absorption spectroscopic technique.

This method is capable of determining down to 4 ng/L of mercury.

Fitzgerald and Lyons [40] have described flameless atomic absorption methods for determining organic mercury compounds, respectively, in coastal water and seawater. These researchers used ultraviolet light in the presence of nitric acid to decompose organomercury. One set of samples is analysed directly to give inorganic bound mercury and the other set is photooxidised by means of ultraviolet radiation for the destruction of organic material and then analysed to give total mercury. The element is determined by a flameless atomic absorption technique.

Millward and Bihan [41] studied the effect of humic material on the determination of mercury by flameless absorption spectrometry. In both seawater and freshwater, association between inorganic and organic entities takes place within 90 min at pH values of 7 or above, and the organically bound mercury was not detected by an analytical method designed for inorganic mercury. The amount of detectable mercury was related to the amount of humic material added to the solutions.

Agemian and Chau [42] have described an automated method for the determination of total dissolved mercury in freshwater and saline water by ultraviolet digestion and cold vapour atomic absorption spectrometry. A flow-through ultraviolet digester is used to carry out photooxidation in the automated cold vapour absorption spectrometric system. This removes the chloride interference. Work was carried out to check the ability of the technique to degrade seven particular organomercury compounds. The precision of the method at levels of 0.07, 0.28, and 0.55 µg/L Hg was ±6.0%, ±3.8%, and ±1.00%, respectively. The detection limit of the system is 0.02 µg/L. Recoveries between 70% and 96% were attained for phenylmercury, phenylmercury nitrite, diphenylmercury, methylmercury chloride, ethylmercury chloride, methyl ethyl mercury chloride, and ethoxy ethyl mercury chloride.

6.2.2   Microwave-Induced Plasma Emission Spectrometry

Chiba et al. [43] used atmospheric pressure helium microwave–induced plasma emission spectrometry with cold vapour generation technique combined with gas chromatography for the determination of methylmercury chloride, ethylmercury chloride, and dimethylmercury in seawater following a 500-fold preconcentration using a benzene–cysteine extraction technique. Detection limits of this system are 6.09 µL for methylmercury chloride, 0.12 µg/L for ethylmercury chloride, and 0.4 µg/L for dimethylmercury.

Yamamoto et al. [44] described a procedure for the determination of ethylmercury compounds in seawater involving extraction with benzene, amalgamation of the mercury on to gold foil, and then the release of mercury from the foil followed by atomic absorption spectrometry. Down to 20 µg/L of organomercury compounds can be determined by this technique.

Researchers at the Department of Environment (UK) [45] have developed a similar technique which involves a preliminary extraction of the dimethylmercury compound with a carbon tetrachloride solution of dithizone.

6.2.3   Gas Chromatography

Summarising, in nonsaline water and seawater, organomercury compounds can be speciated and determined in amounts down to 10 ng/L, thereby meeting the present-day requirements. Generally, the levels of organomercury compounds in seawater are less than 0.2 µg/L [57].

Pongratz and Heumann [59] determined the concentration of monomethylmercury and dimethylmercury in surface seawater samples of the Antarctic and Arctic Oceans as well as of the other remote areas (South Atlantic and South Pacific) during expeditions of the German research vessel *Polarstern*. A purge and trap/gas chromatographic system, equipped with an atomic fluorescence detector, was used. For the analysis of $MeHg^+$, conversion into the volatile methylmercury by reaction with tetraethylborate prior to the purging process was carried out. The detection limit for both methylated mercury compounds was 5 pg Hg/L, which allowed their determination in most ocean water samples even in those of the Antarctic Ocean and Arctic Ocean. A north–south concentration profile in the Atlantic Ocean, covering a distance from 51°N to 58°N, was also examined, which resulted in the most extended set of data in the environment for these important heavy metal species. In anthropogenically influenced areas of the North Atlantic from 51°N to about 40°N, concentrations of methylated mercury in the range of 100–3000 pg/L were found. The contents of these species were significantly lower in remote areas, represented by a range <5–150 pg/L. Concentrations of the methylated mercury species were compared with those of substances often used as biomass indicators, for example chlorophyll *a* and adenosine triphosphate. A positive correlation was found, in general, in remote areas between the contents of methylated mercury and these parameters of bioactivity, demonstrating the biogenic origin of $MeHg^+$ and methylmercury, respectively. The concentration of $MeHg^+$ normally exceeded that of methylmercury, except at locations with especially high bioactivities. This result indicated that methylmercury may be the main primary biogenic product.

6.2.4 MISCELLANEOUS

Other techniques that, to a lesser extent, have been applied to the determination of organomercury compounds in saline water samples include differential pulse polarography [60] and radiochemistry [62–65].

Stoeppler and Matthews [60] have made a detailed study of the storage behaviour of methylmercury and mercuric chloride in seawater. They recommend that samples spiked with inorganic and/or methylmercury chloride be stored in carefully cleaned glass containers acidified with hydrochloric acid at pH 2. Brown glass bottles are preferred. Storage of methylmercury chloride should not exceed 10 days.

Stary et al. [66] developed a preconcentration radioanalytical method for determining down to 0.01 µg/L of methyl- and phenylmercury and inorganic mercury using 100–500 mL samples of water. Extraction chromatography and dithizone extraction were the most promising method for the concentration of organomercurials in the concentration range of 0.01–2 µg/L. The dithizone extraction was used for the preconcentration of inorganic mercury.

Shekhovtsova et al. [67] have described an enzymatic method for the determination of organomercury in seawater.

The enzymatic determination of organomercury (methyl-, ethyl-, phenylmercury) is based on their effect on the induction period (τ_{ind}) caused by the introduction of sodium diethyldithiocarbamate to the oxidation of *o*-dianisidine, *o*-phenylenediamine, and 3,3′, 5,5′-tetramethylbenzidine by H_2O_2 catalysed by native horseradish peroxidise; τ_{ind} is inversely proportional to organomercury compounds concentration over a range of 0.05–10 µM. The lowest detection limit (C_{min}) is 0.03 µM, and the relative standard deviation (RSD) is lower than 3%. This method is simple, inexpensive, and does not require the preliminary conversion of organomercury compounds to elemental or ionic mercury. The procedure was applied successfully to methylmercury determination in water of Kara Sea.

Kirk et al. [68] examined the distribution of total mercury, gaseous elemental Hg(0), and monomethylmercury in marine water of the Canadian Arctic

Archipelago, Hudson Strait, and Hudson Bay. Concentrations of the total mercury were low throughout the water column in samples of all regions (mean ± standard deviations: 0.40 ± 0.47 ng/L). Concentrations of monomethylmercury were also generally low at the surface (23.8 ± 9.9 pg/L); however, at mid and bottom depths, monomethylmercury was present at concentrations sufficient to initiate bioaccumulation of monomethylmercury through Arctic marine food webs (maximum 178 pg/L; 70.3 ± 37.3 pg/L). In addition, at mid and bottom depths, the percentage of total mercury that was monomethylmercury was high (maximum 66%, 28% ± 16%), suggesting that active methylation of inorganic Hg(II) occurs in deep Arctic marine waters. Interestingly, there was a constant, near 1:1, ratio between the concentrations of monomethylmercury and dimethylmercury at all sites and depths, suggesting that methylated mercury species are in equilibrium with each other and/or are produced by similar processes throughout the water column. These results also demonstrate that oceanographic processes, such as water regeneration and vertical mixing, affect mercury distribution in marine waters. Vertical mixing, for example, likely transported monomethylmercury and dimethylmercury upward from production zones at some sites, resulting in elevated concentration of these species in surface waters (up to 68.0 pg/L), where primary production and thus uptake of monomethylmercury by biota are potentially the highest. Finally, calculated instantaneous ocean–atmosphere fluxes of gaseous mercury demonstrated that Arctic marine waters are a substantial source of dimethylmercury and gaseous elemental to the atmosphere (27.3 ± 47.8 and 130 ± 138 $ng/m^2/day$, respectively) during the ice-free season.

6.3 ORGANOARSENIC COMPOUNDS

These have been found in seawater and coastal water in the range of 0.001–2.6 µg/L. Gas chromatographic [69] and atomic absorption spectrometric [70] methods have, respectively, detection limits of 0.25 and 0.14 µg/L.

The most sensitive method available involves preconcentration on Dowex AG50 followed by atomic absorption spectrometry [7,72] extends the detection limit down to 0.02 µg/L fairly near to the lowest organoarsenic content of 0.001 µg/L encountered in marine waters.

Some of the values obtained for inorganic arsenic, monomethylarsenic acid, and dimethylarsenic acid in seawater samples were as follows: AsIII (0.02–0.11 µg/L), AsV (15–42.5 µg/L), monomethylarsenic acid (0.003–0.13 µg/L), and dimethyl arsenic acid (0.002–0.31 µg/L).

6.3.1 SPECTROPHOTOMETRIC METHODS

Haywood and Riley [73] have described procedures for the determination of arsenic in seawater. Whilst this method does not include organic arsenic species, these can be rendered reactive either by photolysis with ultraviolet radiation or by oxidation with potassium permanganate or a mixture of nitric acid and sulphuric acid. Arsenic(V) can be determined separately from total inorganic arsenic after extracting As(III) as its pyrrolidine dithiocarbamate complex into chloroform.

In the method for inorganic arsenic, the sample is treated with sodium borohydride added at a controlled rate. The arsine evolved was in a solution of iodine and the resultant arsenate ion determined photometrically by molybdenum blue method. For seawater, the range, standard deviation, and detection limit were 1–4 µg/L, 1.4%, and 0.14 µg/L, respectively. Silver and copper cause serious interference at concentrations of a few tens of milligrams per litre; however, these elements can be removed either by preliminary extraction with a solution of dithizone in chloroform or by ion exchange.

Haywood and Riley [73] showed that arsenic in the form of tetraphenylarsonium chloride 1-(o-arsonophenylazo)-2-napththol-3,6-disulphuric acid and o-arsono-phenylazo-p-methaminobenzene are quantitatively decomposed in seawater by ultraviolet radiation.

6.3.2 ATOMIC ABSORPTION SPECTROMETRY

Persson and Irgum [71] determined sub ppm levels of dimethyl arsinate by preconcentrating the organoarsenic compound in a strong cation-exchange resin (Dowex AG 50 W-XB). By optimising the elution parameters, dimethyl arsinate can be separated from other arsenicals and sample components, such as group I and II metals, which can interfere in the final determination. Graphite furnace atomic absorption spectrometry was used as a sensitive and specific detector for arsenic. The described technique allows dimethyl arsinate to be determined in a sample (20 mL) containing a 10^5-fold excess of inorganic arsenic with a detection limit of 0.02 ng/mL. Good recoveries were obtained from artificial seawaters, even at the 0.05 µg/L level, but for natural seawater samples, the recoveries were lower (75%–85%). This effect could be attributed to organic sample components that eluted from the column together with dimethyl arsinate.

Various researchers [74–79] have described procedures for the determination of dimethyl arsinate, monomethyl arsenite, and trimethyl arsine oxide in seawater based on the conversion of the organoarsenic compounds to hydrides using sodium boro-hydride, cold trapping of hydrides, and then vaporising them by controlled warming of the cold trap and followed by atomic absorption spectroscopy of the vaporised fractions. The detection limit of this method was 0.02 µg/L organoarsenic, that is, 0.004 µg alkylarsenic absolute in a 20 mL seawater sample.

If this procedure were applied to a 20 mL extract of a 1 g sample of sediment, then a detection limit of 0.0004 mg/kg would be achieved (i.e. 0.0004 µg/g, 0.000 0004 mg/g, and 0.0004 mg/kg).

Yamamoto [80] separated organoarsenic compounds from seawater by column chromatography. The organoarsenic compounds were reduced to arsine with sodium borohydride and analysed by atomic absorption spectrometry.

6.4 ORGANOLEAD COMPOUNDS

Bond et al. [81] examined the interferences occurring in the stripping voltammetric determination of trimethyl lead in seawater by polarography and mercury-199 and lead-207 nuclear magnetic resource spectrometry. NMR and electrochemical data show that Hg(II) reacts with $(CH_3)_3Pb^+$ in seawater. Consequently, anodic stripping voltammetric methods for determining $(CH_3)_3Pb^+$ and inorganic Pb(II) may be unreliable. Solvent extraction followed by spectrophotometry has been used to determine trialkyl lead compounds in water.

6.5 ORGANOCADMIUM COMPOUNDS

Until 1996, organocadmium compounds had not been detected in the environment.

Pongratz and Henmann [82] using differential pulse anodic scanning voltammetry found low levels of methyl cadmium compounds in the Atlantic Ocean. Levels in the South Atlantic were approximately 700 pg/L and those in the North Atlantic were below the detection limit of the method, that is, below 470 pg/L. It is believed that these compounds were formed as a result of biomethylation of inorganic cadmium.

REFERENCES

1. R.B. Laughlin, H.E. Guard, and W.M. Coleman, *Environmental Science and Technology*, 1986, 201, 201.
2. G.T. Blunden and A.N. Chapman, *Analyst*, 1978, 103, 1266.
3. R. Lobinski, W.M.R. Dirk, M. Ceulemans, and F.C. Adams, *Analytical Chemistry*, 1992, 64, 159.
4. V.F. Hodge, S.I. Seidel, and E.D. Goldberg, *Analytical Chemistry*, 1979, 51, 1256.
5. A.D. Valkirs, P.F. Seligman, G.I. Olsen, and F.E. Brinkman, *Marine Pollution Bulletin*, 1986, 17, 320.
6. A. Valkirs, P.F. Seligman and P.M. Stang, *Marine Pollution Bulletin*, 1986, 17, 319.
7. R.S. Braman and M.A. Tompkins, *Analytical Chemistry*, 1979, 51, 12.
8. A.D. Valkirs, P.F. Seligman, G. Closey, and F.E. Brinkman, *Marine Pollution Bulletin*, 1986, 17, 30.
9. A.D. Valkirs, P.E. Seligman, and P.G. Stang, *Marine Pollution Bulletin*, 1986, 17, 319.
10. T.M. Dowling and P.J. Uden, *Journal of Chromatography*, 1983, 64, 153.
11. R.S. Braman and M.A. Tompkins, *Analytical Chemistry*, 1979, 51, 12.
12. R. Pinel and H. Madiec, *International Journal of Environmental Analytical Chemistry*, 1986, 27, 1265.
13. F.E. Brinkmann, in *Trace Metals in Seawater. Proceedings of a NATO Advanced Research Institute on Trace Metals in Seawater*, Sicily, Italy, 30 March–3 April 1981 (eds. C.S. Wong et al.), Pentium Press, New York, 1981.
14. W.A. Aue, *Journal of Chromatography*, 1977, 142, 145.
15. J.A. Jackson, F.E. Brinkmann, and W.P. Iveson, *Environmental Science and Technology*, 1982, 16, 110.
16. I. Tolosa, J. Dachs, and J.M. Bayone, *Mikrochimica Acta*, 1992, 109, 87.
17. M. Ali Sheikh, K. Tsuha, X. Wang, K. Savano, S.T. Imo, and T. Oomori, *International Journal of Environmental Analytical Chemistry*, 2007, 87, 847.
18. H. Tao, R.B. Rajendran, O.R. Quetel, M. Nakazato, M. Tominga, and M. Miyazaki, *Analytical Chemistry*, 1999, 71, 4208.
19. T. Nagase, H. Tai, T. Imgawa, and A. Miyazaki, *Shiger to Kankyo*, 1992, 1, 263.
20. P.O. Bartlett, F.J. Craig, and S.E. Moreton, *Nature*, 1977, 207, 606.
21. D.J. Mackey and H.W. Higgins, *Journal of Chromatography*, 1988, 436, 24.
22. L. Ebdon and G. Alonso, *Analyst*, 1987, 112, 55.
23. R.B. Laughlin, H.E. Guard, and W.M. Coleman, *Environmental Science and Technology*, 1986, 20, 201.
24. A. Kenis and A. Zirino, *Analyst*, 1983, 142, 157.

25. J.D. Smith, *Nature*, 1970, 225, 103.
26. K. Duhamel, G. Blanchard, G. Dorange, and G. Martin, *Applied Organometallic Chemistry*, 1987, 1, 133.
27. P. Tremolada, S. Bristeau, D. Mozzi, M. Sugni, A. Barloglio, T. Dagnac, and M.D.C. Carnevali, *International Journal of Environmental Analytical Chemistry*, 2006, 86, 171.
28. R. Aguliar-Martinez, M.M. Gomez-Gomez, and M.A. Palccios-Carvillo, *International Journal of Environmental Analytical Chemistry*, 2011, 91, 110.
29. H. Ageman and V. Chau, *Analytica Chimica Acta*, 1978, 101, 193.
30. J. Yamamoto, Y. Kandada, and Y. Hisaka, *International Journal of Environmental Analytical Chemistry*, 1983, 16, 1.
31. J. Olaffson, *Analytica Chimica Acta*, 1978, 68, 207.
32. G. Campeau and R. Bartha, *Bulletin of Environmental Contamination and Toxicology*, 1983, 31, 486.
33. I.M. Davies, W.C. Graham, and J. Pirie, *Marine Chemistry*, 1979, 7, 11.
34. K. Olsen, *Analytical Chemistry*, 1977, 42, 23.
35. A. Dean and R.E. Rues, *Analytical Department of Environmental and Natural Water Board, UK*. H.M. Stationery Office, London, UK, 22pp. Ref. 17, 1978.
36. H. Agemian and Y. Chau, *Analytica Chimica Acta*, 1978, 101, 195.
37. E. Yamagami, K. Tateishi, and A. Hashimoto, *Analyst*, 1980, 105, 491.
38. M. Filippelli, *Analyst*, 1984, 109, 515.
39. Department of Environmental and National Water Council, Determination of organo-mercury compounds in waters and effluents, H.M. Stationery Office, London, UK, 23pp., PF 22 ABENV, 1978.
40. W.F. Fitzgerald and W.B. Lyons, *Nature*, 1973, 242, 452.
41. G.E. Millward and A.I. Bihan, *Water Research*, 1978, 12, 979.
42. H. Agemian and A.S.Y. Chau, *Analytical Chemistry*, 1978, 50, 13.
43. K. Chiba, P.T.S. Wong, K. Tanabe, M. Haraguchi, J.D. Winefordner, and K. Fuwa, *Analytical Chemistry*, 1982, 54, 761.
44. J. Yamamoto, Y. Kanada, and Y. Hisaka, *International Journal of Environmental Analytical Chemistry*, 1983, 16, 1.
45. Department of the Environment at National Water Council, UK, Mercury in waters, effluents, soils and sediments. Additional methods (PE-22-AGENW), H.M.S.O., London, UK, 1985.
46. R.J. Pentreath, *Journal Exploration Marine Biology and Ecology*, 1976, 25, 51.
47. R.J. Pentraeth, *Journal Exploration Marine Biology and Ecology*, 1976, 25, 103.
48. A.W. Andren and R.C. Harris, *Nature*, 1973, 245, 256.
49. B.H. Olsen and R.C. Cooper, *Water Research*, 1976, 10, 113.
50. H. Egawa and S. Tajima, In *Proceedings of the Second US Japan Expert Meeting*, Tokyo, Japan, October 1976.
51. E. Shin and P.A. Krenkel, *Journal of Water Pollution Control Federation*, 1976, 48, 473.
52. A. Jeanelov, *Limnology and Oceanography*, 1970, 15, 598.
53. D.G. Langley, *Journal of Water Pollution Control Federation*, 1993, 49, 44.
54. H. Windon, W. Gardner, J. Stephens, and F.T. Taylor, *Estuary Coastal Marine Science*, 1976, 4, 579.
55. I.M. Davies, W.E. Fraham, and S.M. Pirie, *Marine Chemistry*, 1979, 7, 111.
56. K. Chiba, D.T.S. Wong, and K. Tanane, *Analytical Chemistry*, 1983, 55, 1.
57. A.W.P. Jarvic, A.P. Whitmore, R.N. Parkalli, and H.R. Potter, *Environmental Pollution*, 1983, Series B, 6, 69.
58. G. Westhoo, *Analytical Chemistry Scandinavia*, 1978, 22, 2277.
59. R. Pongratz and K.G. Heumann, *International Journal of Environmental Analytical Chemistry*, 1998, 71, 41.

60. M. Stoeppler and W. Matthews, *Analytica Chimica Acta,* 1978, 98, 389.
61. L. Sipos, H.W. Nurnberg, P. Valenta, and M. Brancia, *Analytica Chimica Acta,* 1980, 115, 25.
62. J. Stary and J. Prasilova, *Radiochemistry Radioanalytical Letters,* 1976, 24, 143.
63. J. Stary and J. Prasilova, *Radiochemistry Radioanalytical Letters,* 1976, 26, 33.
64. J. Stary and J. Prasilova, *Radiochemistry Radioanalytical Letters,* 1976, 26, 193.
65. J. Stary and J. Prasilova, *Radiochemistry Radioanalytical Letters,* 1976, 27, 51.
66. J. Stary, B. Hovlick, J. Prasilova, K. Kretzer, and J. Hanusova, *International Journal of Environmental Analytical Chemistry,* 1978, 5, 89.
67. T.N. Shekhovtsova, S.V. Muginova, and I.F. Dolmarova, *International Journal of Environmental Analytical Chemistry,* 1998, 69, 191.
68. J.L. Kirk, V.L. St Louis, H. Hintelmann, I. Hehnhart, B. Else, and L. Poissant, *Environmental Science and Technology,* 2008, 49, 8367.
69. Y. Talmi and B.D. Bostich, *Analytical Chemistry,* 1975, 47, 2145.
70. T.R. Crompton, *The Analysis of Natural Waters,* Vol. 12, Direct Preconcentration, Oxford University Press, Oxford, UK, 1993.
71. J. Persson and K. Irgum, *Analytica Chimica Acta,* 1982, 138, 111.
72. M. Andrea, *Analytical Chemistry,* 1977, 48, 820.
73. M.C. Haywood and J.P. Riley, *Analytical Chemistry,* 1975, 85, 219.
74. W.R. Penrose, *Critical Reviews of Environmental Control,* 1974, 4, 465.
75. T.R. Crompton, *The Analysis of Natural Waters,* Vol. 1, Complex Formation, Oxford University Press, Oxford, UK, 1993.
76. R.S. Braman, D.L. Johnson, and C.O. Forback, *Analytical Chemistry,* 1977, 49, 621.
77. A.G. Haward and M.H. Arbab-Zavor, *Analyst (London),* 1981, 106, 213.
78. R.A. Hinners, *Analyst,* 1980, 106, 213.
79. F.D. Pierce and H.R. Brown, *Analytical Chemistry,* 1977, 49, 1417.
80. M. Yamamoto, *Soil Science America Proceedings,* 1975, 39, 859.
81. M.A. Bond, J.R. Bradbury, P.J. Hanno, G.N. Havell, and H.A. Hulsen, *Analytical Chemistry,* 1984, 56, 2392.
82. R. Pongratz and K.G. Henmann, *Analytical Chemistry,* 1996, 68, 1262.

7 Coastal and Estuary Waters

7.1 METALS

Two factors have a bearing on the concentrations of metals found in coastal and estuary waters: (1) any dilution effects of rivers on discharges into the seawater and (2) the concentrations of metals in such nonsaline water inputs. For example, if a river is heavily polluted with metals, then we would expect the metal concentrations in the river estuary to be higher than that of the open seawater.

Some data on the level of metal accuracy in coastal estuary bay waters are given in Table 7.1, which provides an overall view of the levels of metals found in coastal water, estuary water and, for comparison, seawater and incidentally provides information on the sensitivity requirements for the analytical methods used in their analysis.

7.1.1 SPECIATION OF METALS, MULTIMETAL ANALYSIS

Various researchers have sought information on the speciation of metals in estuary and coastal waters.

Two groups of researchers have studied the speciation of copper, lead, and cadmium in coastal and estuary waters [59]. These elements were fractionated and speciated in coastal waters from the Inner Oslofjord, Norway. They examined the fractions by an operational scheme which involves ultrafiltration followed by the determination of labile, acid-soluble, and total copper, lead, and cadmium by differential pulse anodic stripping voltammetry. It was found that cadmium was present entirely in a low-molecular-weight labile species; lead was mainly in non-labile low-molecular-weight species, with half of the total lead probably occurring in low-molecular-weight organometallic compounds; copper distribution was irregular with extensive organic and colloidal association.

Batley and Gardner [60] studied the speciation of the same three elements in estuarine and coastal waters. They evaluated the potential of a heavy metal speciation scheme to reflect differences in metal distribution within a water mass obtained in a study of soluble copper, lead, and cadmium speciation in water samples from five stations in the Port Hacking Estuary (Australia) and one coastal Pacific Ocean station. The observed metal distribution was found to be consistent with the other measured physical and chemical properties of the sample waters. In all samples, the percentages of metals associated with colloidal matter were high, amounting to 40%–60% of total copper, 45%–70% of total lead, and 15%–35% of total cadmium. The scheme was used to follow changes in metal speciation under different sample storage conditions. Storage at 4°C in polythene containers was shown to prevent losses or changes in the speciation of the metals studied.

TABLE 7.1
Ranges of Metal Concentrations in Coastal Bay and Estuary Waters

Element	Location	Concentration (µg/L)	References
Aluminium	Seto Inland Sea, Japan, and Pacific Ocean	6.4–63	[5]
Antimony	North Sea	0.3–0.82	[6]
Arsenic	North Sea (soluble metals)	1.0	[7]
Arsenic	North Sea coastal waters	1.04	[8]
Barium	Gwangyang Bay, Korea	4.8	[9]
Bismuth	Seawater	0.02–0.11	[10]
	Kattegat	0.0015–0.003	[11]
	North Sea	0.2–0.68	[6]
Cadmium	Sandy Cove, USA	0.04–0.05	[12]
		0.24–0.28	[13]
	Bermuda	0.029	[12]
	Gwangyang Bay, Korea	0.20	[9]
	Coastal seawater	0.05–0.2	[14]
		0.020–0.28	[14]
		0.053	[15]
		0.053–0.07	[16]
		0.2	[17]
	North Sea (soluble metals)	0.02	[7]
	Strait of Gibraltar	<2.8	[18]
	Heligoland Bight	0.02–0.07	[19,20]
	Sea off California	0.015–0.016 (surface)	[21]
		0.94–0.099 (2950 m)	[21]
	German Bight	0.024–0.768	[22]

(Continued)

TABLE 7.1 (Continued)
Ranges of Metal Concentrations in Coastal Bay and Estuary Waters

Element	Location	Cr(III)	Cr(VI)	Organic Cr	Total Cr	Concentration (µg/L)	References
	Danish coastal waters					0.06–0.80	[23]
	Chesapeake Bay					0.05	[24]
	Canadian coastal waters					0.035–0.048	[25]
	Danish coastal waters					0.2–5.0	[23]
	Mediterranean					<5.4	[18]
	Cape San Bay					0.013 (5 m)	[26]
						0.0045 (70 m)	[26]
	North Sea					0.2–0.4	[11]
	Coastal waters					0.3–1.0	[27]
	Estuary water (salinity 10%)					0.5	[28]
	Estuary water (salinity 24.1%)					2.1	[28]
	Coastal waters					0.1	[18]
	Coastal waters					0.05–0.07	[29]
	Gota River Estuary (salinity 0.5%)					0.02	[30]
	Gota River Estuary (salinity 32%)					0.02	[30]
Cerium	Gwangyang Bay, Korea					16.7	[31]
Chromium	North Sea, soluble metals				0.4		[7]
	Sea of Japan	0.057–0.093	0.088–0.15	0.18–0.32	0.37–0.50		[32]
	Japan coast	0.04–0.06			–		[33]
	Sandy Cove, USA				0.84		[73]
	Gwangyang Bay, Korea				2.33		[9]

(Continued)

TABLE 7.1 (Continued)
Ranges of Metal Concentrations in Coastal Bay and Estuary Waters

Element	Location	Concentration (µg/L)			References
	Port Hacking, Australia	0.27	0.49	0.56	[34]
	Drummoyne Bay	0.32	0.05	0.69	[34]
	Australia	0.45	1.26	0.71	[34]
	United Kingdom	0.46	0.60	–	[35]
	Gwangyang Bay, Korea	–	–	–	2.3	[32]
	Canadian coastal waters	–	–	–	0.15–0.5	[25]
	Coastal waters	–	–	–	0.25	[18]
	Coastal waters	–	–	–	0.25–0.29	[29]
	Estuary water (salinity 10%)	–	–	–	0.9	[28]
	Estuary water (salinity 24.1%)	–	–	–	0.5	[28]
Cobalt	Sandy Cove, USA	0.02			[13]
	Coastal seawaters	0.018–0.02			[4]
	Cronulla Beach, Australia	0.21			[24]
	Chesapeake Bay	0.1			[24]
	Canadian coastal waters	0.01			[25]
	Coastal waters	<0.01			[18]
	Coastal waters	0.015–0.028			[29]
Copper	North Sea	0.2			[67]
	Gwangyang Bay, Korea	1.1			[6,9]
	Sandy Cove, USA	0.6–0.7			[13]
	Chirihua, Japan	20			[36]
	Gironde Estuary	3.7			[37]
	Coastal seawater	0.6–0.7			[14]

(*Continued*)

TABLE 7.1 (*Continued*)
Ranges of Metal Concentrations in Coastal Bay and Estuary Waters

Element	Location	Concentration (µg/L)	References
	Sunlace water	0.64	[38]
	North Pacific seawater	0.66–0.72	[39]
		0.2	[38]
	Chesapeake Bay	2.0	[23]
	Osaka Bay	0.89–2.66	[40]
	Delaware Bay	0.83–2.18 (surface)	[41]
		0.73–0.91 (16 m)	[42]
	North Sea	2.82–9.7	[6]
	Canadian coast	1.1–1.2	[25]
	Cape San Blas	0.123 (5 m)	[26]
		0.065 (70 m)	[26]
	Heligoland Bight	0.3–2.04	[19,20]
	Sandy Cove, USA	0.069–0.105 (surface)	[21]
		0.098–0.24 (2950 m)	[21]
	Gwangyang Bay, Korea	1.1	[31]
	Coastal waters	0.5–0.73	[29]
	Coastal waters	1.0	[18]
	Coastal waters	0.6–3.4	[27]
	Gota River Estuary (salinity 0.5%)	1.2	[30]
	Gota River Estuary (salinity 32%)	0.3	[30]
	Sandy Cove, USA	1.4–1.5	[13]
	Gwangyang Bay, Korea	250	[9]
	Coastal seawater	1.4–1.6	[14]

(*Continued*)

TABLE 7.1 (*Continued*)

Ranges of Metal Concentrations in Coastal Bay and Estuary Waters

Element	Location	Concentration (µg/L)	References
	Estuary water	2.1	[2]
	Delaware Bay	2.46–35.1 (surface)	[41]
		2.1–5.2 (10 m)	[41]
	Heligoland Bight	1.13	[20]
	Osaka Bay, Japan	15.4–65.5	[40,41]
	Canadian coastal waters	3.5–4.2	[25]
	Chesapeake Bay	2.1	[24]
	Coastal waters	5.0	[18]
	Coastal waters	3.2–3.7	[29]
	Gota River Estuary (salinity 0.5%)	170	[30]
	Gota River Estuary	16	[30]
Lanthanum	Gwangyang Bay, Korea	0.72	[31]
Lead	Guanabara Bay	0.07–0.55	[43]
	Sandy Cove, USA	0.22–0.35	[13]
	Seawater	0.2–0.3	[39]
		7.1	[44]
		0.038–0.29	[45,46]
		0.51–0.65	[7]
	North Sea soluble metals	0.05	[7]
	Canadian coastal waters	0.34–0.36	[25]
	North Sea	1.8–7.44	[6]
	Chesapeake Bay	0.3	[24]
	Danish coastal waters	0.8–80	[23]
			(*Continued*)

TABLE 7.1 (*Continued*)
Ranges of Metal Concentrations in Coastal Bay and Estuary Waters

Element	Location	Concentration (µg/L)	References
	Heligoland Bight	0.07	[20]
	Coastal waters	0.5–2.4	[27]
	Coastal waters	0.06–0.11	[28]
	Coastal waters	3.1–12	[47]
	Coastal waters	0.25	[18]
	Near shore water	0.22	
	Gota River Estuary (salinity 0.5%)	0.30–0.36	
Manganese	Sandy Cove, USA	1.4–1.8	[13]
	Gwangyang Bay, Korea	1.5	[9]
	Chirihama Bay, Japan	60	[36]
	Tamar Estuary, UK	20–250	[48]
	Coastal seawater	1.4–1.6	[13]
	South Bermuda	1.4–1.8	[49]
	Canadian coastal waters	0.78–0.95	[25]
	Osaka Bay, Japan	11.1–30.6	[40]
	Chesapeake Bay	2.0	[24]
	Heligoland Bight	0.35	[20]
	Coastal waters	4.0	[18]
	Coastal waters	1.9–2.5	[29]
Mercury	Seawater	0.000018–0.000026	[39]
	Seawater	0.01	
	Coastal samples	0.05	
	River Loire Estuary (salinity 20%–30%)	0.06–1.1	[50]

(*Continued*)

TABLE 7.1 (Continued)
Ranges of Metal Concentrations in Coastal Bay and Estuary Waters

Element	Location	Concentration (μg/L)	References
	River Loire, 0–10 km	1.4–11.6	[50]
	River Loire, 10–15 km upstream of estuary (salinity 1‰–10‰)	1.0–7.0	[50]
	River Loire, 15–10 km upstream of estuary (salinity 10‰–20‰)	1–15.1	[50,51]
	River Loire, 10–15 km upstream of estuary (salinity 1‰–10‰)	1.0–7.0	[50]
	River Loire, 15–30 km upstream of estuary (salinity <1‰)	1–15.1	[51]
Molybdenum	North Sea soluble metals	0.002	[7]
	Kagoshima Bay, Japan	8.16–9.7	[5]
	Seawater	2.1–18.8	[52]
	Coastal waters	7–200	[47]
	Coastal waters	10.1–10.3	[8]
Nickel	Sandy Cove, USA	0.033–0.40	[13]
	Profile to 1200 m in Santa Catalina	0.3–0.6	[53]
	Coastal seawater	0.33–0.4	[13]
	Estuary water	1.2–1.3	[13]
	North Sea soluble metals	0.25	[7]
	Chesapeake Bay	1.2	[24]
	Heligoland Bight	0.2–1.2	[19,29]
	Osaka Bay, Japan	2.41–5.33	[40]
	Canadian coastal waters	0.37–0.43	[25]
	Sea off California	0.22–0.3 (surface)	[21]

(Continued)

TABLE 7.1 (Continued)
Ranges of Metal Concentrations in Coastal Bay and Estuary Waters

Element	Location	Concentration (μg/L)	References
	Menai Straits	0.60–0.67 (2950 m)	[21]
Nickel	Coastal waters	1.9	[18]
	Gota River Estuary (salinity 0.5‰)	1.2–1.3	[30]
	Gota River Estuary (salinity 32‰)	0.4	[30]
Thorium	Estuary and seawater	0.0002	[9,31]
	Gwangyang Bay, Korea	1.36–1.86	[8]
	Coastal waters	3.08–3.1	[9]
Uranium	Gwangyang Bay, Korea	0.098	[9]
Rare earths	Gwangyang Bay, Korea	Ce 16.7	[9]
		La 0.72	
Scandium	Gwangyang Bay, Korea	0.098	[9]
Selenium	Seawater	0.4	[54]
Vanadium	Gwangyang Bay, Korea	2.14	[9]
	Osaka Bay, Japan	0.23–0.88	[40]
	Coastal waters	1.22–1.23	[8]
	Coastal waters	<0.01–5.1	[47]
Zinc	North Sea soluble metals	1.0	[13]
	Sandy Cove, USA	1.5–1.9	[13]
	Gwangyang Bay	45.9	[9]
	Cape San Blas	0.055 (5 m)	[26]
		0.030 (70 m)	[26]
	Heligoland Bright	1.3–6.6	[19,20]
			(Continued)

TABLE 7.1 (Continued)
Ranges of Metal Concentrations in Coastal Bay and Estuary Waters

Element	Location	Concentration (µg/L)	References
	Osaka Bay, Japan	5.3–29.1	[41]
	Chesapeake Bay	4.8	[24]
	Danish coastal waters	0.5–250	[23]
	North Sea	7.0–22.0	[6]
	Sea off California	0.007 (surface)	[21]
		0.60–0.65 (2950 m)	[21]
	Coastal	3.28	[18]
	Coastal	1.6–2.0	[29]
	Coastal seawater	1.5–1.9	[14]
		4.1	[44]
	Gota River Estuary (salinity 0.5‰)	0.72–0.84	[30]
	Gota River Estuary (salinity 32‰)	7.6–8.4	[30]
		0.5	[30]
Anions in Seawater	Phosphate Japanese inland waters	12.8–46.0 µg/L	[52–56]
	Fluoride Point Lonsdale, Victoria, Australia (salinity 35‰)	1280–1430	[57]
	Iodate	30–60	[58]
	Iodide	0–20	[58]
	Organic iodine	<5	[1,3,58]

Note: The North Sea and the Mediterranean are included in this list as both are subject to a high degree of metal contamination originating from surrounding coastal areas.

Braunggardt et al. [61] studied the speciation of copper, nickel, and cobalt in a contaminated estuary in southwest Spain.

Four surveys of the Huelva Estuary in southwest Spain and its sources, the Tinto River and the Odiel River, were carried out between 1996 and 1998. The surveys investigated the impact of metalliferous mining of sulphide-rich ores in the catchment area on metal speciation, metal concentrations in a macrophyte, and phytoplankton diversity and abundance. Chemical speciation measurements in the lower Tinto Estuary showed that metals were predominantly electrochemically labile (<99% of total dissolved Cu, Co, and Ni at 10 μm Cu, 424 nM Co, and 500 nM Ni, S = 28). Concentrations of copper-complexing ligands and free cupric ions [Cu^{2+}] in the Gulf of Cadiz ranged between 5.3–38 nM and 0.2–7.9 pM, respectively, with conditional stability constants of the ligands of log k_{Cu} = 11.7–12.6. At enhanced dissolved copper concentrations in the lower Huelva Estuary, copper-complexing ligands were saturated with copper, resulting in nanomolar [Cu^{2+}], which increased upstream. Metal tissue concentrations of the macrophyte *Blidingia marginata* were high, and clear relationship between dissolved labile copper and macrophyte tissue, copper concentrations was observed. A low biodiversity was observed in the Huela system (Shannon–Wiener indices (H) typically <0.2). Nevertheless, the maximum biomass was observed in the lower Tinto Estuary, which showed high labile metal and nutrient concentrations and a low biodiversity ($H < 0.02$), thereby suggesting adaptation through evolutionary processes of the phytoplankton community to the harsh conditions.

7.1.2   Determination of Metals

7.1.2.1   Ammonium

Berg and Abdullah [62] have described a spectrophotometric autoanalyser method based on phenol, sodium hydrochlorite, and sodium nitroprusside for the determination of ammonium in seawater and estuarine water (i.e. the indophenol blue method). The manifold design allows for the determination of ammonium concentration in the range 0.2–20 μg/L over a salinity range of 35%–40% with negligible interference from amino acids.

The interference from amino acids was investigated and found to be negligible. The chloride content of estuary waters can vary over a very wide range from almost zero in rivers entering the estuary to about 18,000 mg/L at the edges of the estuary where the water is virtually pure seawater.

Particularly in analysis that uses autoanalysers, this wide variation in chloride content of the sample can lead to serious 'salt errors' and indeed, in the extreme case, can lead to negative peaks in samples which are known to contain ammonia. Salt errors originate because of changes in pH, ionic strength, and optical properties with salinity. This phenomenon is not limited to ammonia determinations by autoanalyser methods, but it has, as will be discussed later, also been observed in the automated determination of phosphate in estuarine samples by molybdenum blue methods.

In a typical survey carried out in an estuary, the analyst may be presented with several hundred samples with a wide range of chloride contents. Before starting an analysis, it is a good practice to obtain electrical conductivity data for such samples so that they can be grouped into increasing ranges of conductivity and each group analysed under the most appropriate conditions.

In this connection, Mantoura and Woodward [63] have described an indophenol blue method for the automated determination of ammonia in estuarine waters.

Mantoura and Woodward [63] overcame the problem of magnesium precipitation by ensuring a stoichiometric excess of citrate (about 120%). These researchers believe that 'salt errors' occurring with estuarine samples originate from poor pH buffering rather than ionic strength variations. They, in fact, used phenol at a concentration of 0.6 M to make the system self-buffering. Even in the presence of 1 mg NH_3–N per litre, the indophenol blue reaction will consume only 3% of the phenol leaving most of the phenol to act as a pH buffer. Ethanol was used to solubilise the high concentration of phenol used in the system. The salt error of this method was determined by making standard additions of ammonia into water of different salinities. When compared with other methods, this method displays minimal salt error (about 8%), even though the final pH of the river water mixture (pH 10.9) was greater than seawater (pH 9.9).

In addition to the chemical effects of varying salinity, there are optical interferences in colorimetric analysis, which are peculiar to estuarine samples. Saline waters and river waters have, in the absence of colorimetric reagents, an apparent absorbance arising from the following:

1. Refractive bending of light beam by sea salts – 'refractive index blank' [53]
2. Background absorbance by dissolved organics of riverine origin

The former is a function of the optical geometry of the light beam and the flow cell and the latter is related to the organic loading of river water. Both are linearly related to salinity which makes optical blank corrections easy to apply to estuarine samples.

7.1.2.2   Antimony
Both hydride generation atomic absorption spectrometry and emission spectrometry have been applied to the determination of antimony in estuarine waters.

7.1.2.3   Arsenic
Amaukwah and Fasching [64] have discussed the determination of arsenic(V) and arsenic(III) in estuary water by solvent extraction atomic absorption spectrometry using the hydride generation technique.

7.1.2.4   Barium
Epstein and Zander [65] determined barium directly in estuarine water and seawater by graphite furnace atomic absorption spectrometry.

7.1.2.5 Boron

Ball et al. [66] have described a method for determining down to 20 μg/L boron in estuarine waters. A dc argon plasma emission spectrometer was used. Quenching of the plasma by high solute concentrations of easily ionised elements such as alkali metals, as well as high and variable electron density, may be avoided by dilution. The method was found to be more sensitive, equally precise, less subject to interference and with a wider linear analytical range than the carmine spectrophotometric method. Very few interferences were noted when this technique was tested. There is a minor interference from the differential enhancement of tungsten relative to boron in solutions containing high concentrations of alkali metals. The effect of this is to increase the background when estuary water is being analysed, and it can be mitigated by using synthetic estuary water as a blank, by dilution, or by analysis by the method of standard additions.

7.1.2.6 Cadmium

Gardner [67] has reported a detailed statistical study involving 10 laboratories in the determination of cadmium in coastal and estuarine waters by atomic absorption spectrometry. The maximum tolerable error was defined as 0.1 μg/L or 20% of sample concentration, whichever is the larger. Many laboratories participating in this work did not achieve the required accuracy for the determination of cadmium in coastal and estuarine water. Failure to meet targets is attributable to both random and systematic errors.

7.1.2.6.1 Graphite Furnace Atomic Absorption Spectrometry

Inductively coupled plasma atomic emission spectrometry, anodic stripping voltammetry, and isotope dilution mass spectrometry have also been used to determine cadmium in estuarine waters.

7.1.2.7 Chromium

Zhang and Smith [68] preconcentrated various chromium species (total Cr, Cr(VI), Cr(III)) in estuarine and seawater samples by coprecipitation with lead sulphate or lead phosphate prior to determination by neutron activation analysis and gamma spectrometry. Lead phosphate will collect both trivalent and hexavalent chromium while lead sulphate collects hexavalent chromium only. The procedure had a detection limit of 0.1 μg/L for chromium in seawater when 800 mL of samples was used.

Recoveries of both chromium(III) and chromium(VI) were excellent for both sample types with lead phosphate. Chromium(VI) was quantitatively recovered by the lead sulphate procedure from potable water, but its recovery from seawater was incomplete (~87%) because of the considerable amount of competing species, especially sulphate, present. Under the conditions used, the detection limit based on 2 (background)$^{1/2}$ was 0.08 μg/L chromium for 800 mL samples.

Graphite furnace atomic absorption spectrometry, inductively coupled plasma atomic emission spectrometry, anodic stripping voltammetry, and isotope dilution mass spectrometry have been used to determine chromium in estuary waters.

7.1.2.8   Cobalt

Graphite furnace atomic absorption spectrometry, inductively coupled plasma atomic emission spectrometry, anodic stripping voltammetry, cathodic stripping voltammetry, and isotope dilution mass spectrometry have been applied to the determination of cobalt in estuarine waters.

7.1.2.9   Copper

Berger et al. [69] applied a fluorescence quenching titration method to the measurement of the complexation of copper(II) in the Gironde Estuary. Relatively high values of residual fluorescence after titration indicated that much organic fluorescing material does not bind to divalent copper.

Titration was performed in a flow-through system thermostated at 25°C under nitrogen. The pH is adjusted for each step at 8.0 with 0.01 M potassium hydroxide or nitric acid (pH 8.0, this is very close to the natural pH of the Gironde waters). After the addition of copper(II) solution, the sample is circulated through the cuvette for several minutes before measurement. An increase in Rayleigh scattering, which was measured along with fluorescence, signifies aggregation. When Rayleigh scattering doubled its original value before copper(II) was added, the titration was stopped.

Nelson [70] studied voltammetric measurement of copper(II)–organic interactions in estuarine waters. Based on the results of previous studies on the effects of organic matter on the adsorption of copper at mercury surfaces, Nelson developed a method to evaluate the interactions between divalent copper and organic ligands, based on ligand exchange. The copper/organic species competed with glycine, which formed copper glycinate. These two complexes could be distinguished voltammetrically, since copper glycinite gave a higher surface excess of copper at a gelatine-coated hanging drop mercury electrode. The method was applied successfully to both chloride media and estuarine waters. It was demonstrated that estuarine waters contained two types of ligands capable of binding divalent copper: humic material with polyelectrolyte-type binding and discrete ligands with stability constants of about 1000 million. The extent of binding by humic material decreased down the estuary as a result of dilution and increased salinity.

Nelson [71] also examined the role of organic matter in the uptake of copper at a mercury electrode. Experiments were carried out on the induced adsorption of copper on a hanging drop mercury electrode in a stirred solution, using chloride media with added complexing ligand and organic surfactants, and in estuarine water containing added surfactant (gelatine). Copper chloride was the most important copper species adsorbed on the electrode, and adsorption was enhanced by the presence of organic films, which could provide a critical pathway for reducing divalent copper in estuarine waters. The composition of organic monolayers might be determined by utilising the adsorption of divalent and monovalent copper as electroactive probes and determining solution copper-organic binding.

Shuman and Michael [72] applied a rotating disc electrode to the measurement of copper complex dissociation rate constants in marine coastal waters. An operational definition of labile and non-labile metal complexes was established on kinetic criteria. Samples collected off the mid-Atlantic coast of the United States showed varying degrees of copper chelation. It is suggested that the technique should be useful for metal toxicity studies because of its ability to measure both equilibrium concentrations and kinetic availability of soluble metal.

Buerger-Welrich and Sutzberger [73] used solid-phase extraction to measure Cu(II) in estuarine water.

Nduagn [94] carried out model predictions of copper speciation in coastal waters and compared the results obtained with those obtained by voltammetry graphite, furnace atomic absorption spectrometry, inductively coupled plasma atomic emission spectrometry, anodic stripping voltammetry, cathodic stripping voltammetry, and isotope dilution mass spectrometry.

7.1.2.10 Iron

Graphite furnace atomic absorption spectrometry, inductively coupled plasma atomic emission spectrometry, anodic stripping voltammetry, and isotope dilution mass spectrometry have been used to determine iron in estuary water.

7.1.2.11 Lead

Atomic absorption spectrometry, graphite furnace atomic absorption spectrometry, inductively coupled plasma atomic emission spectrometry, anodic stripping voltammetry, and isotope activation mass spectrometry have been used to determine iron in estuary waters.

7.1.2.12 Magnesium

A titration procedure has been applied to the determination of magnesium in estuary waters.

7.1.2.13 Manganese

Knox and Turner [74] have described a polarographic method for determining manganese(II) in estuarine waters which covers the concentration range 10–300 μg/L. The method, which is specific to manganese(II) and its labile complexes, is used in conjunction with a colorimetric technique to compare the levels of manganese(II) and total dissolved manganese in an estuarine system. They showed that the polarographically determined manganese(II) can vary widely from 100% to less than 10% of the total dissolved manganese, determined spectrophotometrically at 450 nm by the formaldoxime method calibrated in saline medium to overcome any salt effects. It is suggested that the manganese not measured by the polarographic method is in colloidal form.

Estuaries in contrast appear to be important sites for manganese redox reactions. Manganese maxima have been observed in several estuaries [75–77], and it has been suggested that these maxima result from a recycling of precipitated manganese [77]. The proposed mechanism is essentially a redox cycle in which dissolved manganese(II) is oxidised into the water column and precipitated. Reduction in anoxic sediments results in the subsequent release of manganese(II) to the water column.

The details of estuarine manganese chemistry are far from clear; however, Sholkovitz [78] notes that while adsorption onto colloidal humic acids or hydrous iron oxides is a major factor controlling the removal of many trace metals from estuarine waters, manganese does not conform to this pattern as it is known to behave independently of iron and associates only weakly with organic matter. Detailed investigation of estuarine manganese reactions requires analytical methods specific to the species involved, a requirement met only by electrochemical methods at natural concentration levels. Davison [79,80] had described the use of direct polarographic methods in the analysis of manganese in lake waters in the concentration range of 0.1–5 mg/L.

Graphite furnace atomic absorption spectrometry, inductively coupled plasma atomic emission spectrometry, anodic stripping voltammetry, and isotope dilution mass spectrometry have all been coupled to the determination of manganese in estuary waters.

7.1.2.14 Mercury

Han et al. [81] have studied speciation in Chinese coastal surface water.

Various other researchers [82–85] have reported on the levels of total mercury in seawater. Generally, the levels are less than 0.2 μg/L with the exception of some parts of the Mediterranean, where additional contributions due to man-made pollution are found [85–89].

7.1.2.15 Nickel

Graphite furnace atomic absorption spectrometry, inductively coupled plasma atomic emission spectrometry, anodic stripping voltammetry, cathodic stripping voltammetry, and isotope dilution mass spectrometry have been applied to the determination of nickel in estuary waters. See Section 7.1.3.22.

7.1.2.16 Selenium

Willie et al. [91] used the hydride generation graphite furnace atomic absorption spectrometry technique to determine selenium in saline estuary waters and seawaters. A Pyrex cell was used to generate selenium hydride which was carried to a quartz tube and then to a preheated furnace operated at 400°C. Pyrolytic graphite tubes were used. Selenium could be determined down to 20 ng/L. No interference was found due to iron, copper, nickel, and arsenic.

Cutter [92] has studied the application of the hydride generation method to the determination of selenium in saline waters.

7.1.2.17 Tin

Tributyltin has been determined in estuarine waters by high-performance liquid chromatography with fluorometric detection according to the method described by Ebdon and Alonso [93] and others [85–90]. Bu_3Sn^+ is quantitatively retained from 100 to 500 mL of sample on a 4 cm long ODS column. The ODS column was back-flushed with methanol–water containing ammonium acetate onto a Partisil SCX analytical column. The eluent was mixed with acetic acid. Morin and Triton X100 was used for Florometric detection.

TABLE 7.2
Multimetal Analysis of Estuarine and Coastal Waters

Element	Detection Limit	References
Atomic Absorption Spectrometry		
Cu, Fe	Cu 0.3 µg/L	
Cu, Ni, Pb, Cd	Ni 0.02 µg/L	[122,123,128,129]
	Pb, 0.7 µg/L	
	Cd 0.5 µg/L	
	–	[124]
Fe, Mn, Cd, Zn, Cu, Ni, Pb, Co	0.1–0.4 µg/L	[125]
Fe, Mn, Cu, Ni	–	[126]
Cd, Pb, Zn	Cd 0.1 µg/L	
	Pb, 0.4 µg/L	[127]
Cd, Pb, Cr	Cr 0.2 µg/L	
Hydride Generation Atomic Absorption Spectrometry		
As, Sb	–	[125]

7.1.2.18 Vanadium
Anodic stripping voltammetry has been applied to the determination of vanadium in estuary waters.

7.1.2.19 Uranium
Cathodic stripping voltammetry has been used to determine uranium in estuary waters.

7.1.2.20 Zinc
Graphite furnace atomic absorption spectrometry, inductively coupled plasma atomic emission spectrometry, anodic stripping voltammetry, and isotope dilution mass spectrometry have been used to determine zinc in estuary waters.

7.1.2.21 Multimetal Analysis
In addition to the aforementioned elements, several elements have been determined by methods discussed in Table 7.2.

7.2 ORGANIC COMPOUNDS

7.2.1 HYDROCARBONS

Walker et al. [95] examined several methods and solvents for use in the extraction of petroleum hydrocarbons from estuarine water and sediments during an *in situ* study of petroleum degradation in seawater. The use of hexane, benzene, and chloroform as solvents is discussed and compared, and quantitative and qualitative differences were determined by analysis using low-resolution computerised mass spectrometry.

Using these data, and data obtained following the total recovery of petroleum hydro-carbons, it is concluded that benzene or benzene–methanol azeotrope is the most effective solvent.

7.2.2 POLYBROMINATED DIPHENYL ETHERS AND OXYORGANOCHLORINE COMPOUNDS

Ramu et al. [96] used mussel samples to measure the levels of polybrominated diphenyl ethers and organochlorines in the coastal waters of Asian countries such as Cambodia, China, Hong Kong, Indonesia, Japan, Korea, Malaysia, the Philippines, and Vietnam. Polybrominated diphenyl ethers were detected in all the samples anal-ysed, and the concentrations ranged from 0.66 to 440 ng/g lipid wt. Apparently higher concentrations of polybrominated diphenyl ethers were found in mussels from the coastal waters of Korea, Hong King, China, and the Philippines, which suggests that significant sources of these chemicals exist in and around this region. With regard to the composition of polybrominated diphenyl ethers congeners, BDE-47, BDE-99, and BDE-100 were the dominant congeners in most of the samples. Among the organo-chlorine compounds analysed, concentrations of DDTs were the highest followed by PCBs > CHLs > HCH > HCB. Total concentrations of DDTs, PCBs, CHLs, and HCHs in mussel samples ranged from 21 to 58,000 to 3.8 to, 0.93 to 900, and 0.90 to 230 ng/g lipid wt., respectively. High levels of DDTs were found in mussels from Hong Kong Vietnam, and China. Polychlorobiphenyls were found in Japan and Hong Kong and in industrialised/urbanised locations of Korea, Indonesia, the Philippines, and India; CHLs were found in Japan and Hong Kong; HCHs were found in India and China. These countries seem to play a role as probable emission sources of cor-responding contaminants in Asia and, in turn, may influence their global distribution.

7.2.3 INSECTICIDES

Jia et al. [97] determined dechloranes, including Dechlorane Plus (DP) Mirex (Dechlorane), Dechlorane 602 (Dec 602), Dechlorane 603 (Dec 603), and Dechlorane 604 (Dec 604) using gas chromatography with mass spectrometric detection for waters, sediment, and oyster samples were collected at 15 sampling sites near the Bohai and Huanghai Sea shore area of northern China in 2008. DP and Mirex were detected in most water, sediment, and oyster samples, which indicated the wide-spread distribution of these two compounds. The mean concentrations in water, sedi-ment, and oyster samples, respectively, were 1.8 ng/L, 2.9 ng/g dry weight (dw), and 4.1 ng/g wet weight (ww) for total DP, and 0.29, 0.9, and 2 ng/L for Mirex. Dec 602 and Dec 603 were not detected in water but in small portions of the sediment and oyster samples, showing a low level of contamination by these two chemicals in the region. Strong and significant correlations were found between total DP and Mirex concentrations in water, sediment, and oyster samples, probably suggesting simi-lar local sources of these two chemicals. Dec 604 was not found in any samples. The biota–sediment accumulation factor of DP, Mirex, and Dec 602 declined along with the increase in their logarithm of octanol–water partition coefficients (log k_{ow}), possibly indicating that compounds with lower log k_{ow} (like Mirex and Dec 602) accumulated more readily in biota. The mean fractional abundance of *syn*-DP (f_{syn}) was 0.34 in water samples, a value lower than that in Chinese commercial mixture (0.41), while the mean f_{syn} for surface sediment (0.44) and oyster (0.45) samples were higher than technical values.

Enrichment of *syn*-Declorane Plus in oyster was in agreement with previously reported findings in Great Lakes fish. There was an enrichment of *syn*-DP in marine surface sediments, however, in contrary to data reported for freshwater sediments.

Ahel et al. [98] have reported a gas chromatographic method for the determination of atrazine and simazine in estuary waters in amounts down to 0.1 µg/L.

Chiron et al. [99] have developed a liquid chromatography–mass (LC/MS) spectrometric method for the determination of five acidic herbicides and two bentazon transformation products in estuarine waters. This method used negative-ion detection and SIM (*X79*). The limit of detection (LOD) was between 5 and 100 ng/L. Several multiresidue particle-beam LC/MS methods have been described. A particle-beam interface allows greater structural information to be obtained from an LC/MS experiment.

7.2.4 MISCELLANEOUS

O'Brian et al. [100] have described a passive phosphate sample for use when sampling estuarine in marine waters. The passive sample was calibrated in the laboratory over a range of flow velocities (0–27 cm/s) and ionic strengths (0–0.62 mol/kg). The observed sampling rates were between 0.006 and 0.20 L/day. An empirical model allowed the estimation of these sampling rates with a precision of 8.5%. Passive flow monitors based on gypsum dissolution rates were calibrated for the same range of flow velocities and ionic strength. Mass loss rates of passive flow monitors increased with increasing ionic strength. It was demonstrated that this increase is quantitatively accounted for by the increased gypsum solubility at higher ionic strengths. O'Brian et al. [100] provided a calculation scheme for these solubilities for an environmentally relevant range of temperature and salinities. The results imply that co-deployed passive flow monitors can be used for estimating the flow effect on the *in situ* sampling rates of the phosphate samplers, and it is to be expected that the same may hold for other passive samplers.

Russak and Sivan [101] have described a hydrochemical tool to identify salinisation of coastal aquifers by the measurement of chemical and isotope data.

Carstensen et al. [102] studied the responses of coastal ecosystems to changing nutrient concentrations.

Sunda and Cas [103] measured the interactive effects of temperature, salinity, and atmospheric pressure of carbon dioxide on entrophication-induced carbon dioxide acidification of subsurface coastal waters.

Measurements were carried out on high salinity (35%) and low salinity (10%) in deep Baltic Sea waters.

7.2.5 OXYGEN DEMAND CHARACTERISTICS

7.2.5.1 Biochemical Oxygen Demand

Whilst the dilution bottle method for biochemical oxygen demand (BOD) yields satisfactory results, on fresh and low saline waters, a discrepancy exists when the test is performed on waters containing elevated levels of sodium chloride and other salts.

Seymour [104] found that an increase in salinity resulted in decreased BOD values. The degradation rate decreased as the salinity level increased, yet only the particulate fraction seemed to be affected. Degradation of the soluble portion was not affected as adversely by salt content. Lysis of bacteria will occur in a growth medium when sodium chloride is added. Cellular constituents released following lysis are metabolised by the remaining microbial population in preference to the substrate present, thus yielding erratic BOD results (Kincannon and Gaudy [105,106]).

Davies et al. [107] set out to quantify the effects of fresh and saline dilution waters on the BOD of saltwater and to quantitatively establish the role of bacterial seed numbers and species associated with changes in BOD results. Significant BOD differences were found when saline samples were diluted with standard (nonsaline) BOD dilution water. Bacterial populations to genera were monitored and it was shown that equivalent numbers of bacteria did not have the same capability to degrade a given amount of organic matter with increases in salt concentrations to the 3% level. Seeding of hypersaline samples with known salt-tolerant species is recommended for consistent BOD results. The BOD values obtained in sewage seed organic standards showed a significant trend. BOD values decreased as salt concentrations increased.

Dilution of sewage-seeded saline organic standards with standard dilution water resulted in BOD values higher than those of corresponding nonsaline organic standard due to increases in bacterial populations and increased organic removal (BOD values) in the presence of low levels of salt. Bacterial populations of sewage seed correlated with corresponding BOD values and concentrations of salt in organic standard solutions show that the addition of 1% and 3% salt resulted in decreased initial bacterial population, decreased growth rates, and decreased ability of an equivalent number of bacteria to degrade an equivalent amount of organic matter. Each salt-tolerant bacterial species yielded comparable BOD values for each organic standard conducted at three salt concentrations. The population increased and the rates of biological oxidation were similar at the three salt concentrations, when using either the standard dilution water or the saline dilution water. Since sewage seed does not compare favourably with salt-tolerant bacterial seed in determining the BOD of saline water, a salt-tolerant bacterial seed should be used to ensure an accurate and reproducible BOD value when saline waste waters are being tested.

7.2.5.2 Chemical Oxygen Demand

The chemical method for the determination of the chemical oxygen demand (COD) of nonsaline waters involves the oxidation of the organic matter with an excess of standard acidic potassium dichromate in the presence of silver sulphate catalyst followed by the estimation of unused dichromate by titration with ferrous ammonium sulphate. Unfortunately, in this method, the high concentrations of sodium chloride present in seawater react with potassium dichromate producing chlorine.

$$6C^- + Cr_2O_7^{2+} + 14H \rightarrow 2Cr^{3+} + 3Cl_2 + 7H_2O$$

Consequently, the consumption of dichromate is many times higher than that due to organic material in the sample. To complicate matters, any amines in the sample consume and release chlorine in a cyclic process leading to high COD.

$$RNH_2 \rightarrow NH_2 + CO_2 + H_2O$$

$$NH_4^+ + 3Cl_2 \rightarrow NCl_3 + 3Cl^- + 4H^-$$

$$2NCl_3 \rightarrow N_2 + 3Cl_2$$

Also, the addition of silver sulphate causes the precipitation of silver chloride, which in the presence of organic compounds is neither completely nor reproducibly oxidised. This method, whilst being applicable to estuarine waters of relatively low chloride content, would present difficulties when applied to highly saline estuarine water and seawater of low organic content.

Southway and Bark [108] discussed the problem of chloride interference in the determination of COD of estuarine waters. They investigated three processes for removing chloride from the test solution: removal by chemical precipitation and filtration, removal as chlorine in an oxidation stage, and removal by a ligand-exchange process. This last process was found to produce the most satisfactory results.

Takimoto et al. [109] conducted a survey on the monthly variations of COD and total organic carbon in coastal marine waters of the northern part of Hiroshima Bay. The COD/TOC ratio varied from 0.13 to 2.2; the smaller value was observed when chlorophyll a concentration was high and vice versa. This variation was found to be due to the imperfect degradation of carbonaceous compounds in the COD method determinations. COD values for samples with high concentration of particulate organic materials tended to give smaller COD/TOC ratio.

7.2.5.3 Dissolved Organic Carbon

Abdullah et al. [110] used 2D correlations of ^{13}C NMR and FTIR to investigate changes in the chemical composition of dissolved organic matter along the Elizabeth River Chesapeake Bay estuarine transient.

This study showed that high-molecular-weight/dissolved organic method consists of three major components that have different biochemical reactivities. The first appears to be a heteropolysaccharide component and its contribution to carbon increases as the marine offshore is approached. The second appears to be composed of carboxyl-rich compounds; its carbon percentage decreases. The third component contains the major functional group of amide/amino sugar and its carbon percentage remains almost constant along the salinity transect. It seems that the heteropolysaccharide and carbon-rich compounds are present in many aquatic environments at different relative ratios. The 2D correlation maps reveal that each of these components is composed of dynamic mixtures of compounds that share similar backbone structures but have significant functional group differences. Two-dimensional correlation spectroscopy is a powerful new biogeochemical tool to track the changes in complex organic matter as a function of space, time, or environmental effects.

Mill et al. [111] carried out reversed-phase liquid chromatographic studies of dissolved organic matter and copper-organic complexes isolated from estuarine waters.

Other studies [112,113] of dissolved organic matter in estuarine water include a study of the chemical composition of dissolved organic compounds in coastal sea surface micro layers in the Baltic Sea and the use of 2D correlations ^{13}C NMR and Fourier transform infrared spectroscopy to investigate changes in the chemical composition of dissolved organic matter along an estuarine transect.

7.3 ORGANOMETALLIC COMPOUNDS

7.3.1 ORGANOMERCURY COMPOUNDS

Davies et al. [114] set out to determine the concentrations of methylmercury in seawater samples much less polluted than Minamata Bay, viz the Firth of Forth, Scotland. They described a tentative bioassay method for determining methylmercury at the 0.06 µg/g level. Mussels from a clean environment were suspended in cages. A small number of them were removed periodically, homogenised, and analysed for methylmercury. The rate of accumulation of methylmercury was determined, and by dividing this by mussel filtration rate, the total concentration of methylmercury in the seawater was calculated. The mussels were extracted with benzene, and the extract was analysed by gas chromatography.

The methylmercury concentration in caged mussels increased from low levels (less than 0.01 µg/g) to 0.06–0.08 µg/g in 150 days giving a mean uptake rate of 0.4 ng/g/day. The average percentage of total mercury in the form of methylmercury increased from less than 10% after 20 days to 33% after 150 days.

Davies et al. [114] calculated the total methylmercury concentration in the seawater as 60 pg/L (i.e. 0.1%–0.3% of total mercury concentration) opposed to 32 ng/L methylmercury found in Minamata Bay, Japan.

Kirke et al. [115] examined the distribution of total mercury (THg), gaseous elemental Hg (GEM), monomethylmercury, and dimethylmercury in marine waters of the Canadian Arctic Archipelago Hudson Strait and Hudson Bay. Concentrations of total mercury were low throughout the water column in samples of all regions (mean ± standard deviation: 0.40 ± 0.47 ng/L). Concentrations of monomethylmercury were also generally low at the surface (23.8 ± 9.9 pg/L); however, at mid and bottom depths, monomethylmercury was present at concentrations sufficient to initiate bioaccumulations of monomethylmercury through Arctic marine food webs (maximum 178 pg/L; 70.3 ± 37.3 pg/L). In addition, at mid and bottom depths, the percent of total mercury that was monomethylmercury was high (maximum 66%, 28% ± 16%), suggesting that active methylation of inorganic mercury II occurs in deep Arctic marine waters. Interestingly, there was a constant, near 1:1, ratio between the concentrations of monomethylmercury and dimethylmercury at all sites and depths, suggesting that methylated mercury species are in equilibrium with each other and/or are produced by similar processes throughout the water column. These results also demonstrate that oceanographic processes, such as water regeneration and vertical mixing, affect mercury distribution in marine waters. Vertical mixing, for example, likely transports monomethylmercury and dimethylmercury upward from production zones at some sites, resulting in elevated concentrations of these species in surface waters (up to 68.0 pg/L), where primary production and thus the uptake of monomethylmercury by biota is potentially the highest. Calculated instantaneous ocean–atmosphere fluxes of gaseous mercury species demonstrate that Arctic marine waters are a substantial source of dimethylmercury and gaseous mercury to the atmosphere (27.3 ± 47.8 and 130 ± 138 ng/m^2/day, respectively) during the ice-free season.

7.3.2 ORGANOTIN COMPOUNDS

Valkirs et al. [116,117] have conducted an inter-laboratory comparison on determinations of di- and tributyltin species in marine and estuarine waters using two methods, namely hydride generation with atomic absorption detection and gas chromatography with flame photometric detection. A good agreement was obtained between the results of the two methods. Studies on the effect of storing frozen sample prior to analysis showed that samples could be stored in polycarbonate containers at $-20°C$ for 2–3 months without significant loss of tributyltin.

Braman and Tompkins [118] found that stannane (SnH_4) and methylstannanes (CH_3SnH_3, $(CH_3)_2SNH_2$, and $(CH_3)_3SnH$) could be separated well on a column comprising silicone oil OV-3 (20% w/w) supported on Chromosorb W. Average tin recoveries from seawater are in the range of 96%–109%.

A number of estuarial waters from in and around the Tampa Bay (Florida) area were analysed by this method for tin content. The results of these analyses show that the average total tin content of estuarine waters was 12 ng/L. Approximately, 17%–60% of the total tin present was found to be in methylated forms. Any monobutyltin present escapes detection.

Ebdon and Alonso [119] have determined tributyltin ions in estuarine waters by high-performance liquid chromatography and fluorimetric detection using morin as a micellar solution. Tributyltin ions were quantitatively retained from 100 to 500 mL of sample on a 4 cm long ODS column. After washing off the salts with 20 mL of distilled water, the ODS column was back-flushed with methanol–water (80 + 20) containing 0.5 M ammonium acetate, on a 25 cm long Partisil SCX analytical column. The eluent from the column (1 mL/min) was mixed with fluorimetric reagent (acetic acid, 0.01 M; morin, 0.0025% m/v; Triton X-100, (0.7% m/v 2.5 mL/min)) for detection at 524 nm with excitation at 408 nm. The detection limit (2σ) is 16 ng of tributyltin (as Sn).

Ali Shiek et al. [120] used gas chromatography to study the speciation and seasonal behaviour of butyl and phenyl organotin compounds in protected subtropical estuarine water in Okinawa, Japan, between February and October 2006. Butyltin compounds were frequently detected in all seasons, while phenyltin was found in the winter and early spring. In the Manko Estuary, the total mean concentrations of butyltin and phenyltin were 22.78 ± 30.85 (mean $\pm$ SD, $n = 53$) and 0.08 ± 0.27 ng (Sn)/L, respectively. In the Okukubi Estuary, the concentrations of butyltin compounds and phenyltin compounds were 12.58 ± 23.96 and 0.47 ± 16.7 ng (Sn)/L, respectively.

Tao et al. [121] have described a sensitive method for the determination of ultra-trace organotin species in seawater. The merits and demerits of derivatisation methods using Grignard reagent or sodium tetraethylborate were evaluated in terms of derivatisation efficiency, applicability to the programmed temperature vaporisation method, and procedural blanks. The sensitivity of gas chromatography/inductively coupled plasma mass spectrometry was improved by more than 100-fold by operating shield torch at normal plasma conditions compared to that obtained without using it. The absolute detection limit as tin reached sub-femtogram (fg) levels. Furthermore, the detection limit in terms of relative concentration was improved 100-fold by using the programmed temperature vaporisation method, which enabled the injection of a large sample volume of as much as 100 μL without loss of analyte. When the organotin species in seawater were extracted into hexane with a preconcentration factor of 1000 after ethylation with sodium tetraethylborate was developed. This method was successfully applied to open ocean seawater samples containing organotin species at a level of 1–100 pg L.

REFERENCES

1. J. Olafsson and C.S. Wong (eds.), Trace metals in seawater, In *Proceedings of a NATO Institute on Trace Metals in Seawater*, 1981, Sicily, Italy, Plenum Press, New York, 1981, pp. 301–314.
2. M.B. Coallela, S. Siggia, and R.M. Barnes, *Analytical Chemistry*, 1980, 52, 2347.
3. B. Pihlar, P. Valenta, and H.W. Nurnberg, *Fresenius Zeitschrift für Analytische Chemie*, 1981, 307, 337.
4. W.F. Fitzgerald, Mercury analysis in seawater using cold trap preconcentrations and gas phase detection, In *Analytical Methods in Oceanography* (ed. T.R.P. Gibb), American Chemical Society, Washington, DC, 1975.
5. T. Korenaga, S. Motomizu, and K. Toei, *Analyst*, 1980, 105, 328.
6. G. Gillam, G. Duyckaens, and A. Disteche, *Analytica Chimica Acta*, 1979, 106, 23.
7. J.M. Hill, A.R. O'Donnell, and G. Mance, The quantities of some heavy metals entering the North Sea technical Report TR205, Water Research Centre, Stevenage, UK, 1984.
8. R.S. Schreadhara Murthy and D.E. Ryan, *Analytical Chemistry*, 1983, 55, 682.
9. C. Lee, N.B. Kim, I.C. Lee, and K.S. Chung, *Talanta*, 1977, 24, 241.
10. T.M. Florence, *Journal of Electroanalytical Chemistry*, 1974, 49, 255.
11. H. Eskillson and D. Jaguer, *Analytica Chimica Acta*, 1982, 138, 27.
12. E. Peuszkowska, G.R. Carnrick, and W. Slavin, *Analytical Chemistry*, 1983, 55, 182.
13. S.S. Berman, J.W. McLaren, and S.N. Willie, *Analytical Chemistry*, 1980, 52, 488.
14. R.E. Sturgeon, S.S. Berman, and I.A.H. De Sauliniers, *Analytical Chemistry*, 1980, 52, 1585.
15. R. Guevremont, R.E. Sturgeon, and S.S. Berman, *Analytica Chimica Acta*, 1980, 115, 1633.
16. A.P. Mykytiak, D.S. Russell, and R.E. Sturgeon, *Analytical Chemistry*, 1980, 52, 1251.
17. D. Jagner, *Analytical Chemistry*, 1978, 50, 1924.
18. R.E. Sturgeon, S.S. Berman, I.A.H. Sauliniers, and D.S. Russell, *Analytical Chemistry*, 1979, 51, 2364.
19. D. Schmidt, *Heligolander Meeresunter-suchungen*, 1980, 33, 576.
20. Private Communication.
21. J.M. Lo, J.C. Yun, I. Hutchinson, and C.M. Wal, *Analytical Chemistry*, 1982, 54, 2536.

22. K. Bruland, R.P. Franks, G.A. Kanaeur, and J.H. Martin, *Analytica Chimica Acta*, 1979, 105, 233.
23. R.K. Sperling, *Fresenius Zeitschrift für Analytische Chemie*, 1982, 310, 254.
24. I. Drabach, P. Pfeiffer Madsen, and J. Sorensen, *International Journal of Environmental Analytical* Chemistry, 1983, 15, 153.
25. H.M. Kingston, I.L. Barnes, T.J. Brady, and R.C. Rains, *Analytical Chemistry*, 1978, 50, 2064.
26. C.C. Won, S. Chiang, and A. Corsini, *Analytical Chemistry*, 1985, 57, 719.
27. S.R. Peotrowicz, Traces metal in seawater, In *Proceedings of a NATO Advanced Research Institute of Trace Metals in Seawater* (eds. C.S. Wong et al.), 30 March–3 April 1981, Sicily, Italy, Plenum Press, New York, 1981.
28. G. Scarponi, G. Capodaglio, P. Cescou, B. Cosma, and R. Frache, *Analytica Chimica Acta*, 1982, 135, 263.
29. V.B. Stein, B. Canelli, and A.H. Richards, *International Journal of Environmental Analytical Chemistry*, 1980, 8, 99.
30. W. Slavin, *Atomic Spectroscopy*, May–June 1980, 1, 66.
31. L.G. Danielsson, I. Magnusson, I. Westerlund, and J. Zhang, *Analytica Chimica Acta*, 1982, 144, 183.
32. C. Lee, N.B. Kim, I.C. Lee, and K.S. Chung, *Talanta*, 1977, 24, 241.
33. E. Nakayama, T. Kuwamoto, H. Tokoro, and T. Fujinaka, *Analytica Chimica Acta*, 1981, 131, 247.
34. M. Ishibashi and T. Shigenmatsu, *Bulletin of the Institute of Chemical Research, Kyoto University*, 1950, 23, 59.
35. T.L. Mullins, *Analytica Chimica Acta*, 1984, 165, 97.
36. L. Cheukas and J.P. Riley, *Analytica Chimica Acta*, 1966, 35, 240.
37. M. Murata, M. Omatsu, and S. Muskimoto, *X-Ray Spectroscopy*, 1984, 13, 83.
38. P. Berger, M. Ewald, D. Liu, and J.H. Weber, *Marine Chemistry*, 1984, 14, 289.
39. K. Yoshimura, S. Nigo, and T. Tarutani, *Talanta*, 1982, 29, 173.
40. R.S.S. Murthy and D.E. Ryan, *Analytica Chimica Acta*, 1982, 140, 163.
41. A. Sugimae, *Analytica Chimica Acta*, 1980, 121, 331.
42. R.E. Pellenberg and T.M. Church, *Analytica Chimica Acta*, 1978, 97, 81.
43. S.S. Berman, J.W. McLaren, and S.N. Willie, *Analytica Chimica Acta*, 1980, 52, 488.
44. S.A. Acebal, O. De Luca, and A. Robello, *Analytica Chimica Acta*, 1983, 148, 78.
45. D. Jagner, *Analytical Chemistry*, 1978, 50, 1924.
46. G. Torsi, *Analytica Chimica (Rome)*, 1977, 67, 557.
47. G. Torsi, E. Oesimoni, F. Palimisano, and L. Sabbatini, *Analytica Chimica Acta*, 1981, 124, 143.
48. H. Tominaga, K. Bansho, and Y. Umezaki, *Analytica Chimica Acta*, 1985, 169, 171.
49. S. Knox and D.R. Turner, *Estuarine and Coastal Marine Science*, 1980, 10, 317.
50. G.R. Carnrick, W. Slavin, and D.C. Manning, *Analytical Chemistry*, 1981, 53, 1866.
51. M. Fresnet-Robin and F. Offmann, *Estuarine and Coastal Marine Science*, 1978, 7, 425.
52. T. Kiriyama and R. Kuroda, *Talanta*, 1984, 31, 472.
53. T. Nakahara and C.L. Chakahorti, *Analytica Chimica Acta*, 1979, 104, 99.
54. D.S. Lee, *Analytical Chemistry*, 1982, 54, 1182.
55. J.H. Tseng and H. Zeitlin, *Analytica Chimica Acta*, 1978, 101, 71.
56. A.J. Paulson, *Analytical Chemistry*, 1986, 58, 183.
57. S. Motomizu, T. Wakimoto, and K. Toei, *Analytical Chemistry*, 1982, 138, 329.
58. C.J. Rix, A.H. Bond, and J.D. Smith, *Analytical Chemistry*, 1976, 48, 1236.
59. V.W. Truesdale, *Marine Chemistry*, 1978, 6, 253.
60. G.E. Batley and D. Gardner, *Estuarine and Coastal Marine Science*, 1978, 7, 59.

61. C.R. Braunggardt, E.P. Achleeberg, M. Gledhill, M. Nimmi, F. Elbaz-Poulihet, A. Ceuzado, and Z. Velazquez, *Environmental Science and Technology*, 2007, 41, 4214.
62. B.R. Berg and M.J. Abdullah, *Water Research*, 1977, 11, 637.
63. R.F.C. Mantoura and E.M.S. Woodward, *Estuary and Coastal Shelf Science*, 1983, 17, 219.
64. O. Amaukwah and P. Fashing, *Analytical Chemistry*, 1979, 51, 1101.
65. H.S. Epstein and A.J. Zander, *Analytical Chemistry*, 1979, 51, 915.
66. J.W. Ball, J.M. Thompson, and E.A. Jenner, *Analytica Chimica Acta*, 1978, 98, 67.
67. M.J. Gardner, Analytical quality control for trace metals in the coastal and marine environment, the determination of cadmium, Water Research Centre Environment, Medmenham Laboratory, Bucks, UK Report PRS 1516-M, May 1987.
68. O. Zhang and R.S. Smith, *Analytical Chemistry*, 1983, 55, 100.
69. P. Berger, M. Ewald, D. Liu, and J.H. Weber, *Marine Chemistry*, 1984, 14, 289.
70. A. Nelson, *Analytica Chimica Acta*, 1985, 109, 287.
71. A. Nelson, *Analytica Chimica Acta*, 1985, 169, 273.
72. M.S. Shuman and L.C. Michael, *Environmental Science and Technology*, 1978, 12, 1069.
73. D. Buerger-Welrich and B. Sutzberger, *Environmental Science and Technology*, 2004, 38, 1843.
74. S. Knox and D.R. Turner, *Estuarine and Coastal Marine Science*, 1980, 10, 317.
75. G.W. Bryan and I.G. Hummerstone, *Journal of the Marine Biological Association of the UK*, 1973, 53, 705.
76. L.M. Holliday and P.S. Liss, *Estuarine and Coastal Marine Science*, 1976, 4, 349.
77. D.W. Evans, N.H. Cutshall, F.A. Cross, and D.A. Wolfe, *Estuarine and Coastal Marine Science*, 1977, 5, 71.
78. E.R. Sholkovitz, *Earth and Planetary Science Letters*, 1978, 41, 77.
79. W. Davison, *Journal of Electroanalytical Chemistry*, 1976, 72, 229.
80. W. Davison, *Limnology and Oceanography*, 1977, 22, 746.
81. F. Han, X.Q. Shang, S.Z. Zhang, and B. Wen, *International Journal of Environmental Analytical Chemistry*, 2004, 84, 583.
82. J. Olafsson, K. May, and M. Steopller, *Fresenius Zeitschrift für Analytische Chemie*, 1974, 66, 68.
83. R. Chester, D. Gardner, J.P. Riley, and J. Stoner, *Marine Pollution Bulletin*, 1973, 2, 28.
84. A. Renzoni, E. Bacci, and L. Falciai, *Review of International Oceanographic Medicine*, 1973, 32, 31.
85. R.A. Fitzgerald, D.C. Gordon Jr., and R.E. Cranston, *Deep Sea Research* 1974, 21, 139.
86. Y. Thibaud, *Science et Peche, Bull. Inst. Peche Marit*, 1971, 209, 1.
87. G. Cumon, G. Vialex, H. Lelievre, and P. Bobenrieth, *Review of International Oceanographic Medicine*, 1972, 26, 95.
88. A. Renzoni and F. Baldi, *Accua e Aria* 1975, 8, 597.
89. M. Stoeppler, F. Backhaus, W. Matthes, M. Bernhard, and E. Shulte, In *Proceedings of Verb. XXVth Congress and Plenary Assembly of ICSEM*, Split, Croatia, 1976.
90. M. Stoeppler, M. Bernhard, F. Bakhaus, and E. Schulte, *Marine Pollution Bulletin*, in press.
91. SN. Willie, D.E. Sturgeon, and S.S. Berman, *Analytical Chemistry*, 1986, 58, 1140.
92. G.A. Cutter, *Analytica Chimica Acta*, 1978, 98, 59.
93. L. Ebdon and J.L. Alonso, *Analyst*, 1987, 112, 1151.
94. K. Nduagn, *Environmental Science and Technology*, 2012, 46, 7640.
95. J.D. Walker, R.P. Cowell, M.C. Hamming, and H.I. Ford, *Bulleting of Environmental Contamination and Toxicology*, 1975, 13, 245.
96. K. Ramu, N. Kagiwara, A. Sudaryante et al., *Environmental Science and Technology*, 2007, 41, 4580.

97. H. Jia, Y. Sun, X Liu, M. Yang, D. Wang, H. Qi, S. Le Chen, E.S. Verko, E.J. Reiner, and Y. Fhi, *Environmental Science and Technology*, 2011, 45, 2613.

98. M. Ahel, K. Evans, T.W. Fileman, and R.A.C. Mantoura, *Analytica Chimica Acta*, 1992, 268, 195.

99. S. Chiron, E. Martinez, and D.J. Barelo, *Journal of Chromatography A*, 1994, 665, 283.

100. D.S. O'Brian, T.K. Booij, D.W. Hawker, and J.F. Mueller, *Environmental Science and Technology*, 2011, 45, 2871.

101. A. Russak and O. Sivan, *Environmental Science and Technology*, 2010, 44, 4096.

102. J. Carstensen, M. Sanchez-Camacho, C.M. Duate, D. Krause-Jensen, and M. Marle, *Environmental Science and Technology*, 2011, 45, 9122.

103. W.G. Sunda and W.-J. Cas, *Environmental Science and Technology*, 2012, 46, 10651.

104. M.P. Seymour, The biodegradation of particulate organic matter in seawater medium, Thesis Rice University, 1977.

105. D.F. Kincannon and A.F. Gaudy, *Biotechnology Engineering*, 1966, 8, 371.

106. D.F. Kincannon and A. Gaudy, *Water Pollution Control Federation*, 1966, 38, 1148.

107. B.M. Davies, J.R. Bishop, R.K. Gurhrie, and R. Forthafers, *Water Research*, 1978, 12, 917.

108. C. Southway and L.S. Bark, *Analytical Proceedings, London*, 1981, 18, 7.

109. K. Takimoto, H. Kuratoni, A. Eto, T. Mukai, and M. Konuro, *Water Wastes*, 1980, 22, 187.

110. H.A.N. Abdullah, E.C. Minor, and P.G. Hatcher, *Environmental Science and Technology*, 2010, 44, 8044.

111. G.I. Mill, E. McFadden, and J. Quinn, *Marine Chemistry*, 1987, 20, 31.

112. M. Van Pinxstera, C. Muller, Y. Huma, C. Statie, and H. Hermann, *Environmental Science and Technology*, 2012, 46, 10455.

113. H.A.N. Abdullah, E.C. Minor, and P.G. Hatcher, *Environmental Science and Technology*, 2010, 44, 8044.

114. I.M. Davies, W.C. Graham, and J. Pirie, *Marine Chemistry*, 1979, 7, 111.

115. J.L. Kirke, V.L. St. Louis, H. Hintemann, I. Lehnhart, B. Else, and L. Poissant, *Environmental Science and Technology*, 2008, 49, 8369.

116. A.D. Valkirs, P.E. Seligman, and P.M. Strong, *Marine Pollution Bulletin*, 1986, 17, 319.

117. A.O. Valkir, P.E. Seligan, G.J. Olsen, F.E. Brinkman, C. Matthias, and J.M. Bellamy, *Analyst*, 1987, 112, 17.

118. R.S. Braman and M.H. Tompkins, *Analytical Chemistry*, 1979, 51, 12.

119. L. Ebdon and G. Alonso, *Analyst*, 1987, 112, 1551.

120. M. Ali Shiek, K. Tsuha, X. Wang, K. Samrono, S.T. Imo, and T. Omori, *Journal of International Environmental Analytical Chemistry*, 2007, 87, 847.

121. H. Tao, R.B. Rajendran, C.R. Quetel, T. Nakazaot, M. Tominge, and A. Miyazaki, *Analytical Chemistry*, 1999, 71, 4208.

122. R.E. Pellenberg and T.M. Church, *Analytica Chimica Acta*, 1978, 97, 81.

123. S.C. Apte and A.M. Gunn, *Analytica Chimica Acta*, 1987, 193, 147.

124. R.E. Sturgeon, S.S. Berman, J.A.H. Desauleniers, J.A.H. Mykytiuk, A.Q. McLaren, and D.S. Russell, *Analytical Chemistry*, 1980, 52, 1585.

125. M. Yamamoto, K. Urata, K. Murashige, and Y. Yamamoto, *Spectrochimica Acta*, 1981, 36B, 671.

126. A.J. Paulson, *Analytical Chemistry*, 1986, 58, 183.

127. V.B. Stein, B. Canelli, and A.H. Richards, *International Journal of Environmental Analytical Chemistry*, 1980, 8, 99.

128. J. Gardner and J. Yates, The determination of dissolved cadmium and lead in estuarine water samples, CEP Consultants Ltd., Edinburgh, UK, 1979, pp. 427–430.

129. R.G. Clem and A.T. Hodgson, *Analytical Chemistry*, 1978, 50, 102.

8 Determination of Metals in Marine Sediments

8.1 SEPARATION AND FRACTIONATION OF SEDIMENTS FROM SEAWATER

Sedimentary solid matter is of two main classes: one is settled on the sea bed and the other is suspended in the water column. The former can be collected by hand in shallow waters or by using a suitable solid sampler in deeper waters, as discussed by Butman [1]. Particulates present in samples of the water column can be isolated by a variety of means, including filtration and centrifugation.

In the first place, it is generally agreed that the distribution of particle sizes in the oceans is continuous from the whale to the simple single molecule [5]. The size at which one calls an aggregate of molecules a particle is, therefore, arbitrary. In the case of seawater, the dividing line between dissolved and particulates has been chosen as 0.45 µm, largely because of the first commercially available membrane filters, which had the same pore size.

8.1.1 SEPARATION BY FILTRATION

It should be mentioned at this point that the acceptance of 0.45 µm as the dividing line is purely nominal since few researchers in the field actually use filters with this pore size. The glass fibre filters used by many researchers have considerably larger pore sizes, ranging from 0.7 µm for Whatman GF F to 1.32 µm for GF C. With these filters, all particles larger than the normal pore size are retained, but many smaller particles are also trapped. The silver filters and most particularly the 0.4 µm size contain relatively large and variable amounts of carbon, which must be removed by combustion. After this combustion, the pore sizes are considerably enlarged, with the 0.45 µm filter approaching 0.8 µm in pore size. The normal 0.8 µm pore size filter is used by many investigators because the pore size changes very little under heat treatment. Thus, although 0.45 µm has been accepted as the minimum size particulate matter by definition, the filters actually used have a somewhat larger pore size and retain particles which are considerably smaller than the normal cut-off size.

The choice of filter can be determined by the amount of material considered as particulate, sometimes with unexpected results. Thus, the Whatman filters with large pore size actually retain about three times as much particulate organic carbon as does the 0.8 µm silver filters. Presumably, the difference results from the larger number of small particles retained by the glass fibre filters.

The method of calculation of the blank can also influence the determined sediment content. If surface seawater is filtered through a pad consisting of two or more filters, glass fibre or silver, the bottom filter will often contain a small amount of sediment above the blank value. Some researchers have maintained that this is due to the adsorption of dissolved organic matter on the filter and that this value should, therefore, be subtracted from the weight of the sediment found on the top filter [2]. Other researchers feel that the material caught in the second filter is largely composed of similar particles passing through the first filter. Depending on the way in which the particulate fraction is defined, the material caught by the second filter should either be added to that collected on the first filter [3] or ignored [4–6]. It can easily be seen that the choice of blank calculation can cause a considerable difference to the final values given for sediment content at least in surface waters. As far as the particulate fraction is defined, not in terms of particle size, but in terms of material caught on a specific filter, it is recommended that only one filter, rather than a pad of two or more, be used, since the material caught on subsequent filters is irrelevant by definition.

When uniform methods for collection and analysis are used, the deeper layers of the oceans give remarkably consistent results. Replicate samples, taken with a Niskin rosette sampler rigged to close six 5 L bottles simultaneously, displayed a standard deviation of ± 1.3 µg of carbon per litre [6].

Methods which collect a greater proportion of the smaller particles have also been employed. For example, a layer of fine inorganic particulate matter deposited on a filter of coarser porosity separates the particulate from the dissolved fraction. Thus, Fox et al. [7] used layers of calcium hydroxide and magnesium hydroxide while Ostapenya and Kovalevskaya [8] used alumina. These filters suffered from three disadvantages: they were troublesome to construct, the nominal pore size was irreproducible, and absorption of truly dissolved material was possible. The techniques were abandoned with the advent of the first membrane filters having a graduated pore size.

Many sediment collection methods used are biased towards those particles falling very slowly. If the residence time of a particle in the water column is only a few days, the probability of being caught in a 5 L Niskin bottle is small. This has been pointed out by the work of Bishop et al. [9]. These investigations used an *in situ* pump and filtration apparatus to filter very large quantities (5–30 m³) of seawater. This caught many classes of particles never seen in Niskin bottle samples. Their results are not comparable to those obtained in other filtration methods.

8.1.2 Separation by Continuous Centrifugation

A method for removing particles, which is not limited in volume sampled and which suffers less from the problems of overlapping classification, is continuous-flow centrifugation. Separation into density classes can be achieved by high-speed centrifugation. Jacobs and Ewing [10] used continuous centrifugation to collect total suspended matter in the oceans, and Lammers [11] discussed the possible uses of the method. The biggest drawback seems to be that the separation is governed by particle density rather than particle size. However, the method holds considerable promise for the collection of colloidal material as a separate fraction.

8.1.3 FRACTIONATION BY SIZE FRACTIONATION

Some work has been done on the size fractionation of particulate matter in water samples by the use of graduated filters. Since the filters in common use do not display a sharp cutoff in particle size retention, interpretation of the results is difficult. Repeated filtration of a single sample through filters of different pore sizes does not divide the particulate matter into definite size classes, since each filter retains particles smaller than the normal pore size. The results of the filtration of separate aliquots through a series of filters can only be reported in terms of 'particles smaller than' the nominal pore size and are equally difficult to interpret. Although such size fractionation has been reported [12], these conclusions can only be accepted in the broadest possible sense. Particle size distributions based on filtration should be supported by Coulter counter data before any conclusions can be drawn.

Once a sample of dissolved organic matter has been isolated, it is seldom in a form which permits simple analysis. In most cases, there are far too many compounds present, and some form of fractionation must take place in order to remove interferences and simplify analytical procedures.

One could devise many different bases for the fractionation of organic materials, and functional groups, degree of saturation, presence or absence of aromatic groups, and degree of polarity have all been used. The approach most often used is fractionation by size. At the upper end of the size range, we deal with particles consisting of many discrete molecules. Fractionation is accomplished by differential filtration using filters and screens of decreasing pore size [13].

Particles of smaller sizes, from the colloidal to the micromolecular, are separated by membrane filters. The most familiar of these is the Amicon Diflo filter, although several other companies manufacture similar products. Separations in the same size range can also be achieved with hollow polymeric fibres. At the upper end of the size range, these filters can be used to separate different size classes of material that we would normally consider colloidal. At the smaller end, the separation is made on the basis of molecular size. The results are presented in terms of molecular weight, but the molecular weight calibration is done with spherical molecules. The results are, therefore, given as equivalent spheres rather than as true molecular weights. The techniques have been applied to coastal seawater.

Ultrafiltration as a fractionation method gives recoveries of 80%–100% when the carbon present in each fraction is summed.

Ultrafiltration techniques employing membrane filters and those using hollow fibres both work well for the concentration and desalting of humic and fulvic acids. However, the high priming volume needed for the hollow-fibre apparatus restricts it to large-volume applications. This is not likely to be a problem in marine work, where large volumes are required because of the low concentrations of organic materials.

Both membranes and fibres retain material well below the expected molecular weight cutoff.

These techniques are just coming into use in marine organic chemistry. The apparatus is now available for processing large quantities of seawater at pilot plant levels to yield gram quantities of dissolved organic materials in specified molecular size classes. This should be one of the most fruitful methods for accumulation, separa-tion, and rough fractionation of dissolved organic materials.

8.1.4 FRACTIONATION BY COLUMN CHROMATOGRAPHY

Separation into molecular size classes by ultrafiltration is necessarily discontinuous; the fractions resulting are composed of mixtures of compounds within a given band of molecular sizes. We would often prefer a continuous separation by molecular size, particularly if it is suspected that the material in question might naturally fall into only a few fractions, each consisting of a tight grouping of molecular sizes or weights. This kind of fractionation is best carried out by some form of column chro-matography. If molecular size is to be the criterion for separation, then materials such as Sephadex can be used as column packing. Sephadex separates compounds by exclusion, holding the smaller molecules within the particles and rejecting those which will not fit within the pores of the resin. Thus, with a Sephadex column, the large components come off the column first. The system is not perfect; some charged compounds, such as phenols, can be bound irreversibly to some of the resinsomitt [14,15].

XAD resins have been used to collect and concentrate organic materials from seawater. They can also be used as packing for fractionation by column chromatog-raphy. While they have been used in simple gravity flow column chromatography, high-pressure liquid chromatography has also used them [16,17]. This technique in marine chemistry offers many possibilities and a few major drawbacks. The first drawback is the lack of sensitivity of the detectors. If the sought compounds happen to be fluorescent, or can be made into fluorescent derivatives, the inherent sensitiv-ity of fluorescence can be used. Otherwise, the technique is limited to the much less sensitive refractive index and ultraviolet absorbance detectors.

There is also the possibility of combining liquid chromatography with mass spec-trometry as a standard technique for the identification of at least compounds of lower molecular weight.

Researchers who have examined the applicability of high-performance liquid chromatography to water analysis include Waggot et al. [18], Scott and Kucera [19], Snyder [20], and Englehardt [21].

Reversed-phase chromatography is a variant of high-performance liquid chroma-tography where a non-polar organic phase is immobilised and a polar solvent is used as eluent. This variant may also be applied when non-polar compounds are sorbed from a polar solvent [22]. This situation is encountered in the attempt to accumulate non-polar organic substances from seawater by liquid–solid adsorption.

Activated charcoal was one of the first absorbents used to accumulate dis-solved organic matter [23]. For more specific applications, activated charcoal has been replaced by synthetic absorbants such as macroreticular resins, for example Amberlite XAD [24,25] or polyurethane foam [26]. These absorbants suffer from the drawback of being difficult to clean and of retaining traces of the materials col-lected rather tenaciously. Therefore, these absorbants have to be Soxhlet extracted in order to remove the sorbate which, for all practical purposes, eliminates the possibil-ity of fractionated desorption.

Applying the principle of reversed-phase chromatography in the accumulation of dissolved organic material from water, Ahling and Jensen [27] used a mixture of n-undecane and Carbowax 4000 monostearate on chromosorb W as the collecting medium. Uthe and Reinke [28] tested porous polyurethane coated with liquid phases such as SE 3, DC 200, QF-1, DEGS, OV-25, and OV-225 for the same purposes. In each case, the coating is achieved easily and may be modified to the desired adsorption properties. However, the coating is not chemically bonded to the support and may thus be removed together with the sorbate. Aue et al. [29] demonstrated the potential of support-bonded polysiloxanes for a simple, fast, and sensitive analysis of organochlorine compounds in natural aqueous systems.

Derenbach et al. [30] tested a technique for the accumulation of certain fractions of dissolved organic material from seawater and, subsequently, for the fractionated desorption of the collected material.

The handling of water extracts and possible sources of contamination would thus be reduced to a minimum. Furthermore, fractioned desorption of the accumulated material under mild conditions should result in less complex mixtures with little risk of denaturation.

These researchers investigated the suitability of numerous support materials used in reversed-phase performance liquid chromatography for the recovery of non-polar organic compounds from seawater. Porous glass treated with trichloro-n-octadecyl silane was found to permit at least a semi-quantitative recovery of test compounds. This silanized glass support was found to be easy to keep free from contamination and, in addition, had a relatively high adsorption capacity, permitting fractional desorption of the test compounds. Results obtained with this column were compared with those obtained using Amberlite XAD-2.

Derenbach et al. [30] gave full details of the preparation of this support material. They used ^{14}C-labelled spike compounds, ^{14}C n-hexadecane, and di(2-ethylhexyl) (carboxyl-^{14}C) phthalate) and also non-labelled compounds (n-C_{16}–n-C_{24} alkanes, di-n-butyl, butylbenzyl, and various phthalic acid esters, p,p' DDE, DDMU, Dieldrin, Endrin pesticides) in recovery experiments.

Some 25 L samples of natural seawater were spiked and 5 L subsamples of these were extracted with 20 mL and then 10 mL of pure n-hexane. The hexane phases were allowed to separate for 30–60 min. The combined extracts were dried over anhydrous sodium sulphate, reduced in volume with a rotary evaporator at 40°C and tap water vacuum, and taken for silica gel cleanup followed by gas chromatography. The remaining 20 L of the sample was drawn through the adsorption system at a pumping rate of 2–5 bed volumes per minute. The system consisted of a pre-combusted glass fibre filter (same type as earlier: diameter 140 mm), and the adsorption column (length 90 mm; i.d. 23 mm) was packed with either silanized porous glass or silanized glass beads or Amberlite XAD-2. Columns and filters were then Soxhlet extracted for approximately 8 h with methanol–water or acetone–water (v/v 3:2). After evaporating a major portion of the organic solvent (as earlier), the remaining extract was partitioned into hexane and dried over anhydrous sodium sulphate.

Table 8.1 shows spiking recoveries from 25 L seawater samples obtained by liquid–liquid extraction followed by gas chromatography (a) and by liquid–solid adsorption onto silanized porous glass (b), silanized glass beads (2) or XAD-2 resin (c).

TABLE 8.1
Separation and Fractionation of Sediments from Seawater

Spike Compound	Recovery by Liquid–Liquid Extraction	Recovery by Liquid–Solid Adsorption Onto:			Recovery by Glass Fibre Filters
		XAD-2	Silanized Porous Glass	Silanized Glass Beads	
n-alkanes	71 ± 8	7 ± 2	6 ± 2	~5	13 ± 2
Phthalates	103 ± 5	67 ± 28	73 ± 36	14 ± 12	5 ± 4
DDT, DDE, DDMU	74 ± 30	47 ± 16	50 ± 21	47 ± 35	8 ± 12
Dieldrin, Endrin	71 ± 13	65 ± 4	68 ± 15	26 ± 15	1 ± 2

Source: Derenbach, J.R. et al., *Mar. Chem.*, 6, 351, 1978.

Note: Average spike recovery from 25 L samples given in percent (spike concentration added to the sample is taken as 100%).

The recovery was measured from 25 L samples for a range of spike compounds: n-alkanes and phthalates at concentrations of 0.5 μg/L and pesticides at concentrations of 40 and 20 μg/L.

The liquid–liquid extraction is superior to any liquid–solid adsorption technique. The adsorption materials XAD-2 and silanized porous glass gave poor recoveries for alkanes varying between 3% and 20% for different sets of water samples. Both adsorption materials were equally inefficient. However, the recovery of phthalates is much better. On the other hand, the recovery of phthalates drops with increasing alkane character, for example diisobutyl-, di-n-butyl, benzylbutyl-, dicyclohexyl–phthalate acid esters are still recovered at about the average value diethyl–phthalate was detected at below average values due to evaporation in the workup procedure, while for bis(ethylhexyl)–phthalate, 28% was found on XAD-2 and 29% on silanized porous glass. The same effect is less pronounced for alkanes with increasing chain lengths, which can be taken as a further indication of a micelle formation for nonaromatic hydrocarbons. The recoveries of pesticides are in the range of 30%–100% and are distinguished by the scattering of the results over a wide range, sometimes exceeding 100% and by more than 50% on average, DDE and DDMU giving the highest values. This might be explained by the degradation of DDT. Dieldrin and eldrin were equally well recovered by all sampling techniques. When the column capacity was increased by a factor of 2 (two columns connected in series, either XAD-2 or silanized porous glass), an additional 3%–8% of alkanes and phthalates were recovered. No pesticides could be eluted from the second column.

Spiked phthalates and n-alkanes easily desorb from a silanized porous glass column in fractions according to the polarity of solvents and compounds.

For best recoveries, therefore, the sampling unit of the liquid–solid adsorption system has to be kept within certain proportions to the concentration of compounds to be sampled.

A disadvantage of the Derenbach et al. [30] reversed-phase high-performance liquid chromatography technique is that it is only semi-quantitative. Outweighed against this is the fact that the technique using silanized porous glass is easy to keep free of contamination, has a comparatively high adsorption capacity, and permits the fractioned desorption of accumulated material.

Hayase et al. [31] applied reversed-phase liquid chromatography with double detector (fluorescence and absorption) in the determination of dissolved organic matter in estuarine seawater. The dissolved organic matter was extracted into chloroform at pH 3 or 8. The hydrophilic–hydrophobic balance and aromatic character of the seawater-dissolved organic matter were represented on the chromatograms. The results indicated that reversed-phase liquid chromatography with double detectors was an effective technique in the characterisation of dissolved organic matter in seawater.

8.1.5 FRACTIONATION BY CHEMICAL LEACHING

Different chemical reagents dissolve different proportions of a sediment sample, and the concentration of inorganic or organic substances will differ in each fraction. Thus, as shown in Table 8.2, the following reagents dissolve different percentages of a sediment, and the stated percentage of the total cadmium content of the unfractionated samples is found in each fraction.

8.1.6 FRACTIONATION BY SEDIMENTATION

A dispersion of dry sediment in seawater can be used to carry out sedimentation measurements of the particle size distribution of sediment [32]. In this procedure,

TABLE 8.2
Effect of Choice of Extraction Agents on Efficiency

Extractant	Species Extracted	Percentage of Total Cadmium Content of Unfractionated Sample Found in the Fraction
LiCl–CsCl MeOH at 20°C	Readily exchanged ions	17
CH_3COONa pH 5.0 at 20°C	Carbonate bound surface oxide bound ions	31
$NH_2OH–HCl–CH_3COOH$ at 20°C	Ions bound to Fe and Mn oxides	34
H_2O_2, pH 2.0 at 90°C	Organically and sulphide bound ions	12
Aqua regia–HF–CH1–H_2O_2	Ions bound to residual phase	6

the sediment to be analysed is introduced as a 5% suspension in water into the sedimentation cell by means of a peristaltic pump. A finely collimated beam of X-rays is passed through the suspended medium. Radiation is detected as pulses by a scintillation detector on the opposing side of the cell to the source. The concentration of sediment in the beam is proportional to the X-ray intensity. The sedimentation cell is continuously moved across the beam to reduce the analysis time. A computer solves Stokes, law and presents the results linearly as a cumulative mass percent distribution.

8.1.7 FRACTIONATION BY CENTRIFUGING

Centrifuging a dry sediment in tetrabromoethane provided several fractions based on density (Table 8.3). The mass balance of lead, cadmium, and zinc in each fraction is also given. It is seen that the majority of the heavy metals occur in the organic and conglomerate fractions obtained from this method.

8.1.8 OTHER FRACTIONATION METHODS

Kiff [33] and Dines and Wharfe [34] have described procedures for sampling and particle size analysis of estuarine sediments. For the determination of particle size down to 45 μm, sieving is used; for the sub-sieve range, gravitational and centrifugal sedimentation methods and the Coulter counter are applied. By means of disc centrifuge, sizes down to 0.1 μm can be determined. The scheme is designed to be accurate and reproducible for long periods of operation and does not require elaborate calibration procedures.

Hakanson et al. [35] investigated the relation between physical and chemical characteristics of sediments in marine sediments and lakes for determining how representative sediment data can be established for recent sediments. He developed a mathematical model for predicting the distribution of physical and chemical parameters in sediments.

TABLE 8.3
Centrifuging Dry Sediment in Tetrabromoethane

Density	Fraction	%	Mass Balance			Concentration in Fraction (µg/kg)		
			Pb	Cd	Zn	Pb	Cd	Zn
<2.4	Organics	9	20	1	22	2220	110	2460
2.4–2.55	Conglomerates	6	6	0.3	7	1000	50	1160
2.55–2.66	Quartz and calcite	46	11	0.5	18	240	11	390
2.66–2.75	Magnesium and calcite	20	6	0.6	9	300	30	450
2.75–2.95	Aragonite	19	2	0.2	4	105	10	210
>2.95	Heavy minerals	0.1	<	<	<	<	<	<

8.2 DETERMINATION OF METALS

8.2.1 ARSENIC

Maher [36] has described a procedure for the determination of total arsenic in marine sediments. The sample is first digested with a mixture of nitric, sulphuric, and perchloric acids. Then arsenic is converted into arsine using a zinc reductor column; the evolved arsine is trapped in a potassium iodide–iodine solution; and the arsenic is determined spectrophotometrically at 866 nm as the arseno-molybdenum blue complex. The detection limit is 0.05 mg/kg dry sediment and the coefficient of variation is 5.5% at this level. The method is free from interferences by other elements at levels normally found in marine sediments. Values of 9.7 ± 0.3 and 13.2 ± 0.4 mg/kg obtained from NBS reference waters SRM 1S71 and SRM 1566 were in good agreement, respectively, with the nominal values of $10.2 \pm$ and $13.4 \pm$ mg/kg.

Siu et al. [37] determined arsenic in marine sediments down to 1 µg/kg by digestion with concentrated acid and derivatisation with 2.3 mercapto-propanol and electron capture chromatography.

Brzezinska-Paudyn et al. [38] compared five methods for the determination of arsenic in marine and river sediments. The results obtained by graphite furnace atomic absorption spectrometry, neutron activation analysis, combined furnace flame atomic absorption ICPAES, and flow injection/hydride generation ICPAES showed that all methods were appropriate for arsenic determination higher than 5 mg/kg in sediments. Graphite furnace atomic absorption spectrometry had a detection limit of 0.5–1 mg/kg.

Charkraborty et al. [39] have described a simple and sensitive spectrofluorometric method for the determination of arsenic using rhodamine-B as a fluorescent agent. This method is based on the reaction of As(III) with potassium iodate in acid medium to liberate iodine, which decreased the fluorescence intensity of rhodamine-B. This decrease in intensity was used to quantify As(III). A linear decrease in the response was observed with the increasing As(III) concentrations. An R^2 value of 0.995 was obtained for As(III) in the concentration range of 0.4–12.5 µg/mL; this showed linearity and reproducibility by this method. The limit of quantitation was found to be 0.4 µg/mL of As(III). This method was successfully used to determine the total concentration of arsenic in coastal estuarine sediments. This study suggests that the estuarine sediments were more contaminated with arsenic than the coastal sediments, and the probable source of high arsenic content in estuarine sediment is agricultural sewage in the study area. The concentrations of dynamic (non-residual) arsenic species in sediments which can be subsequently released to the overlying water column as a result of either physical or biochemical disturbances were determined by using this method. It was observed that total arsenic content did not correlate with the dynamic fraction of arsenic in the sediments. Total organic carbon present in the sediments played a crucial role in controlling its bioavailability. Dissociation rate constants of arsenic-sediment complexes were successfully determined by using this spectrofluorometric method.

8.2.2 BISMUTH

Lee [40] has used flameless atomic absorption spectrometry with hydride generation to determine down to 6 µg/kg of bismuth in marine sediments. The precision
is 6.7% at the 14 µg/kg bismuth level. The sediment (0.5 g) is completely digested
with nitric acid, perchloric acid, hydrochloric acid, and hydrofluoric acid on a hot
plate. The residue is dissolved in 50 mL 1 N hydrochloric acid. Bismuth is reduced in
solution by sodium borohydride to bismuthine, stripped with helium, and collected
in situ in a modified carbon rod atomiser. The collected bismuth is subsequently
atomised by increasing the atomiser temperature and detected by an atomic absorption spectrometer. High concentrations of cobalt, copper, gold, molybdenum, nickel,
palladium, platinum, selenium, silver, and tellurium interfere in this procedure. The
results obtained in bismuth determinations on sediments taken in Narragansett Bay
ranged from 0.27 to 0.64 mg/kg and those obtained from the North Pacific Ocean
ranged from 0.10 to 0.12 mg/kg.

8.2.3 CADMIUM

Wavelength modulation inductively coupled plasma echelle spectrometry has been
used to determine cadmium [41]. This method was applied to two standard marine
sediments, one (U.S. Geological Survey MAG-1) from a relatively unpolluted marine
sediment and the other (Environment Canada WQB-2) from a freshwater harbour of an
industrial city. A correction was necessary to overcome interference by the 214.445 nm
and 226.505 nm lines of iron, which interfere at both the cadmium 214.438 nm and
226.502 nm lines. Cadmium determinations were in good agreement with nominal
values for the two standard sediments: WQS-2, expected 1.5 ± 0.2 mg/kg, found $2.0 \pm$
0.5 mg/kg; and MAG-1, expected 0.17 ± 0.1 mg/kg, found less than 0.2 mg/kg.

8.2.4 COBALT

Spectrophotometric methods involving the use of dimethylglyoxime and 2-nitroso-1
naphthanol have been used to determine cobalt and nickel, respectively, in marine
sediments [42].

8.2.5 COPPER

Differential pulse anodic stripping voltammetry using *in situ* plating techniques has
been used to determine down to 0.1 µg/kg copper in acid extracts of marine sediments [43].

Madsen et al. [44] used potentiometric stripping analysis to determine down to
3 µg/kg of copper (and lead) in estuarine sediments with a precision of 3.9%–4.5%.

Soil contamination by heavy metals has become a widespread dangerous problems in many parts of the world, including the Mediterranean. This is closely
related to the increased irrigation of land by waste waters, the uncontrolled application of sewage sludge and industrial effluents, the atmospheric deposition of dust
and aerosols, the vehicular emissions, and many other negative human activities.

In this context, Kheir et al. [45] have predicted the spatial distribution and concentration levels of copper in the 195 km^2 of Naht el-Jawz watershed situated in northern Lebanon using a geographic information system and regression tree analysis. The chosen area represents a typical case study of Mediterranean coastal landscape with deteriorating environment. Fifteen environmental parameters (parent material, soil type, pH, hydraulic conductivity, organic matter, stoniness ratio, soil depth, slope gradient, slope aspect, slope curvature, land cover/use, distance to drainage line, proximity to roads, nearness to cities, and waste area surroundings) were generated from satellite imageries, digital elevation models, ancillary data, and/or field observations to statistically explain copper laboratory measurements. A large number of tree-based regression models were developed using (1) all parameters, (2) all soil parameters only, and (3) selected pairs of parameters. The best regression tree model (with the lowest number of terminal nodes) combined soil pH and surroundings to waste areas and explained 77% of the variability in copper measurements made into laboratory measurements. The overall accuracy of the predictive quantitative copper map produced using this model (at 1:50,000 cartographic scale) was estimated to be ca. 80%. Applying the proposed tree model is relatively simple and may be used in other coastal areas. It is certainly of significant interest to local governments and municipalities. It will serve several development projects concerned with improving the environmental conditions and the quality of living in coastal areas.

8.2.6 INDIUM AND IRIDIUM

Hodge et al. [46] determined picogram quantities of indium and iridium in marine sediments by the isolation of anionic forms on a single ion-exchange bead followed by graphite furnace atomic absorption spectrometry.

8.2.7 IRON

Lucotte and D'Anglejan [47] compared five different methods for the extraction and determination of iron hydroxides and/or associated phosphorus in estuarine particulate matter from the estuaries of St. Lawrence and Eastmain Rivers. The extraction reagents examined were citrate–dithionite–bicarbonate, ammonium oxalate–oxalic acid, hydroxylamine hydrochloride, calcium nitrilotriacetic acid, and acetate-tartrate. The procedure using the citrate tartrate dithionite bicarbonate extraction reagent was generally more specific and reproducible than the others. This extraction procedure was included in an extraction sequence which differentiated the exchangeable phosphorus and iron fraction, the iron hydroxides and associated phosphorous, and the organic phosphorus and lithogenous iron.

8.2.8 LEAD

Bettoli et al. [48] used a ^{210}Pb-based method to determine lead in marine sediments and snow samples from the Ross Sea region, Antarctica.

^{210}Pb is widely used to determine accumulation rates in order to obtain a timescale in environmental samples. The most accurate method uses the determination of ^{210}Pb via its granddaughter ^{210}Po by alpha spectrometry. Unfortunately, this method requires a complex wet-chemistry procedure to achieve the separation of ^{210}Po from its matrix. Bettoli et al. [48] described this work in a simplified procedure for the chemical separation of ^{210}Po and applied it to three marine sediment cores and a 10 m snow core. The calculated sedimentation rates for marine sediments range from 0.053 to 0.071 cm/year. The mean annual accumulation rate for the snow is 16.6 cm/year w.e.

The wavelength modulation inductively coupled plasma echelle spectrometric technique referred to under cadmium [41] has also been applied in the determination of lead in marine sediments. The 220.353 nm lead line was used. Only aluminium at concentrations higher than that usually encountered in marine sediments is expected to interfere in the determination of lead at 220.353 nm. Lead determinations in standard reference samples were in good agreement with expected values.

Harivan et al. [49] studied the natural and anthropogenic sources of lead in surface sediments offshore of the Israeli Mediterranean coast using the isotopic composition of lead in diluted acid sediment extracts. Surface sediments were collected at the lower reaches of coastal streams, along a south–north shore transect and at selected monitoring stations of the Dan Region Wastewater Plant (DRWP) outfall pipe. The background values of the lead isotopic composition were determined from the deepest part of the two representative cores collected offshore and were found to have a narrow range dominated mainly by clays derived from both inland soils and the Nilotic cell and to a lesser extent from the Saharan dust. The impact of the Dan Region Wastewater Plant activated sludge can be traced to a distance of ca. 2 km from the outfall pipe. Enrichment factors of zinc, copper, and lead were up to 25 and are strongly correlated with each other and with the lead isotopic composition, thus demonstrating the sludge to be their common source. The isotope composition of lead in stream sediments has the widest range of values and indicates a strong anthropogenic contribution, probably from both post-1992 aerosols and point sources.

8.2.9 Mercury

The microbial methylation of inorganic mercury to methylmercury observed in lake water sediments has also been observed in marine sediments. Mercury has been found in some sediments but at very low concentrations, mainly from the areas of known mercury pollution. It represents usually less than 1% of the total mercury in the sediment and frequently less than 0.1% [50–53]. Microorganisms within the sediments are considered responsible for the methylation [50,54], and it has been suggested that methylmercury may be released by the sediments to the seawater, either in dissolved form or attached to particulate material and thereafter rapidly taken up by organisms [55–57].

Abo-Rady et al. [58] have described a method for the determination of total inorganic plus organic mercury in nanogram quantities in sediments. This method is based on the decomposition of organic and inorganic mercury compounds with acid permanganate, removal of excess permanganate with hydroxylamine hydrochloride, reduction to metallic mercury with tin and hydrochloric acid, and transfer of the liberated mercury in a stream of air to the spectrometer. Mercury was determined by using a closed, recirculating air stream. Sensitivity and reproducibility of the 'closed system' were better (it is claimed) than those of the 'open system'. The coefficient of variation was 5.6% for sediment samples.

Hutton and Preston [59] have described a non-dispersive atomic fluorescence method using cold vapour generation for the determination of down to 0.4 mg/kg mercury in marine sediments. In this procedure, mercury in the sediment was reduced to its element form with acidic stannous chloride solution and swept with argon into the fluorimeter. Marine estuarine sediments examined by this procedure were found to contain between 1 and 16 μg mercury per kilogram.

8.2.10 MOLYBDENUM

Pavlova and Yasimirskii [60] employed a spectrophotometric method based on the catalytic acceleration by molybdenum of the oxidation of dithiooxamide by hydrogen peroxide. This method was used for the determination of down to 2 ng of molybdenum in seawater sediments.

8.2.11 NICKEL AND COPPER

Burzminskii and Fernando [61] studied methods for the extraction of copper and nickel as their oxalate complexes from deep-sea ferromanganese nodules.

8.2.12 PLATINUM

Hodge et al. [46] have determined picogram quantities of platinum in marine sediments by the isolation of ionic forms on a single ion-exchange bead, followed by graphite furnace atomic absorption spectrometry.

Cobelo-Garcia et al. [62] have described a method for determining platinum in marine sediments using stripping voltammetry. Results obtained by this method may be seriously affected by the presence of intensive matrix background or interfering peaks, leading to poorer detection limits and/or inaccurate quantitative results. In their work, Cobelo-Garcia et al. [62] tested the use of signal transformation (e.g. second derivative) in the analysis of platinum in seawater and sediment digests by means of catalytic adsorptive stripping voltammetry. In natural waters, the time of detection of platinum is affected by a broad background wave due to the formazone complex used in the sample matrix for its determination, while in sediment digests, the platinum peak may be interfered by elevated concentrations of zinc, affecting the accuracy of the determination. Results applying second derivative signal transformation revealed a significant improvement (two- to threefold) of the detection limit in water due to the minimisation of background effects, therefore allowing shorter accumulation times and faster determinations. In the presence of interfering peaks, the inaccuracy resulting from erroneous baseline selection in the original signal is eliminated when the second derivative is used.

Signal processing should be considered a useful tool for other voltammetric methodologies where more accurate or faster determinations were needed.

8.2.13 PLUTONIUM

Wu et al. [63] determined the isotopic composition and distribution of plutonium in the Northern South China Sea. This revealed continuous release and transport of plutonium from the Marshall Islands.

The 239 to 240 plutonium activities and the $^{240}Pu/^{239}Pu$ atom ratios in sediments in the Northern South China Sea and its adjacent Pearl River Estuary were determined to examine the spatial and temporal variation of plutonium inputs. It was clarified that plutonium was sourced from a composition of global fallout and close in fallout from the Pacific Proving Grounds in the Marshall Islands where above-ground nuclear weapons testing was carried out during 1952–1958. The latter source dominated the plutonium input in the 1950s, as proved by elevated $^{240}Pu/^{239}Pu$ atom ratios (70:30) in a dated sediment core. Even after the 1950s, the Pacific Proving Grounds was still a dominant plutonium source due to continuous transport of remobilise plutonium from the Marshall Islands about 4500 miles away along the North Equatorial Current followed by the transport of the Kuroshio Current and its extension into the South China Sea through the Hudson Strait using a simple two end-member model. Wu et al. [63] quantified the contribution of plutonium for the Pacific Proving Grounds to the Northern South China Sea Shelf and the Pearl River Estuary, which were 60% ± 1% and 30% ± 3%, respectively. This study confirmed that there were no clear signals of plutonium from the Fukushima Daiichi Nuclear Power Plant accident impacting the South China Sea.

8.2.14 SCANDIUM

A method has been described [63] to determine scandium in marine sediments, which involves digestion of the sample with hydrochloric acid–hydrofluoric acid and then separation of scandium from other elements by successive cation and anion chromatography prior to coprecipitation of scandium with ferric hydroxide. Finally, scandium is determined spectrophotometrically at 610 nm with bromopyrogallol red.

8.2.15 SELENIUM

Siu and Berman [64] determined down to 0.2 pg selenium in marine sediments by a procedure involving acid digestion in a PTFE bomb, conversion to 50-nitropiazsele-nol, and electron capture gas chromatography of the toluene extract.

Willie et al. [65] have described a method for the determination of down to 1.4 µg/kg total selenium in marine sediments based on the generation of SeH_2 using sodium borohydride with its subsequent trapping in graphite furnace at 600°C. Typical precisions of 5%–10% are claimed for this method. Copper, nickel, iron, and arsenic did not interfere at the concentrations likely to occur in sediments. Using this method, a selenium content of 2.3 ± 0.10 mg/kg was obtained on reference marine sediment NBS 1566 compared to an accepted value of 2.1 ± 0.5 mg/kg.

Selenium speciation in sediments and soil affects its bioavailability to crops. Stroud et al. [66] have described an analytical procedure for the determination of inorganic selenium species (selenite and selenate) in soil extracts by anion-exchange liquid chromatography with ICP-MS detection. This method has 10-fold higher sensitivity than existing HGAAS-based soil selenium measurements. A comparison of phosphate extraction solutions on agricultural soils amended with 20 µg/kg selenate or selenite was carried out, and a 0.016 M KH_2PO_4 extraction solution is recommended. Recovery of selenate was greater than 91%; however, selenite recovery ranged between 18.5% and 46.1% due to rapid binding to the sample soil. Soil preparation did not have a significant ($p > 0.5$) effect on the extractability of the selenate or selenite amendments. The stability of selenium species in the phosphate extracts was variable, depending on temperature and storage time. Therefore, immediate (<1 h) analysis of the soil extracts is preferable. The method developed was applied for determining extractable selenium from six arable soils in the United Kingdom. Extractable selenium levels in these soils ranged between 92% and 96%.

8.2.16 SILVER

Bloom [67] has used Zeeman-corrected graphite furnace atomic absorption spectrometry to determine silver in marine sediments.

8.2.17 THALLIUM

Riley and Siddique [68] described a procedure for the determination of thallium in deep-sea sediments. It involves preliminary concentration by adsorption on a strongly basic anion-exchange resin as the tetrachlorothallate ion, followed by elution with sulphur dioxide, and evaporation, before determination by graphite furnace atomic absorption spectrometry or differential pulse anodic stripping voltammetry. There was good agreement between the results obtained by the two methods.

8.2.18 ZINC

Bai et al. [69] determined the effectiveness of Pd/Md chemical modifier for the accurate direct determination of zinc in marine/lacustrine sediments by graphite furnace atomic absorption spectrometry using slurry samples. A calibration curve prepared by aqueous zinc standard solution with the addition of Pd/Mg chemical modifier is used to determine the zinc concentration in the sediment. The accuracy of the method was confirmed using Certified Reference Materials, NMIJ CRM 7303-a (lacustrine sediment) from the National Metrology Institute of Japan, National Institute of Advanced Industrial Science and Technology, Japan, and MESS-3 (marine sediment) and PACS-2 (marine sediment) from the National Research Council, Canada. The analytical results obtained employing Pd/Mg modifier are in good agreement with the certified values of all the reference sediment materials. Although for NRC MESS-3 an accurate determination of zinc is achieved even without the chemical modifier, the use of Pd/Mg chemical modifier is recommended as it leads to the establishment of a reliable and accurate direct analytical method. One quantitative analysis takes less than 15 min when performed on dried sediment samples. This is several ten times faster than conventional analytical methods using acid-digested sample solutions. The detection limits are 0.13 µg/g (213.9 nm) and 16 µg/g (307.6 nm), respectively, in sediment samples, when 40 mg of dried powdered samples are suspended in 20 mL of 0.1 mol/L nitric acid and a 10 μL portion of the slurry sample is measured. The precision of the proposed method is 8%–15% (RSD).

8.2.19 ZIRCONIUM

Sastry et al. [70] separated zirconium from marine sediments by digestion with hydrochloric acid. They then coprecipitated zirconium with cerium iodide prior to spectrophotometric determination as the Alizarin R-Mordant Red-S chromophore.

8.2.20 HEAVY METALS

This is a subject of considerable recent interest. Various methods have been used to analyse sediments for these elements [71,72].

Hickey and Kittrick [73] used sequential extraction to investigate the chemical partitioning of cadmium, copper, nickel, and zinc in marine sediments.

8.2.20.1 Atomic Absorption Spectrometry

Belhomme et al. [74] have carried out a comparative study of methods of digesting a number of metals in marine sediments prior to their determination. The metals were those usually found in sediments, such as aluminium, iron, manganese, and titanium, and some which occur less frequently and at very low concentrations, such as zinc, copper, lead, cadmium, chromium, and arsenic. The analytical methods used were visible spectrometry for aluminium and titanium, flame atomic absorption spectrometry for iron, manganese, and zinc, and the graphite furnace for other metals. Digestion techniques investigated were acid digestion in hydrochloric, nitric, hydrofluoric, and perchloric acids and a dry technique involving calcining the sediments and taking up the ashes in nitric acid. The results show that the selection of a particular digestion technique is dependent on the particular metal to be analysed, but it is indicated, in general terms, that the simplest, most rapid, most free from contamination and most effective is digestion with nitric acid, in a bomb, at 150°C.

Microwave digestion techniques have also been used in the sequential extraction of calcium, iron, chromium, manganese, lead, and zinc in sediments. The extraction rates of metals in each of the five binding fractions of marine sediments (metal exchangeable, carbonate bound, iron–manganese oxide bound, organic bound, and residual) were determined for both conventional and microwave heating techniques. Microwave digested sediment samples (20 mL) were centrifuged at 10,000 rpm for 30 min. Metals were determined by flame atomic absorption spectrometry. To prepare material for the determination of extraction rates for one particular fraction, a quantity of material was taken through the conventional sequence of steps up to the fraction for which the rate was determined. Sequential microwave extraction procedures were then established from the results of the rate experiments. Recoveries of total metals from standard sediment samples were comparable for both conventional and microwave techniques except for iron (62% by microwave and 76% by conventional method).

Pai et al. [75] studied the volatilisation behaviours of Cd, Cu, Ni, and Pb in different extracting media using graphite furnace atomic absorption spectrometry to determine metal concentrations in natural sediments. Considerable interference was found for Cd, Ni, and Pb in the extracts of high matrix concentrations and could be largely reduced by carefully selecting the ashing and atomisation temperatures of graphite furnace atomic absorption spectrometry. Optimal heating programs are proposed to suit a variety of extracts.

8.2.20.2 Inductively Coupled Plasma Atomic Emission Spectrometry

McLaren et al. [76] have described a procedure which permits the simultaneous determination of six major and minor elements (aluminium, iron, calcium, magnesium, sodium, and phosphorus) and eight trace elements (beryllium, cobalt, copper, manganese, nickel, lead, vanadium, and zinc) in near-shore marine sediments by inductively coupled plasma atomic emission spectrometry. Dissolution of the samples is achieved with a mixture of nitric, perchloric, and hydrofluoric acids in sealed Teflon vessels. Accurate calibration for all elements can be achieved with simple aqueous standards, provided that proper correction for various spectroscopic interferences is made. The method is not suitable for the determination of arsenic, cadmium, and molybdenum in these materials because of inadequate sensitivity but not for chromium and thallium because of incomplete dissolution. Detection limits varied from 0.5 mg/kg (beryllium) to 0–5 mg/kg (arsenic).

To determine arsenic and antimony in marine sediments, McLaren et al. [76] digested marine sediment with 4 g potassium hydroxide in a nickel crucible. The crucible was placed in a furnace and is then heated to 500°C for 30 min. The crucible was then cooled, and the contents were dissolved in 50 mL of 1 mol/L hydrochloric acid. The precipitated silicic acid was allowed to settle or was filtered before analysis. A similar digestion procedure was used in the determination of selenium but using solid sodium hydroxide or nitric–perchloric–hydrofluoric acids [76]. The addition of potassium bromide and concentrated hydrochloric acid and heating to 50°C for 50 min convert selenium to the SeIV state, leaving arsenic and antimony in the pentavalent state. Reduction with sodium borohydride then converts all three metalloids to their hydrides.

Good results were obtained by this procedure for NBS and NRCC (National Research Council of Canada) reference marine sediments.

Petronio et al. [77] described a multi-method experiment to study humic compounds and metal speciation in a marine sediment sample. The humic and fulvic acid structures were investigated by infrared spectrometry, ^{13}C NMR, thermogravimetry, and elemental analyses. Trace metals (Cu, Pb, Cd, Zn, Fe, Cr, Ni, Co, Mn, and Al) were determined in a digest of the sample and in solutions obtained by selective extraction procedures. The distribution of metals associated with humic substances was also investigated by high-pressure liquid chromatography using inductively coupled plasma atomic emission spectrometer as detector.

The results show that manganese, iron, and aluminium are present in inorganic form, whereas 37% of nickel and 8.8% of copper are present in the metal-humic compound form.

Palanquez et al. [78] used inductively coupled plasma atomic emission spectrometry and graphite furnace atomic absorption spectrometry to determine heavy metals in suspended matter in the Gulf of Cadiz.

8.2.20.3   Neutron Activation Analysis

This technique has been applied in the determination of cobalt, chromium, caesium, iron, rubidium, antimony, scandium, and strontium in suspended matter isolated from seawater.

Detection limits achieved by Ackerman [79] ranged from 0.06 mg/kg (cobalt) to 70 mg/kg (barium, iron, tin).

8.2.20.4   Photon Activation Analysis

This technique has been used to determine 12 trace elements (arsenic, barium, antimony, chromium, cobalt, lead, manganese, nickel, strontium, uranium, zinc, and zirconium) and 6 minor elements (sodium, manganese, potassium, calcium, titanium, and iron) in a certified marine sediment Standard Reference Material (BCSS-1). Photon activation analysis was carried out at the National Research Council of Canada electron linear accelerator using the Bremsstrahlung produced by the impact of a focussed electron beam on a tungsten converter.

Landsberger and Davidson [80,81] used photon activation analysis to determine a wide range of elements in marine sediments.

Eleven trace metals (As, Ba, Cr, Co, Mn, Ni, Pb, Sb, Sr, Li, and Zr) and given minor elements (Na, Mg, Cl, K, Ca, Tl, and Fe) were determined in a certified marine sediment Standard Reference Material. The precision and accuracy of the results when compared to the accepted values clearly demonstrated the reliability of this technique.

After an irradiation period of 6 h and a cooling-off period of 15 h, the y-ray spectrum was accumulated of 20,000 s. Peaks are labelled with y-ray energy in KeV and assigned a parent radionuclide.

An excellent agreement was obtained between results obtained by photon activation analysis on a standard marine sediment ranging from 0.63 ± 0.01 to 0.59 ± 0.06 mg/kg for antimony, 122–119 mg/kg for zinc, and 347 ± 36 mg/kg to 330 mg/kg for barium. Detection limits for these three elements were, respectively, 02, 7.0, and 0.4 mg/kg.

8.2.20.5   Heavy Metal Surveys

Gregory et al. [82] studied the concentration of heavy metals in coastal sediments using energy dispersive X-ray fluorescence.

Morillo et al. [83] examined the heavy metal fraction in 17 sediment samples from the Tinto and its main tributaries. For each of these samples, the association of metals (Cu, Zn, Cd, Pb, Fe, Ni, Cr, and Co) was determined in four fractions: acid soluble, reducible, oxidizable, and residual. The total metal content was also determined. Results showed high mean concentrations of Fe (109,000 mg/kg), Pb (2330 mg/kg), Zn (901 mg/kg), Cu (805 mg/kg), and Cd (2.7 mg/kg) in the sediments studied. However, the mean values found for Co (21 mg/kg), Cr (56 mg/kg), and Ni (17 mg/kg) are comparable to those in unpolluted areas. Heavy metal fractionation of the Tinto River sediments showed that the metals with the greatest mobility are Cd and Zn. These are the metals that showed the highest percentages in the first two fractions (the most labile) and the lowest percentages in the residual fraction. However, the percentage of metal present in the fourth fraction (residual) was high for Cr (78%) and Co (66%), which implies that these metals are strongly linked to the sediments.

In an exposure assessment of heavy metals (Pb, Cr, Cu, Ni, and Pb) in water and sediments from a Western Mediterranean basin (Rio Guadalhorse Region of Analusie, Southern Spain), Alonso-Castillo et al. [84] measured concentrations of heavy metals in sediment and water from the Guadalhorce River. In the later twentieth century, cities such as Málaga (capital of Costa del Sol) have suffered the impact of mass summer tourism. The ancient industrial activities, abandoned mine sites and the actual urbanisation and coastal development, recreation and tourism, and wastewater treatment facilities have been sources of pollution. This river has been heavily modified, with three dams for volume regulation purposes due to the climate cycles, with some years very dry and others with torrential rains. In this study, different indices for the assessment of sediment contaminations, statistical tools (Kruskal–Wallis test, conglomerate analysis), sequential extraction methods, and environmental quality guidelines have been employed to assess the possible contamination of this basin. Other physical–chemical parameters such as chloride concentrations, pH, and conductivity were also measured. The results indicated that Ni and Cu were the most troublesome metals because they were more easily mobilisable than Cr and Pb; Ni exceeds the SQGs guidelines and Cu presents considerable contamination. These metals were derived from lithogenic and anthropogenic sources, respectively, according to the values of enrichment factors. Ni was the most dangerous because in 96.6% of the samples analysed, Ni concentrations exceeded the threshold effect concentration below which harmful effects are likely to be observed, and even the probable effect concentration (PEC) above which harmful effects are likely to be observed in 56.6%. The cause of this pollution was postulated to be by abandoned Ni mines, which indicates that the pollution from mining persists after several decades.

Taymaz et al. [85] determined mercury, cadmium, and lead levels in water, sediment, and fish samples from Izmit Bay, Turkey. Sampling and analysis methods are described. Variations of heavy metal concentrations from different sampling stations are discussed. Results indicate that the levels of mercury and cadmium were highest in the vicinity of a chlor-alkali plant, while the highest concentration of lead was near a metallic pipe factory. The amounts of heavy metals found in the shoreline sediment samples were similar to those found in fish species from the Bay. Angelidis and Aloupi [86] carried out an assessment of heavy metal contamination of shallow coastal sediments around Mytipene, Greece.

Normalisation procedures were applied on a data set of metal concentrations of sediments from 28 stations in the harbour and coastal area of Mytilene, Lesvos Islands, Greece. Due to the great granulometric variability of the sediments, the normalisation for grain size was not applicable. Also the normalisation to a carbonate-free basis did not provide additional information. On the other hand, the normalisation to Al and the calculation of enrichment factors of metals made it possible to distinguish between contaminated and non-contaminated areas. The harbour sediments were highly enriched in Cd, Cu, and Zn and also slightly enriched in Pb. No enrichment was found in the sediments of the coastal area outside the harbour, indicating that the metal-rich deposits of that confined area are not affecting the quality of the neighbouring coastal environment.

8.3 DETERMINATION OF NONMETALS

8.3.1 BORON

Kiss [87] examined various techniques for the efficient separation and preconcentration of boron from marine sediments. Alkaline fusion with potassium carbonate was used to render boron reactive, even in the most resistant silicate minerals. Fusion cakes were extracted with water, and borate was isolated by Amberlite XE-243 boron-selective resin. Borate was determined spectrophotometrically, following elution with 2 mol/L hydrochloric acid. Either the carminic acid complex (620 nm), formed in sulphuric acid (94%) or sulphuric: acetic acid (1:4), or the azomethine hydrogen ion association complex (415 nm) formed at pH 5.2 was used for borate measurement.

8.3.2 SULPHIDE AND ELEMENTARY SULPHUR

Valkov and Zhabina [88] reduced elemental sulphur in marine sediments to hydrogen sulphide using metallic chromium. The liberated hydrogen sulphide was absorbed in cadmium solution and estimated iodimetrically.

Chau et al. [89] extracted elemental sulphur from marine sediments with aromatic solvents and determined S_4, S_6, and S_8 in the extract by gas chromatography.

Morse and Cornwell [90] investigated methods for determining acid volatile sulphides and pyrites in marine sediments from several typical depositional environments, which were permanently or seasonally anoxic. Extractants gave similar results under different acid conditions, with the exception of stannous chloride in hot hydrochloric acid. The use of scanning microscopy combined with simultaneous elemental analysis showed that the identifiable iron sulphides were almost always pyrite.

8.3.3 ORGANIC CARBON

Mills and Quinn [91] used persulphate oxidation at 130°C with subsequent measurement of carbon dioxide produced by a non-dispersive infrared detector to determine organic carbon in high carbonate oceanic sediments and estuarine sediments. Sediment samples were pre-dried for 4 h at 110°C to remove moisture and, if carbonate was present, were pretreated with hydrochloric acid and phosphoric acid to decompose carbonates, then washed and dried. The results or organic carbon determinations in some marine sediments by this method, and by high-temperature combustion using the Carlo Erba elemental analyser, show that there appears to be no significant bias between the methods. The average relative standard deviation obtained by the persulphate method is 11.4%, which is higher than the value of 5.1% obtained by high-temperature oxidation. The overall precision of the persulphate method was 6.4%.

Weliky et al. [92] have developed a procedure for the determination of both organic and inorganic carbon in a single sample of marine deposit, thus avoiding errors occurring in the method most commonly used. Carbonate carbon is determined from the carbon dioxide evolved by the treatment of the sample with phosphoric acid, and the residue is then treated with a concentrated solution of dichromate and sulphuric acid to release carbon dioxide from the organic matter.

8.3.4 PHOSPHORUS

Waldeback et al. [93] have described an accelerated solvent extract method for the determination of phosphorus in sediment. The bulk of phosphorus compounds stored in lake sediments contributes to a large extent to this process. It is, therefore, of great interest to get an adequate estimate of the amount of potentially mobile phosphorus for future release from sediment. The recently developed accelerated solvent extraction technique has, in this study, been applied to effectively extract the mobile forms of phosphorus in sediments. By using a buffered dithionite solution at 25°C followed by water at 100°C as solvents in sequential extractions, the total extraction process was completed within 90 min. The feasibility of using aqueous solutions and water as solvents with this technique is demonstrated here for the first time. The results obtained have been comparable to extractions resulted from the traditional techniques used to define and quantify the different amounts and forms of phosphorus in sediment.

8.4 RADIOELEMENTS

Published work on the determination of radioelements in marine sediments is reviewed in Table 8.4.

TABLE 8.4
Radioelements in Marine Sediments

Element	Sample Workup	Analytical Finish	Detection Limit	References
Ac227, Th228, Th230, Th232, Th234, Pa231, U234, α emitting plutonium isotopes	Acid solution, iron–aluminium preconcentration, ion-exchange chromatography		–	[94]
U235	–	Neutron activation analysis	1.4 mg/kg	[95]
Te204	Digestion, absorption of TeCl$_4$ on anion-exchange column, desorption with H$_2$SO$_3$	Neutron activation analysis	–	[96]
K, U, Th, Ra	Inter-laboratory comparison		–	[97]
Pb210, Ra	Inter-laboratory comparison		–	[98]

REFERENCES

1. A. Butman, *Journal of Marine Research*, 1986, 44, 645.
2. W. Banoub and P.J. Williams, 1972, 19; D.W. Menzel, *Deep Sea Research*, 1967, 14, 220.
3. J.K.B. Bishop and J.M. Edmond, *Journal of Marine Research*, 1976, 34, 181.
4. D.C. Gordon Jr. and W.H. Sutcliffe Jr., *Limnology and Oceanography*, 1974, 19, 989.
5. J.H. Sharp, *Limnology and Oceanography*, 1974, 19, 984.
6. P.K. Wangersky, *Limnology and Oceanography*, 1974, 19, 980.
7. D.L. Fox, C.H. Oppenheimer, and J.S. Kittredge, *The Journal of Marine Research*, 1953, 12, 233.
8. J.P. Ostapenya and R.Z. Kovalevskaya, *Okeanologiya*, 1965, 4, 694.
9. J.K.B. Bishop, J.M. Edmond, D.R. Ketten, M.P. Bacon, and W.B. Silker, *Deep Sea Research*, 1977, 24, 511.
10. M.B.M. Jacobs and M. Ewing, *Science*, 1969, 163, 180.
11. W.G. Lammers, *Environmental Science and Technology*, 1967, 1, 52.
12. M.M. Mullin, *Limnology and Oceanography*, 1965, 10, 459.
13. Private communication.
14. E. Gjessing and F.G. Lee, *Environmental Science and Technology*, 1967, 1, 631.
15. I.S. Siroka, G.M. Varthel, Y.Y. Lurr, and N.P. Stepanova, *Zhurnal Analiticheskoi Khimii*, 1974, 29, 1626.
16. D.J. Pietrzyk and C.H. Chu, *Analytical Chemistry*, 1977, 49, 757.
17. D.J. Pietrzyk and C.H. Chu, *Analytical Chemistry*, 1977, 49, 860.
18. A. Waggott, *Proceedings of the Analytical Division of the Chemical Society* (*London*), 1978, 15, 232.
19. R.W.P. Scott and P. Kucera, *Analytical Chemistry*, 1973, 45, 749.
20. L.R. Snyder, *Analytical Chemistry*, 1974, 46, 1384.
21. H. Engelhardt, *Hochdruck Flüssigkeits-Chromatographische*, Springer, Berlin, Germany, 1975, 213 pp.
22. H. Braus, F.M. Middleton, and G. Walton, *Analytical Chemistry*, 1951, 23, 1160.
23. J.W. Eichelberger and J.J. Lichtenberg, *Journal of the American Water Works Association*, 1971, 63, 25.
24. J.P. Riley and D. Taylor, *Analytica Chimica Acta*, 1969, 46, 307.
25. G.R. Harvey, Absorption of chlorinated hydrocarbons from sea water by a cross linked polymer, Woods Hole Oceanographic Institution Technical Report WHOK-72–86, Woods Hole, MA, pp. 9 (unpubl).
26. H.D. Gesser, A. Chow, F.C. Davis, J.F. Uthe, and J. Reinke, *Analytical Letters*, 1971, 4, 883.
27. B. Asling and S. Jensen, *Analytical Chemistry*, 1970, 42, 1483.
28. J.F. Uthe and J. Reinke, *Environmental Letters*, 1972, 3, 117.
29. W.A. Aue, S. Kapila, and C.R. Hastings, *Journal of Chromatography*, 1972, 69, 73.
30. J.R. Derenbach, M. Ehrhardt, C. Osterrahf, and G. Petrick, *Marine Chemistry*, 1978, 6, 351.
31. K. Hayase, K. Shimoshima, and H. Tsuibota, *Journal of Chromatography*, 1985, 322, 350.
32. A.B. Weaver and D.C. Cirohler, *Water South Africa*, 1981, 7, 79.
33. P.R. Kiff, *Laboratory Practice*, 1973, 22, 259.
34. E. Dines and J. Wharfe, *Water Bulletin*, 1987, 250, 6.
35. L. Hakanson, *Water Resources Research*, 1981, 17, 1625.
36. W. A. Maher, *Analyst*, 1983, 108, 939.
37. K.W.M. Siu, S.Y. Roberts, and S.S. Berman, *Chromatographia*, 1984, 19, 398.
38. A. Brzezinska-Paudyn, A. Van Hoon, and R. Hancock, *Atomic Spectroscopy*, 1986, 7, 72.

39. P. Charkraborty, P.V. Raghunadh, and V.V. Sarma, *International Journal of Environmental Analytical Chemistry*, 2012, 92, 133.
40. D.S. Lee, *Analytical Chemistry*, 1982, 54, 1682.
41. J.W. McLaren and S.S. Berman, *Applied Spectroscopy*, 1981, 35, 403.
42. K. Byasimirskii, E.M. Enid Yambo, V.K. Poblova, and Y.A. Gevinarenco, *Okeandogiyc*, 1990, 10, 1111.
43. D.J. Heggle, Study of reservoir fluxes and pathways in an Alaskon Fjord, PhD dissertation, University of Alaska, Fairbanks, AK, 217 pp., 1997.
44. P.P. Madsen, I. Brabach, and J. Sorenson, *Analytica Chimica Acta*, 1983, 151, 479.
45. R.B. Kheir, M.H. Greve, J.D. Deroin, and N. Rabai, *International Journal of Environmental Analytical Chemistry*, 2013, 93, 75.
46. V. Hodge, M. Stallard, M. Kiode, and E.D. Goldberg, *Analytical Chemistry*, 1978, 58, 616.
47. M. Lucotte and B. D'Anglejan, *Chemical Geology*, 1985, 48, 257.
48. M.G. Bettoli, L. Cantelli, D. Quarin, M. Ravanelli, L. Tositi, and O. Takartini, *International Journal of Environmental Analytical Chemistry*, 1998, 71, 321.
49. Y. Harivan, A. Almogi-Labin, and B. Herut, *Environmental Sciences and Technology*, 2010, 44, 6576.
50. B.H. Olsen and R.C. Cooper, *Water Research*, 1976, 10, 113.
51. A.W. Andreu and R.C. Harriss, *Nature*, 1973, 245, 256.
52. P.O. Bartlett, P.J. Craig, S.F. Morton, *Analyst* 1980, 267, 606.
53. H. Windon, W. Gardner, J. Stephens, F. Taylor, *East Coast Marine Station*, 1976, 4, 579.
54. E. Shin and P.A. Krenkel, *Water Pollution Control Federation*, 1976, 48, 44.
55. D.C. Gillespie, *Fisheries Research Board Canada*, 1972, 29, 1035.
56. A. Jernelow, *Limnology and Oceanography*, 1970, 15, 958.
57. D.G. Langley, *Water Pollution Control Federation*, 1973, 48, 473.
58. M.D.K. Abo-Rady, *Fresenius Zeitschrift für Analytische Chemie*, 1979, 299, 187.
59. R.C. Hutton and B. Preston, *Analyst*, 1980, 105, 981.
60. U.K. Pavlova and K.R. Yasimirskii, *Zhurnal Analiticheskoi Khimii*, 1969, 24, 1347.
61. M.J. Burzminskii and Q. Fernando, *Analytical Chemistry*, 1978, 50, 1177.
62. A. Cobelo-Garcia, J. Santos-Echeandia, D.E. Lopez-Sanchez, C. Almercija, and D. Omanovic, *Analytical Chemistry*, 2014, 85, 2308.
63. J. Wu, L. Zheng, C. Welfang Chen, K. Tagami, and S. Uchido, *Environmental Science and Technology*, 2014, 48, 3136.
64. K.W. Siu and S.S. Berman, *Analytical Chemistry*, 1983, 55, 1603.
65. S.N. Willie, R.R. Sturgeon, and S.S. Berman, *Analytical Chemistry*, 1986, 58, 1140.
66. J.L. Stroud, S.P. McGrath, and F.J. Zhao, *International Journal of Environmental Analytical Chemistry*, 2012, 92, 222.
67. N. Bloom, *Atomic Spectroscopy*, 1983, 4, 204.
68. J.P. Riley and S.A. Siddique, *Analytica Chimica Acta*, 1986, 181, 117.
69. J. Bai, T. Nakatani, Y. Sasaki, H. Minami, S. Inoue, and N. Takahashi, *International Journal of Environmental Analytical Chemistry*, 2011, 91, 856.
70. V.N. Sastry, T.M. Krislmamourthy, and T.P. Sarma, *Current Science, India*, 1969, 38, 279.
71. E.S. Pilkinton and L.T. Warren, *Environmental Science and Technology*, 1979, 13, 295.
72. D.T. Piper and G.G. Gales, *Analytica Chimica Acta*, 1969, 47, 560.
73. M.A. Hickey and J.A. Kittrick, *Journal of Environmental Quality*, 1984, 13, 372.
74. J.M. Belhomme, R. Erb, J. Dequidt, and A. Phillip, *Review Francail des Sciences de C'ean*, 1982, 1, 305.
75. S. Pai, F. Lin, C. Tseng, and D. Shen, *International Journal of Environmental Analytical Chemistry*, 1993, 50, 193.
76. J.W. McLaren, S.S. Berman, V.I. Buyku, and D.S. Russell, *Analytical Chemistry*, 1986, 58, 1802.

77. B.M. Petronio, B. Cosme, A. Marz, and P. Rivaro, *International Journal of Environmental Analytical Chemistry*, 1993, 54, 45.
78. A. Palanquez, F. Plana, M. Baucelle, G. Lacorte, and M. Roura, *International Journal of Environmental Analytical Chemistry*, 1989, 36, 85.
79. E. Ackerman, *Deutsche Gewassa Kundliche Mittelunga*, 1977, 21, 53.
80. S. Landsberger and W.F. Davidson, *Analytical Chemistry*, 1985, 57, 197.
81. S. Landsberger and W.F. Davidson, *Analytical Chemistry*, 1985, 57, 196.
82. M.A. Gregory, T.M. McClurg, and C.J. Bronckaert, *International Journal of Environmental Analytical Chemistry*, 2003, 83, 65.
83. J. Morillo, J. Usero, and I. Gracie, *International Journal of Environmental Analytical Chemistry*, 2002, 82, 245.
84. M.I. Alonso-Castillo, E.V. Alonso, G. Garcia-de-Torez, and J.M. Cano Pavon, *International Journal of Environmental Analytical Chemistry*, 2013, 94, 441.
85. K. Taymaz, Y. Vigit, H. Ozbal, A. Ceritioglu, and N. Muftugil, *International Journal of Environmental Analytical Chemistry*, 1984, 16, 253.
86. M.A. Angelidis and M. Aloupi, *International Journal of Environmental Analytical Chemistry*, 1997, 68, 281.
87. E. Kiss, *Analytica Chimica Acta*, 1988, 211, 342.
88. I.I. Valkov and N.N. Zhabina, *Zhurnal Analiticheskoi Khimii*, 1971, 21, 359.
89. K.Y. Chau, M. Monanavi, and A. Sycip, *Environmental Science and Technology*, 1973, 7, 948.
90. J.W. Morse and J.C. Cornwell, *Marine Chemistry*, 1987, 22, 55.
91. G.I. Mills and I.G. Quinn, *Chemical Geology*, 1979, 25, 165.
92. C. Weliky, E. Suess, P.J. Muller, and K. Fischer, *Limnology and Oceanography*, 1983, 28, 1252.
93. M. Waldeback, T. Rydin, and K. Markides, *International Journal of Environmental Analytical Chemistry*, 1998, 79, 257.
94. R.S. Anderson and A. Fleer, *Analytical Chemistry*, 1982, 54, 1240.
95. A.D. Suttle, B.D. O'Brien, and D.W. Mueller, *Analytical Chemistry*, 1969, 41, 1265.
96. A.D. Matthews and J.P. Riley, *Analytica Chimica Acta*, 1969, 48, 25.
97. R. Bojanoski, R. Fukai, and E. Holm, *Journal of Radioanalytical and Nuclear Chemistry*, 1987, 110, 113.
98. B.R. Harvey and A.K. Young, *Science of the Total Environment*, 1988, 69, 13.

9 Determination of Organic Compounds in Marine Sediments

A large number of organic compounds are being detected in marine sediment samples. Some of these are reviewed as follows.

9.1 ALIPHATIC HYDROCARBONS

Walker et al. [1] studied the profiles of hydrocarbons in sediments according to the depth of sediment cores collected from Baltimore Harbour in Chesapeake Bay, Maryland. Gas–liquid chromatography was used to detect hydrocarbons present at different depths in the sediment, while low-resolution spectrometry was employed to measure the concentrations of paraffins, cycloparaffins, and aromatic and poly-nuclear aromatics. Their data showed that the concentrations of total and saturated hydrocarbons decreased with increased depth. It was noted that hydrocarbons in oil-contaminated sediments should be identified and quantified to determine the fate of these compounds in dredge spills.

May et al. [2] described a gas chromatographic method for analysing hydrocarbons in marine sediments and seawater, which is sensitive at the sub-microgram per kilogram level. Dynamic headspace sampling for volatile hydrocarbon components requires minimal sample handling, thus reducing the risk of sample component loss and/or sample contamination. The volatile components are concentrated on a Tenax gas chromatographic precolumn and determined by gas chromatography or gas chromatography–mass spectrometry.

Brown et al. [3] described a gas chromatography–mass spectrometry technique for fingerprinting petrogenic hydrocarbons. The technique identified and quantified n-alkanes, isoprenoids pristine and phytane, pentacyclic triterpanes, unresolved complex mixture, and total hydrocarbon content. Results obtained using sediments preserved with chloroform during sediment trap collection were compared with those for unpreserved anoxic sediments and anoxic bottom surface sediment. Petrogenic hydrocarbons were detected at all stations, whose concentrations decreased with increasing distance from an urban area. Carbon preference index values increased along this transect, indicating a great dominance of biogenic hydrocarbons further out in the archipelago. The compositions of preserved and unpreserved anoxic samples were very similar. The results indicated that the sediment trap technique was a useful method for collecting and preserving material for fingerprinting petrogenic hydrocarbons.

Takada and Ishimatari [4] extracted alkylbenzenes with normal C_{10}–C_{14} and branched C_{11}–C_{13} alkyl chains from marine and coastal sediments and suspended matter in benzene methanol extracts. The extract in benzene was then applied to a Florisil column for the removal of copper sulphide and polar materials and subjected to silica gel column chromatography. Alkylbenzenes were quantified and identified using gas chromatography with flame ionisation detection. The recoveries of alkylbenzenes were 81%–84%.

Whittle [5] described a thin-layer chromatographic method for the identification of hydrocarbon marker dyes in oil-polluted waters.

McLeod et al. [6] conducted inter-laboratory comparisons of method for determining traces of aliphatic and aromatic hydrocarbons in marine sediments. Agreement within a factor of 2–3 was obtained between the 12 participating laboratories.

Mark [7] described an infrared method for the determination of the oil content of marine sediments. He showed that the magnitude of the CH_2 stretching band at 2925 cm^{-1}, normally used to determine oil in a sediment, was enhanced in the presence of biological matter. The concentration of this material can generally be estimated from the magnitude of the protein–NH band at 1650 cm^{-1} with the use of a calculated correction to the total absorption at 2925 cm^{-1}, but the oil must contribute less than 10% to the total absorption at 2925 cm^{-1}. It is desirable, however, that the nature of the organic matter be determined by studying the complete infrared spectrum.

Fluorescence spectroscopy has been adapted as an alternative analytical method for estimating oil in sediments [8–12].

Interlocutory comparisons have been performed for determining selected trace aliphatic and aromatic hydrocarbons in marine sediments [13–15].

Wakeham [16] discussed the application of synchronous fluorescence spectroscopy in the characterisation of indigenous and petroleum-derived hydrocarbons in lacustrine sediments. The author reported a comparison, using standard oils, of conventional fluorescence emission spectra and spectra produced by synchronously scanning both excitation and emission monochromators.

9.1.1 OIL SPILLS

Jones et al. [17] examined the hydrocarbon composition of surface sediments at two lithologically different sites in the Humber Estuary (United Kingdom) by gas chromatography and gas chromatography–mass spectrometry. The sediments were samples taken 5, 7, and 12 months after the accidental spillage of 6000 tonnes of Nigerian light crude oil into the estuary from the tanker *Sivand*. The distinctive marker compounds in the Nigerian oil facilitated in identifying residues of the 'Sivand' cargo in the sediments up to 12 months after the spill.

Interpretation of chromatograms revealed several other sources of hydrocarbons contributing to the sediments, including those from algae, higher plants, and fossil fuel combustion products. Field observations and laboratory studies indicated that the latter compounds were less readily biodegraded in sediments than the oil-derived hydrocarbons.

Morel and Courtol [18] developed an extraction procedure for determining trace levels of hydrocarbons in marine sediments. Excellent accuracy (94%) was obtained by this method and down to ppm quantities could be determined.

Gas chromatography has also been used to distinguish between fossil added to sediments through oil pollution and those hydrocarbons present in low concentrations as natural biogenic products [19,20].

Page et al. [21] used capillary gas chromatography and capillary gas chromatography–mass spectrometry to determine aliphatic hydrocarbons in interstitial sediments collected from the French coastline following the Amoco Cadiz disaster.

Brown et al. [22] described a rapid field method for detecting down to 2 μg of oil in sediments associated with marine spills. The method was employed in connection with the *Argo Merchant* oil spill off Nantuckett in December 1976.

In this method, the sediment is mixed with sodium sulphate and extracted with *n*-hexane. A portion of the extract is applied to a paper strip, which is then eluted with petroleum ether: benzene (35:65) for 60 s. Under ultraviolet light, the strip reveals a blue fluorescent spot indicating the presence of oil in the sediment.

9.2   AROMATIC HYDROCARBONS

Henning [23] applied ultraviolet spectroscopy for determining the aromatic constituents of residual fuel oil in hexane extracts of marine sediment samples. Examination of the ultraviolet spectra of samples of an oil pollutant from a beach and crude oil, at various concentrations, revealed strong absorption maxima at approximately 228 and 256 nm. The ratio of the peak heights at this wavelength is constant for a particular oil and is independent of concentration. These permit quantitative analysis of sediment samples many months after an oil spill.

Hargrave and Phillips [24] used fluorescence spectroscopy to evaluate concentrations of aromatic constituents an aquatic sediments. The oil concerned, a Venezuelan crude, contained about 35% by weight or aromatic constituents. Aromatic substances were extracted with *n*-hexane, and fluorescence spectroscopy was used to produce a series of contour diagrams of fluorescence intensity at various excitation and emission wavelengths in order to compare the fluorescence spectral patterns of sample extracts and standard oils. Petroleum residues were determined, and it was found that total oil concentrations ranged from 10 to 3000 μg/g wet sediment, with the highest concentrations occurring in sedimented particles.

Krahn et al. [25] described a high-performance liquid chromatographic method for the determination of 127 aromatic hydrocarbons and 21 chlorinated hydrocarbons in solvent extracts of marine sediments.

Vowles and Mantoura [26] determined sediment–water partition coefficients and the high-performance liquid chromatography capacity factors for 14 alkylbenzene and polyaromatic hydrocarbons. The partition coefficient correlated well with the alkyl-cyano capacity factors, and it was concluded that this phase gave a better indication of sorption on sediment than either the octanol or octadecylsilane phases.

Mazeas and Budinski [27] described a rapid and simple analytical procedure that allowed for accurate quantification of aliphatic and aromatic hydrocarbons in sediments and petroleum. Sediments were collected with a Soxhlet extractor using methylene chloride for 48 h. Oil maltene fractions were isolated by asphaltene precipitation in pentane. Sediment and oil extracts were first purified using alumina microcolumns. Saturated fractions were then separated from aromatic ones by fractionation on silica gel. Alkanes and polycyclic aromatic hydrocarbons were analysed separately by gas chromatography coupled to mass spectrometry (GC-MS). Polycyclic hydrocarbons and alkanes were quantified relative to perdeuterated polycyclic hydrocarbons and alkanes introduced prior to extraction. For accurate quantification, it is important to use several perdeuterated alkanes that cover the volatility range of alkanes of interest and at least one perdeuterated polycyclic hydrocarbons per class of aromaticity. This analytical procedure was validated on a standard reference material on sediment (SRM 1941a) and a standard reference material on crude oil (SRM 1582). This analytical procedure was then applied to the study of the contamination of marine sediments collected along the Atlantic coast after the Erika oil spill in December 1999.

McAlister et al. [28] used aryl hydrocarbon receptor-based, real-time polymerase chain reaction assay (Procept®) to determine dioxin-like compounds in soil and sediment samples from sites known to be contaminated by dioxin and related compounds. The biological equivalent quotient generated by the Procept assay was compared with toxicity equivalent measurements by isotope dilution gas chromatography–mass spectrometry. This research demonstrates that the Procept assay has several limitations, when compared with GC-MS, including poorer detection limits and precision, and increased sensitivity to a wide range of aryl hydrocarbon receptor agonists and antagonists, such as polycyclic aromatic hydrocarbons and polychlorinated biphenyls. However, results indicate that the Procept assay can be used as an effective screening technique for dioxin-like compounds as the EPA draft recommended interim preliminary remediation level for dioxin in soils of 72 pg/g toxicity equivalent.

Saber et al. [29] used high-resolution Shpol'skii spectrofluorimetry at 10 K to quantitatively determine polyaromatic hydrocarbons in lacustral sediments. Polyaromatic hydrocarbons incorporated into n-alkane matrix at low temperature yielded high-resolution fluorescence spectra of quasi lines with a multiplet structure related to several insertion sites. Samples required extraction and purification, and the choice of sample treatment, which depended on the total organic pollution levels, is discussed.

A high-performance liquid chromatographic procedure has been applied in the determination of polyaromatic hydrocarbons in saline sediments.

Dunn and Stich [30] and Dunn [11] described a monitoring procedure for polyaromatic hydrocarbons, particularly benzo[a]pyrene in marine sediments. The procedures involved extraction and purification of hydrocarbon fractions from the sediments and determination of compounds by thin-layer chromatography and fluorometry or gas chromatography. In this procedure, the sediment was refluxed with ethanolic potassium hydroxide, then filtered, and the filtrate was extracted with isooctane. The isooctane extract was cleaned up on a Florisil column, and then the polycyclic aromatic hydrocarbons were extracted from the isoactive extract with pure dimethyl sulphoxide. The latter phase was contracted with water and then extracted with isooctane to recover polycyclic aromatic hydrocarbons. The overall recovery of polycyclic aromatic hydrocarbons in this extract by fluorescence spectroscopy was 50%–70%.

Karakas and Pekey [12] carried out a source apportionment of polycyclic aromatic hydrocarbons in the surface sediments of Izmit Bay, Turkey. Thirty-five surface sediment samples were collected from the northeastern coast of the Izmit Bay to apportion the sources of polycyclic aromatic hydrocarbons entering the Bay. Samples were collected in February and June 2002, and they were analysed for 16 polycyclic aromatic hydrocarbon compounds using HPLC-UV. Total polycyclic aromatic hydrocarbons ranged from 1.1 to 68.4 µg/g dry weight. Both the factor analysis and the factor analysis absolute factor score multiple linear regression analysis were applied to the results of 11 polycyclic aromatic hydrocarbon compounds, which were observed in more than 80% of the samples. From the factor analysis, two factors explaining 91.3% of the total variance were identified. The first factor was petrogenic and explained 76% of the variance. Except for anthracene, 57%–85% of the lower-molecular-mass polycyclic aromatic hydrocarbon compounds (from fluorine to chrysene) were contributed by this factor.

The percentage recoveries of a range of polycyclic aromatic hydrocarbons lie in the range from 86.4% (benzo (ghi) perylene) to 12.0% (acenaphthylene).

Readman et al. [31] used flame ionisation capillary gas chromatography to determine polyaromatic hydrocarbons in extracts of sediments from Mersey, Dee, and Tamar rivers.

Hyotylainen et al. [32] used pressurised hot water extraction with an LC–GC system for the determination of polycyclic aromatic hydrocarbons in the sediment. The sediment was first extracted with pressurised hot water, and the analytes were absorbed into a solid-phase trap. The trap also functioned as an LC column, which removed most of the interfering matrix components. The 780 µL liquid chromatographic fraction containing the analytes was directly transferred to the gas chromatograph using an on-column interface. The whole pressurised hot water extraction–LC–GC analysis took place in a closed system, and no sample pretreatment was required. The sensitivity of the method was excellent due to the efficient concentration of the LC–GC system. Sensitivity was ~800 times better than in traditional systems. In addition, only a small amount of sample (10 mg) was required for the analysis. The pressurised hot water extraction–LC–GC method proved to be linear in the concentration range of 0.01–2 µg/g; the limits of quantification were below 0.01 µg/g for all the analytes, and the relative standard deviations were between 3% and 28%. Liquid chromatographic cleanup and the improved sensitivity made detection with FID sufficient for the determination of analytes. The results were comparable to those obtained in an inter-laboratory comparison study as well as to the results obtained with off-line super fluid–gas chromatography–mass spectrometry.

Noura et al. [33] assessed the contamination by polycyclic aromatic hydrocarbons in sediments collected at Manastir Bay, Turkey in a chromatographic–mass spectrum method.

The total polycyclic aromatic hydrocarbon concentrations were in the range of 25.6–576.8 ng/g dry weight in the winter and 44.9–395.8 ng/g dry weight in the summer. A comparison of the results with the Sediment Quality Guidelines suggested no ecotoxicological risks for benthic organisms. The use of molecular indices has shown that polycyclic aromatic hydrocarbons in surface sediments originate from pyrolytic sources.

9.3 OXYGEN-CONTAINING COMPOUNDS

9.3.1 SURFACTANTS

Electrospray MS has been used to determine nonylphenol polyethoxylated surfactants in marine sediments [34].

Martin et al. [35] used pressurised liquid extraction followed by liquid chromatography–mass spectrometry for the determination of major surfactants in marine sediments.

9.3.2 UNSUBSTITUTED AND HYDROXY-SUBSTITUTED FATTY ACIDS

Mendoza et al. [36] determined unsubstituted and hydroxy-substituted fatty acids in a 5 mol/L lacustrine sediment core taken from Leman Lake. Unbound and tightly bound compounds were not converted from one form to another. The abundance profile below 30 cm was not only similar but showed no decreasing trend, suggesting a common origin in three forms. The presence of unsubstituted monounsaturated acids in the C_{20}–C_{32} range suggested a possible origin for long-chain fatty acids other than from higher plants. Nothing was known of the origins of (omega-1)-hydroxy acids longer than C_{20} or those of 2-methylnonacosanoic acid.

9.3.3 NAPHTHENIC ACIDS

Wan et al. [37] studied the occurrence of naphthenic acids in North Sea coastal sediments after the Hebei Spirit oil spill.

Naphthenic acids are highly toxic. Their concentration was found to be in the range of 7.8–130, 3–4.4, and 0.8–20 µg/L in sediments from the Taean area, which were much greater than those measured in the reference sites of Manlipo and Anmyundo beaches. Concentrations of naphthenic acids were 50–100 times greater than those (0.077–2.5 mg/kg dry weight) of polycyclic atomic hydrocarbons in the same sediment samples. Thus, the ecological risk of naphthenic acids in oil spill–affected areas deserves more attention.

The sedimentary profiles of oil-derived naphthenic acids and background naphthenic acids centred around compounds with 21–35 and 12–21 carbons, respectively, indicating that the crude-derived naphthenic acid mixtures originating from the 2007 oil spill were persistent. Acyclic $NAs_{n=5-20}$ were easily degraded compared to cyclic $NAs_{n=21-41}$ during the oil weathering processes, and the ratio of oxy-$NAs_{n=21-41}$ relative to $NAs_{n=21-41}$ could be a novel index to estimate the degree of oil weathering in sediments. Altogether the persistent oil-derived $NAs_{n=21-41}$ could be used as a potential indicator for oil-specific contamination, as such compounds would not be much affected by the properties of coastal sediments possibly due to the high sorption of the negatively charged compounds, including naphthenic acids.

9.3.4 REACTIVE OXYGEN SPECIES

The photochemical reduction of Fe(III) complexes to Fe(II) is a well-known initiation step for the production of reactive oxygen species in sunlit waters. Murphy et al. [38] showed a geochemical mechanism for the same in dark environments based on the tidally driven, episodic movement of anoxic groundwaters through oxidised Fe(III)-rich sediments. Sediment samples were collected from the top 5 cm of sediment in a saline tidal creek in the estuary at Murrells Inlet, South Carolina and characterised with respect to total iron, acid volatile sulphides, and organic carbon content. These sediments were air dried, resuspended in aerated solution, and then exposed to aqueous sulphide at a range of concentrations chosen to replicate the conditions characteristic of a tidal cycle, beginning with low tide. No detectable reactive oxygen species production occurred from this process in the dark until sulphide was added. Sulphide addition resulted in the rapid production of hydrogen peroxide, with maximum concentrations of 3.85 μM. The mechanism of hydrogen peroxide production was tested using a simplified three-factor representation of the system based on hydrogen sulphide, Fe(II), and Fe(III). The resulting predictive model for maximum hydrogen peroxide agreed with the measured hydrogen peroxide in field-derived samples at the 95% level of confidence, although with a persistent negative bias suggesting a minor undiscovered peroxide source in sediments.

9.3.5 CARBOHYDRATES

Cowie and Hedges [39] described a flame ionisation gas chromatographic method for the determination of equilibrated isomeric mixtures of monosaccharides (galactose, glucose, xylose, mannose, rhamnose, fucose, arabinose, and lyxose) in saline sediments. Acid hydrolysis yields monomeric carbohydrates, which may exist in up to five isomeric forms when in solution. Lithium perchlorate was used to catalytically equilibrate carbohydrate mixtures in pyridine prior to conversion to the trimethylsilyl ether derivatives. Analysis was carried out by the use of gas–liquid chromatography on fused-silica capillary columns. Quantification on the basis of a single clearly resolved peak for each carbohydrate was made possible by the equilibration step. Carbohydrate losses and optimal conditions for maximum reproducible carbohydrate recovery were determined for each extra stage.

Carbohydrates recovered through the analytical procedures were in the range of 101% (lyxose, xylose, and galactose) to 108% (glucose) with reproducibility between 2.8% and 18.9%. Down to 0.1 μg of each monosaccharide can be determined in a sample hydrolysate.

9.4 HALOGEN-CONTAINING COMPOUNDS

9.4.1 CHLOROPHENOLS

Xie [40] determined trace amounts of chlorophenols and chloroguaiacols in marine
sediments collected off the Swedish coast. The compounds were desorbed from
sediment surfaces by a mixture of acetic anhydride and hexane, after buffering with
0.1 mol/L sodium carbonate. The optimal pH was achieved by a 1:4 ratio of buffer to
acetic anhydride. The acetylated extracts were analysed by glass capillary gas chromatography with electron capture detection. The recoveries (in µg/kg) ranged from
85% to 100% with standard deviations of 4%–11%.

9.4.2 CHLOROBENZENES

An isooctane extraction–gas chromatographic procedure [41] has been applied in the
determination of various (0.003–0.07 mg/kg) chlorobenzenes in estuarine sediments.

9.4.3 POLYCHLORINATED BIPHENYLS

Three different detection methods (gas chromatography with electron capture, mass
spectrometry, and atomic emission detectors) have been compared for the determination of polychlorobiphenyls in highly contaminated marine sediments [42]. Only
atomic emission detection in the chlorine-selective mode provided excellent polychlorobiphenyl profiles without interference. However, the lower sensitivity of the
atomic emission detector compared to the other two detectors required a 10–20 g
sample size for analysis.

A method has been described [43] for separating polychlorinated biphenyls from
chlorinated insecticides. This procedure involves adsorption chromatography on
alumina and charcoal columns, elution with increasing fractional amounts of hexane
on alumina columns, and with acetone–diethyl ether on charcoal columns. The polychlorinated biphenyls and chlorinated pesticides are then determined by gas chromatography–mass spectrometry on the separate eluates without interference.

Jensen et al. [44] described a method applicable to marine sediments for the
determination of polychlorobiphenyls and organochlorine insecticides, in which
the sample is extracted with acetone and then *n*-hexane, together with 1% aqueous
ammonium chloride. The extracts are then concentrated for purification with concentrated sulphuric acid and aqueous sodium sulphate in the presence of tetrabutyl
ammonium sulphate and finally analysed by gas chromatography.

Japenga et al. [45] determined polychlorinated biphenyls and chlorinated insecticides in River Elbe estuary sediments by a procedure in which the sediments were
pretreated with acetic acid, mixed with silica, and Soxhlet extracted with benzene/
hexane. Humic material and elemental sulphur were removed by passing the extract
through a chromatographic column containing basic alumina, on which sodium sulphite and sodium hydroxide were adsorbed. Silica fractionation was followed by
gas chromatography to analyse chlorinated pesticides, polychlorinated biphenyls,
and polyaromatic hydrocarbons. Recovery experiments with standard solutions gave
recoveries of 90%–102%.

Mascolo et al. [46] have described a method for the direct analysis of polychloro-biphenyls in heavily contaminated sediments and soils using thermal desorption/gas chromatography/mass spectrometry. Thermal desorption and analysis are performed in-line employing a limited amount of sample (2 mg), which eliminates the need for any solvent and time-consuming extraction. The analytical procedure was optimised using a sample spiked with Aroclor 1254 and Aroclor 1260 and validated with a certified industrial soil sample for which the concentrations of 13 polychlorobiphenyl congeners are known. Limits of detection were sensitive to matrix effects and varied substantially among analytes.

Mascolo et al. [47] employed gas chromatography and gas chromatography in electron capture detection for determining polychlorinated biphenyls in the ng/g range in marine sediments. Several reference sediment samples were used as comparison standards.

9.4.4 POLYBROMINATED DIPHENYL ETHERS

Desis et al. [48] determined polybrominated diphenyl ether concentrations in marine samples collected from locations around Thermaikos Gulf in North Greece. Polybrominated biphenyls were detected in all sampling sites, and their average total concentration ranged from 0.26 to 4.92 ng/g dry weight. Concentrations were an order of magnitude higher in locations outlining the inner part of the Gulf, which were also closer to industrial areas, sewage treatment plant discharges, city harbour, and landfill area. These findings suggest pollution in the aquatic ecosystem from industrial and urban activities in the area. Congener profiles exhibit predominance of BDE-209, while concentrations of other polybrominated biphenyl congeners were usually lower, when compared to similar studies from other countries globally, indicating that the Thermaikos Gulf is among the low-polluted areas.

9.5 NITROGEN-CONTAINING COMPOUNDS

Krone et al. [49] used capillary column gas chromatography with nitrogen-specific detection and gas chromatography–mass spectrometry to determine nitrogen-containing aromatics originating from creosote oil in solvent extracts of sediments taken from Eagle Harbour, Puget Sound and from uncontaminated areas. Organic sediment extracts and the commercial creosote oil were fractioned by silica/alumina column chromatography. No nitrogen-containing aromatics were detected in sediments from a pristine reference area. Over 90 nitrogen-containing aromatics were identified in the sediments from Eagle Harbour and in the creosote oil. The total concentration of nitrogen-containing aromatics in Eagle Harbour sediments ranged from 200 μg to 1200 mg/kg sediment (dry weight). Primarily, three-ring and four-ring nitrogen-containing aromatics were identified from a wood-creosoting facility on the shores of the harbour.

Li and Brownwell [50] presented a sensitive and robust method of analysis for quaternary ammonium compounds in marine sediments. Methods for extraction, sample purification, and HPLC-time-of-flight MS analysis were optimised, providing solutions to problems associated with the analysis of quaternary ammonium compounds, such as dialkyldimethylammonium and benzalkonium compounds. The exceptionally high positive mass defects characteristic of alkylammonium or protonated alkylamine ions are discussed in this study. The ability to resolve masses of alkylamine fragment ions is much greater than that for the molecular ions of benzalkonium compounds and many other chemicals, opening up a range of potential applications. The power of utilising a combination of approaches is illustrated with the identification of non-targeted dialkyldimethylammonium $C_8:C_8$ and $C_8:C_{10}$, two biocides previously unreported in environmental samples. Concentrations of quaternary ammonium in sewage-impacted estuarine sediments (up to 74 μg/g) were higher than concentrations of other organic contaminants measured in the same or nearby samples.

Kido et al. [51] determined basic organic compounds such as quinoline, acridine, aza-fluorene, and their N-oxides in marine sediments from an industrial area. The sediments were extracted with benzene by using a continuous extractor for 12 h. Hydrochloric acid (1 N) was added to the benzene extracts, and the mixture was shaken for 5 min; the acid layer separated from the benzene layer was made alkaline by the addition of sodium hydroxide and the alkaline aqueous solution was extracted with diethyl ether; the ether extracts were then dehydrated with anhydrous sodium sulphate and concentrated with a Kuderna–Danish evaporator. The concentrations were separated and analysed by gas chromatography–mass spectrometry and gas chromatography–high-resolution mass spectrometry.

9.6   SULPHUR-CONTAINING COMPOUNDS

In a method described by Bates and Carpenter [52] for the characterisation of organosulphur compounds in the lipophilic extracts of marine sediments, it was shown that the main interference was elemental sulphur (S_8). Techniques for its elimination were discussed. Saponification of the initial extracts was shown to create organosulphur compounds. Activated copper removed S_8 from an extract and appeared to neither create nor alter organosulphur compounds. However, mercaptans and most disulphides were removed by the copper column. The extraction efficiency of several other classes of sulphur compounds was 80%–90%. Extracts were analysed with a glass capillary chromatograph equipped with a flame photometric detector. Detection limit was in the nanogram range, and precision was ±10%. The recovery of 100 ng of individual sulphur standards added the pre-extracted sediment varied from 80% to 90% for sulphides, sulphonates, and aromatic sulphur compounds. Mercaptans and disulphides were not recovered since these classes of compounds were retained in the copper column.

9.7   PHOSPHORUS-CONTAINING COMPOUNDS

Perfluoro octene sulphonamide ethanol-based phosphate has the empirical formula $(F_{17}O_8SO_2)_2PO_4$.

The environmental occurrence of perfluorooctane sulphonate (PDFS) can arise from its direct use as well as from the transformation of precursors (N-alkyl-substituted perfluorooctane sulphonamides). Perfluorooctane sulphonated ethanol–based phosphate (PAP) esters are among the numerous potential perfluorooctane sulphonate precursors which have not been previously detected in the environment and for which little is known about their stability based on their high production volume during 1970 to 2002 and widespread use in food contact paper in packaging applications. Perfluorooctane sulphamide ethanol–based phosphate esters may be potentially significant sources of perfluorooctane sulphanate.

Benskin et al. [53] reported the first environmental occurrence of perfluorooctane sulphamide ethanol–based phosphate diesters, SAM PAD, in marine sediments from an urbanised marine harbour in Vancouver, Canada. Diester concentration in sediments (40–200 ng/g of clay) was similar to those of perfluorooctane sulphanate (71–180 pg/g dry weight). A significant ($p = 0.05$) correlation was observed between perfluorooctane sulphamide ethanol–based phosphate diesters and N-ethylperfluorooctane sulphamide acetate on anticipated degradation products of perfluorooctane sulphamide ethanol–based phosphate diester. Perfluorooctane sulphonate (PFDS) precursor (FUSAM) concentration sediment (120–110 pg/g) was 1.6–24 times greater than that of perfluorooctane sulphonate in sediment.

Although perfluorooctane sulphamide ethanol–based diester was not detected in water, perfluorooctane sulphonate was observed at concentrations up to 110 pg/g. Among the per- and polyfluoroalkyl substances monitored in this work, transformed sediment–water distribution coefficients ranged from 2.3 to 4.3 and increased with the number of CF_3 units and N-alkyl substitutions in the case of (N-alkyl-substituted) perfluorooctane sulphonamides. Overall, these results highlight the importance of (N-alkyl-substituted) perfluorooctane sulphonamides as potentially significant sources of perfluorooctane sulphonate in particular urban marine environments.

9.8 INSECTICIDES

9.8.1 CHLORINATED INSECTICIDES

Picer et al. [54] described a method for measuring the radioactivity of labelled DDT-contaminated sediments in relatively large volumes of water, using a liquid scintillation spectrometer. Various marine sediments, limestone, and quartz in seawater were also investigated. External standard ratios and counting efficiencies of the system investigated were obtained, as well as the relation of efficiency factor for external standard ratios for each system was studied.

Jensen et al. [44] have described a method applicable to marine sediments for the determination of polychlorobiphenyls and organochlorine insecticides, in which the sample is extracted with acetone, then n-hexane together with 1% aqueous ammonium chloride. The extracts are then concentrated for purification with concentrated sulphuric acid and aqueous sulphate in the presence of tetrabutyl ammonium sulphate and finally analysed by gas chromatography.

Various researchers have discussed methods for separating chlorinated insecticides from interfering levels of polychlorobiphenyls [43,45]. In one method, polychlorinated biphenyls and chlorinated insecticides in River Elbe estuary sediments were determined by a procedure in which the sediments were pretreated with acetic acid, mixed with silica, and Soxhlet extracted with benzene/hexane. Humic material and elemental sulphur were removed by passing the extract through a chromatographic column containing basic alumina, on which sodium sulphite and sodium hydroxide were adsorbed. Silica fractionation was followed by gas chromatography to analyse chlorinated pesticides, polychlorinated biphenyls, and polyaromatic hydrocarbons. Recovery experiments with standard solutions gave recoveries of 90%–102%.

Japenga et al. [45] separated polychlorobiphenyls from chlorinated insecticides by a procedure involving adsorption chromatography on alumina and charcoal columns and elution with increasing fractional amounts of hexane on alumina columns and with acetone–diethyl ether on charcoal columns. The polychlorinated biphenyls and chlorinated pesticides were then determined by gas chromatography–mass spectrometry on the separate eluates without interference.

Erkman and Kolankaya [55] determined organochlorine pesticides in sediments. They carried out a study between May 2002 and August 2003 in Meric Delta, a site where the Mericrover River falls into the Agean Sea. Residues of organochlorine pesticides in surface water, sediments, and fish were analysed by gas chromatography. The study showed that all the 20 analysed organochlorine pesticides and their residues were widespread throughout the study area. It was found that the concentrations of these selected organochlorine pesticides in sediments were higher than those in water.

Yamashita et al. [56] measured the amount of organochlorine pesticides discharged into river water and sediments in part samples taken from the Nile, Egypt. Among different organochlorine pesticides analysed, p,p'-DDE was the most predominant in fish (7.6–67 ng/g wet weight), sediments (3.2–432 ng/g dry weight), and suspended solids (5.3–138 pg/L). However, in the dissolved phase of water samples, HCH compounds predominated (α-HCH, 71–2815 pg/L). Concentrations of organochlorine pesticides, except chlordane, were higher in the Manzala Lake than those in the Nile River. Concentrations of organochlorine pesticides in fish corresponded with those in sediments from each location. Comparison of organochlorine concentrations in the Nile River water with those reported in earlier studies suggested a decrease in concentrations during the last decade. However, concentrations of p,p'-DDE have increased in fish. It appears that the release of this metabolite from contaminated sediment is the major source of p,p'-DDE in fish during recent years.

Salvado et al. [57] evaluated different procedures for the extraction of organochlorine pesticides from natural waters and sediments. In the case of extraction from water, a C_{18} disk solid-phase extraction method was employed. Recovery experiments in the range of 40–200 ng/L with selected organochlorine compounds resulted in average recoveries between 80% and 100%. Four different solvents – hexane, ethyl acetate, acetonitrile, and methanol – were tested as eluting agents. Best recoveries were obtained with ethyl acetate and hexane. A comparative study of organochlorine pesticides sediment extraction procedures was performed employing sonication, Soxhlet extraction, and shake-flask methods. The capacity of these methods to recover organochlorine pesticides from a sediment sample fortified at 50 ng/g was evaluated using hexane: acetone (1: 1 v/v), hexane: acetone (8.2 v/v) acetonitrile and dichloromethane. The three extraction techniques gave similar results, and dichloromethane was the most effective solvent. Optimised methods were applied in the analysis of waters and sediments from the 'Aiguamolls de l'Empordà' Nature Park, Girona (Spain).

Mangani et al. [58] reviewed published articles on the levels of chlorinated insecticides, polycyclic aromatic hydrocarbons, and polychlorobiphenyls in the Mediterranean coastal sediments. Analyses were performed by capillary gas chromatography–mass spectrometry after Soxhlet extraction with a petroleum ether–acetone mixture. Samples collected in three different areas along the Italian coast were analysed, and consistently different concentrations were found for the three locations. Possible sources of pollution were investigated and identified.

9.8.2 Miscellaneous Insecticides

Kjolholt et al. [59] determined trace amounts of organophosphorus pesticides and related compounds using capillary gas chromatography and nitrogen-specific detector.

Acetone-n-hexane extracts of the sediment were partitioned between methylene chloride and water and subjected to adsorption chromatography and analysed by gas chromatography. Recoveries were between 55% and 82%, and detection limit was in the range of 95–220 µg depending on the type of organophosphorus insecticide.

Chai et al. [60] developed a simple extraction and cleanup procedure for the analysis of 24 organophosphorus, organochlorine, and pyrethroid pesticides in mineral and peat soils using modified QuEChERS method. The pesticides were extracted from the soil with acidified acetonitrile. The water was removed from the extract by salting out with sodium chloride and addition of magnesium sulphate. For organophosphorus pesticides, the extracts were cleaned up with 0.2 g of primary secondary amine packed in glass Pasteur pipette and determined by gas chromatography with flame photometric detector. For organochlorine and pyrethroid pesticides, the extracts were cleaned up with 0.2 g of silica gel packed in a glass Pasteur pipette and determined by gas chromatography with an electron capture detector. After the cleanup, the extracts had lower colour intensity and reduced matrix interferences. The recovery of the organophosphorus and organochlorine pesticides determined at 0.10–1.0 mg/kg fortified levels ranged from 79.0%–120.0% and 82.2%–117.6%, respectively. The detection limits for the organophosphorus and organochlorine pesticides were 0.001–0.01 and 0.002–0.005 mg/kg, respectively. The recovery of the pyrethroid pesticides ranged from 87.5% to 111.7% at the detection limits of 0.002–0.010 mg/kg. The relative standard deviations for all pesticides studied were below 10.8%. The modified method was simple, fast, and had utilised less reagents than the conventional methods. The method was applied for determining the pesticide residues in mineral and peat soil samples and sediments collected from vegetable farms.

Various method have been described for the determination of insecticides in solid samples. With little or no modification, these methods may be applicable to the analysis of marine sediments. Insecticides discussed include diaphenthiuron [61], friazines [62], cypermethrin [63], difulin [64], and organochlorine, organonitrogen, and organophosphorus [65] pesticides.

9.9 STEROLS

Dryier et al. [66] determined sterols in lacustrine sediments. Samples of wet lacustrine were heated under anoxic conditions at 150°C, 175°C, 200°C, and 250°C for

5 days; at 175°C for 5 days with influx of potassium hydroxide and methanol to remove sterols; and at 175°C for 12, 18, 24, and 48 h, after which extraction was performed. Heating the sediment increased the amounts of extractable sterols provided that the temperature did not exceed 200°C, because degradation became rapid above that temperature. The behaviour of sterol ketones was similar, but the temperature limit was slightly higher. The various levels of the sterols extracted are tabulated; 4-methylsterols had a high stability towards thermal degradation under the conditions used.

Readman et al. [31] selected capillary gas chromatography using a flame ionisation detector as the method for quantifying sterols, in particular coprostanol, as a marker of faecal pollution. The hydrocarbon fraction produced as a by-product of the sterol analysis was used for quantifying 'oil-derived' and polycyclic aromatic hydrocarbons. Analyses of sediments from the estuaries of Mersey, Dee, and Tamar rivers are given as examples of how to interpret results obtained by this method. Petrogenic and biogenic inputs of saturated hydrocarbons could be distinguished.

9.10   HUMIC AND FULVIC ACIDS

Hayase et al. [67] measured the fluorescence and adsorption spectra of humic acid and fulvic acid in sediments collected from Tokyo Bay at 20°C and pH 8. The maximal excitation and emission wavelengths for humic acid were longer than those for fulvic acid, independent of molecular weight, and could, therefore, be used to differentiate between humic acid and fulvic acid in marine deposits. Smaller molecules showed greater fluorescence than larger molecules. Fluorescence intensity per weight concentrations unit increased from humic acid and decreased for fulvic acid, with an increase in adsorption coefficient.

Pontanen and Morris [68] compared the structure of humic acids from marine sediments and degraded diatoms by infrared and C_{13} and proton NMR spectroscopy. Samples of marine sediments taken from the Peru continental shelf were extracted with water, sodium hydroxide (0.05 mol/L), and sodium pyrophosphate (0.05 mol/L) under an atmosphere of nitrogen and fractionated by ultrafiltration. Humic acids of molecular weight 300,000 and above were examined. Diatoms were collected from ecosystem bags in Loch Ewe, United Kingdom. Highly branched aliphatic compounds formed the major fraction of the acids. Carbohydrates and, to a lesser extent, aromatic compounds, and carbonyl, ether, alcohol, and amino groups were found in marine humic acids. The structures of the marine and algal humic acids were similar, and there was evidence that certain sediment humic acids originated from planktonic debris. Raspor et al. [69] examined humic acids isolated from marine deposits in the Adriatic Sea and Norwegian Sea and humic acid in fulvic acid isolated from an estuarine sediment from Borneo. They presented data on their elementary composition, absorption, and infrared spectra, distribution of molecular weight, trace metal content, and adsorption at a hanging mercury drop electrode. Humic acids from the marine environments had a higher nitrogen content than the estuarine humic acid. The most humified material was that isolated from deep-sea sediment in the Norwegian Sea. Fulvic acid had a lower carbon content and a higher oxygen content than the humic acids and was more hydrophilic.

Hayase [70] characterised fulvic acid from sediments collected from Tokyo Bay by reversed-phase liquid chromatography. Sedimentary fulvic acid exhibited increasing hydrophilic character with increasing molecular weight. The method used was effective for the hydrophobic/hydrophilic characterisation of humic substances.

Gregory et al. [71] carried out the mapping of concentrations of various organic compounds originating from industrial effluents in coastal water sediment.

In a study of the distribution of humic substances in sediments taken from the western Ross Sea, Antarctica, Calace et al. [72] sampled the first 40 cm of sediment from three basins in the Ross Sea using a box corer. Site Y1 was located close to the coast in Terra Nova Bay; the sediment of site Y3 was collected in a more distal basin in the central sector of the Ross Sea; finally, site Y5 was sampled in the deepest zone of Joides Basin.

Sediment cores were sliced and analysed with a depth resolution of 2–4.5 cm. The distribution of humic substances and their structural features along the cores were determined and related to the pattern of the total organic carbon (TOC) and sedimentological data. The grain size distribution and the ^{210}Pb inventories allow the sediment of the study sites to be characterised.

The humic substances content in the sediment decreased with a change in slope between 23 and 26 cm at Y1, between 12 and 15 cm at Y3, and constant values with further depth. At Y5, the depth profile of humic substance content shows constant values in the upper 17 cm and values decrease with further depth. The pattern of humic substance yield is similar to that found for TOC. The analysis of the elemental composition of the humic acids extracted from different sediment depths shows an increasing C/N atomic ratio at sites Y1 and Y3 and constant values along the Y5 core. The depth profile of the C/N atomic ratio is confirmed by the changes observed in the structural characterisation and indicates a shift from the freshly deposited organic matter on the sediment surface to more humified material (humin). The results obtained highlight a different sediment rate at the three sites as deduced from sedimentological analysis.

9.11 LIGNIN

Pocklington and Macgregor [73] used alkaline nitrobenzene oxidation to produce vanillin and syringaldehyde from the lignin fraction of marine sediments and particulate matter filtered from seawater. Results indicated that this material constituted a substantial proportion (1%–14%) of the organic matter in the sediments and that a general background level (20–60 mcg/L) of material in the suspended fraction was attributed to lignin. The highest levels were found in the vicinity of forest industrial plants.

9.12 PRIORITY POLLUTANTS

Ozretich and Schroeder [74] developed an extraction procedure, utilising sonication with acetonitrile and cleanup using aminopropyl and/or C-18 bonded phase columns, to prepare marine sediments for priority organic pollutant analysis by gas chromatography. Recoveries from standard reference and inter-laboratory comparison sediments and tissue preparations compared favourably to publish mean values. Mean recoveries of 22 priority organic pollutants from the sediments ranged from 0% to 84% with a median recovery of 71% and an average percent relative standard deviation (%RSD) of 9%. The effects of sediment type and storage method on the spike recoveries are discussed.

Priority pesticides (alachlor, aldrin, γ-chlordane, chlorofenvinphos, chlorpyrifos, Dieldrin, 4,4'-DDT, 4,4'-DDE, α-endosulfan, β-endosulfan, endosulfan sulphate, endrin, α-HCH, β-HCH, γ-HCH, δ-HCH, HCB, HCBD, heptachlor, heptachlor epoxide, isodrin, methoxychlor, mirex, Quintozene, terbuthylazine, and trifluralin) are a group of toxic substances known for their persistency in the aquatic environment. Their screening in marine sediments may provide information on the sources and distribution in the water mass of fresh transitional and coastal waters. Pinto et al. [75] proposed a rapid and reliable method to extract multi-residues of priority pesticides by ultrasound irradiation from marine sediments. Multiple variables were optimised: ultrasound frequency, sonication intensity, signal operation mode, time of extraction, and water bath temperature. After sample cleanup and preconcentration of the pesticides by stir bar sorptive extraction, the compounds were analysed by gas chromatography–mass spectrometry using the selective ion monitoring acquisition mode.

Better performance was found for ultrasonic-assisted extractions at the frequency of 35 kHz and an output intensity of 60% in the sweep mode of operation. An increase in the water bath temperature to 80°C had a significant effect on the extraction of pesticides with high octanol–water partitioning coefficient (K_{ow}). Under optimal conditions, method detection limits and method quantification limits ranged from 0.3 to 4.4 ng/g and from 0.8 to 14 ng/g, respectively. Recoveries between 70% and 111%, at high precision levels, were found at different types of marine sediments with a single extraction cycle. Method performance was in good agreement with quality control guidelines.

9.13   ORGANIC CARBON PARAMETERS

9.13.1   CARBON CONTENT

Electron beam X-ray excitation [76], wet oxidation, and microcombustion techniques [77] have all been employed for the determination of organic carbon in estuarine sediments.

The persulphate oxidation technique [78] has also been applied to estuarine sediments. Organic carbon was determined at 10 g/kg in bay and estuarine sediments with a coefficient of variation of about 10%. Dankers and Leane [77] compared two methods based on wet oxidation with potassium peroxydisulphate and loss on ignition at 560°C for the determination of particulate organic carbon in estuarine suspended matter. They found that in estuarine sediments with high clay content, loss of ignition overestimated organic matter content approximately four times. Discrepancies were attributed to incomplete oxidation in the wet digestion method and loss of carbonate and water in the dry ignition method.

9.13.2 OXYGEN UPTAKE

Nothlich and Reuten [79] devised an apparatus for the measurement of oxygen uptake by sediments from coastal waters and estuaries. The specific oxygen uptake per unit surface area of undisturbed sediment was much smaller (by a factor of about 30) than that of the suspended sediments, which is of considerable significance for an assessment of the oxygen balance in waters transporting dredged solids. In addition, the content of organic degradable matter increases with an increase in the proportion of fines in the suspended sediment, which further enhances the oxygen consumption. Typical silty sediments are characterised by a high oxygen uptake, whereas sandy sediments exhibit much lower values.

REFERENCES

1. J.D. Walker, R.R. Calwell, M.C. Hamming, and H.T. Ford, *Environmental Pollution*, 1975, 9, 231.
2. W.E. May, S.N. Chesler, S.P. Cram, B.H. Gump, H.S. Hertz, D.P. Eragonio, and S.M. Dryzel, *Journal of Chromatographic Science*, 1975, 13, 535.
3. D. Brown, A. Colsimo, B. Ganning, C. Naf, Y. Yabuhr, and C. Ostman, *Marine Pollution Bulletin*, 1987, 18, 380.
4. H. Takada and R. Ishimatari, *Journal of Chromatography*, 1985, 346, 281.
5. P.J. Whittle, *Analyst*, 1977, 107, 976.
6. W.D. McLeod, P.G. Prohaska, D.D. Gennero, and D.W. Brown, *Analytical Chemistry*, 1982, 31, 281.
7. H.B. Mark, *Environmental Science and Technology*, 1972, 6, 833.
8. V. Zitko and W.V. Carson, Technical Report, Fisheries Research Board, Ottawa, Quebec, Canada, 1970, 217, 29pp.
9. D.J. Skarrett and V. Zitko, *Fisheries Research Board, Ottawa, Canada*, 1972, 29, 1347.
10. D.J. Scarratt and V. Zitko, *Journal Fisheries Research Board Canada*, 1972, 29, 1347.
11. B.P. Dunn, *Environmental Science and Technology*, 1976, 10, 1018.
12. D. Karakas and B. Pekey, *International Journal of Environmental Analytical Chemistry*, 2005, 85, 433.
13. W.D. McLeod, P.G. Prohaska, D.D. Gennero, and D.W. Brown, *Analytical Chemistry*, 1982, 54, 386.
14. L.R. Hilpert, W.E. May, S.A. Wise, S.N. Chealer, and H.S. Hertz, *Analytical Chemistry*, 1982, 54, 458.
15. J. Albaiges and J. Grimalt, *International Journal of Environmental Analytical Chemistry*, 1987, 31, 281.
16. S.G. Wakeham, *Environmental Science and Technology*, 1977, 11, 272.
17. M. Jones, S.J. Rowland, A.G. Douglas, and S. Howells, *International Journal of Environmental Analytical Chemistry*, 1986, 24, 227.
18. G. Morel and P. Courtol, *International Journal of Environmental Analytical Chemistry*, 1987, 30, 105.
19. M. Blumer and J. Sass, *Marine Pollution Bulletin*, 1972, 3, 92.
20. J.W. Farrington and J.G. Quinn, *Estuary and Coast Marine Science*, 1973, 1, 71.
21. D.S. Page, C. Foster, P.M. Ficket, and E.S. Gilfillan, *Marine Pollution Bulletin*, 1988, 19, 107.
22. L.R. Brown, G.S. Pabst, and M. Light, *Marine Pollution Bulletin*, 1978, 9, 81.
23. H.F.O. Henning, *Marine Pollution Bulletin*, 1979, 10, 234.
24. B.T. Hargrave and G.A. Phillips, *Environmental Pollution*, 1975, 8, 193.

25. M.M. Krahn, L.K. Moore, R.G. Bogar, C.A. Wigren, S.L. Chau, and D.W. Brown, *Journal of Chromatography*, 1988, 437, 161.

26. F.D. Vowles and R.F. Montoura, *Chemosphere*, 1987, 16, 109.

27. L. Mazeas and H. Budinski, *International Journal of Environmental Analytical Chemistry*, 2002, 82, 157.

28. D.R. McAlister, R. Carion, A. Dindal, and S. Billets, *International Journal of Environmental Analytical Chemistry*, 2013, 93, 35.

29. A. Saber, J. Jazocz, and M. Barin-Bouer, *Journal of Environmental Analytical Chemistry*, 1987, 28, 171.

30. B.P. Dunn and H.F.J. Stitch, *Journal of Fisheries Research Board Canada*, 1976, 33, 2040.

31. J.W. Readman, M.R. Preston, and R.F.C. Mantoure, *Marine Pollution Bulletin*, 1986, 17, 298.

32. T. Hyotylainen, T. Andersson, K. Hartonen, K. Kuosamen, and M.L. Rickkola, *Analytical Chemistry*, 200, 72, 3070.

33. T. Noura, M. A., Li Tagorti, H. Budrinski, H. Elchelbert, and H. Boussetta, *International Journal of Environmental Analytical Chemistry*, 2013, 14, 1470.

34. P.Y. Shang, M.G. Ikowomon, and R.W. McDonald, *Journal of Chromatography*, 199, 849, 467.

35. P.A.L. Martin, A. Gomes Parra, and E. Gonzalez Mazo, *International Journal of Environmental Analytical Chemistry*, 2005, 85, 293.

36. Y.A. Mendoza, F.O. Gulacar, Z.L. Hu, and A. Bucks, *International Journal of Environmental Analytical Chemistry*, 1987, 31, 107.

37. I. Wan, B. Wang, J.S. Khim, S. Hong, and J. Hu, *International Science and Technology*, 2014, 48, 4153.

38. S.A. Murphy, B.M. Solomon, S. Meng, J.M. Copeland, T.J. Shaw, and J.L. Ferry, *Environmental Science and Technology*, 2014, 48, 3815.

39. G.I. Cowie and J.L. Hedges, *Analytical Chemistry*, 1984, 56, 479.

40. T.H. Xie, *Chromosphere*, 1983, 12, 1183.

41. F.I. Onuska and K.A. Terry, 1985, 57, 801.

42. S. Pedersen-Bjergaard, J. Vedde, and E.M. Brevis, *Chromatography*, 1991, 43, 44.

43. J. Teichman, A. Bevenue, and J.W. Hylin, *Journal of Chromatography*, 1978, 151, 155.

44. S. Jensen, L. Renberg, and L. Reutergard, *Analytical Chemistry*, 1977, 49, 316.

45. J. Japenga, N.J. Wapenaar, F. Smedes, and W. Salomons, *Environmental Technology Letters*, 1987, 8, 9.

46. G. Magiolo, G. Baggnuolo, B. de Tommaso, and V. Uricchio, *International Journal of Environmental Analytical Chemistry*, 2013, 93, 1030.

47. A.R. Magcolo, E.J. Reiner, S.N. Liss, and T. Chen, *International Journal of Environmental Analytical Chemistry*, 2010, 90, 197.

48. I. Desis, K. Kamarionos, M. Athanasiadon, I Athanassiadis, and X. Karahanlis, *International Journal of Environmental Analytical Chemistry*, 2011, 91, 1151.

49. C.A. Krone, B.W. Burrows, D.W. Brown, P.A. Rabisch, A.F. Friedman, and D.C. Halins, *Environmental Science and Technology*, 1986, 20, 1144.

50. X. Li and B.J. Brownwell, *Analytical Chemistry*, 2009, 81, 7926.

51. A. Kido, R. Shinohara, and S. Ero, *Japan Journal of Water Pollution Research*, 1979, 2, 245.

52. T.S. Bates and R. Carpenter, *Analytical Chemistry*, 1979, 51, 551.

53. J.P. Benskin, M.G. Ikono, A.P.C. Gobas, M.R. Wondneb, and J.P. Cosgrove, *Environmental Science and Technology*, 2012, 46, 6505.

54. N. Picer, M. Picer, and P. Strohal, *Bulletin of Environmental Contamination and Toxicology*, 1975, 14, 565.

55. B. Erkman and D. Kolankaya, *International Journal of Environmental Analytical Chemistry*, 2006, 86, 161.

56. N. Yamashita, Y. Urushigawa, S. Masunaga, M.I. Walash, and A. Miyazaki, *International Journal of Environmental Analytical Chemistry*, 2000, 77, 289.
57. V. Salvado, A. Alcaide, N. Carandell, and M. Hidalgo, *International Journal of Environmental Analytical Chemistry*, 2001, 81, 243.
58. F. Mangani, G. Crescenltini, E. Sisti, F. Bruner, and S. Cannarsa, *International Journal of Environmental Analytical Chemistry*, 1991, 45, 89.
59. J. Kjolholt, *Journal of Chromatography*, 1985, 325, 231.
60. L.K. Chai, F. Elie, and C. Jinang, *International Journal of Environmental Analytical Chemistry*, 2014, 94, 519.
61. R. Iuan and F. Tekalp, *International Journal of Environmental Analytical Chemistry*, 2012, 92, 85.
62. Y. Wang, A. Chang, X. Zhoni, and X. Zang, *International Journal of Environmental Analytical Chemistry*, 2012, 92, 1503.
63. H. Garoias, B. Hammouti, and A. Rios, *International Journal of Environmental Analytical Chemistry*, 2012, 92, 1378.
64. K. Zhang, X. Meng, D. Jiang, D. Hu, Y. Zhang, P. Lu, Z. Zeng, S. Yang, and B. Song, *International Journal of Environmental Analytical Chemistry*, 2014, 94, 370.
65. I. Ali, *International Journal of Environmental Analytical Chemistry*, 1997, 68, 83.
66. F. Dryier, A. Bucks, and F.O. Gulakar, *Geochimica and Cosmochimica Acta*, 1988, 52, 1662.
67. K. Hayase, *Geochimica and Cosmochimica Acta*, 1985, 49, 159.
68. E.L. Pontanen and R.J. Morris, *Marine Chemistry*, 1985, 17, 115.
69. B. Raspor, H.W. Nurnberg, P. Valentia, and M. Bramica, *Marines Chemistry*, 1984, 15, 217.
70. K. Hayase, *Journal of Chromatography*, 1954, 295, 530.
71. M.A. Gregory, T.P. McClure, and C.A. Brouckaert, *International Journal of Environmental Analytical Chemistry*, 2003, 83, 65.
72. N. Calace, F. Giglio, S. Mirantes, B.M. Petronio, and M. Ravaioli, *International Journal of Environmental Analytical Chemistry*, 2004, 84, 423.
73. R. Pocklington and C.D. Macgregor, *International Journal of Environmental Analytical Chemistry*, 1973, 3, 81.
74. O. Ozretich and W. Schroeder, *Analytical Chemistry*, 1986, 58, 2041.
75. M.L. Pinto, E. Vale, G. Sontag, and J.I. Noronka, *International Journal of Environmental Analytical Chemistry*, 2013, 93, 1638.
76. A.B. Pode, *X-ray Spectrometry*, 1973, 2, 165.
77. N. Dankers and R. Leane, *Environmental Technology Letters*, 1983, 4, 283.
78. N.R. McQuaker, P.D. Khuchner, and G.N. Chang, *Analytical Chemistry*, 1979, 51, 888.
79. I. Nothlick and W. Reuten, *Gewasser Kindlicle Mitteil Tungen*, 1982, 26, 112.

10 Determination of Organometallic Compounds in Marine Sediments

10.1 ORGANOTIN COMPOUNDS

Despite the prohibition on their use as antifoulants on boats and harbour works during the last decade, concentrations of organotin compounds are very persistent in seawater and marine sediments. Hence, work on determining these substances is continuing. The sorption of tributyltin on estuarine sediments under various conditions of alkalinity has been studied by Unger et al. [1]. The resulting estimated equilibrium sorption coefficients were in general agreement with apparent sorption coefficients calculated from the concentrations of tributyltin in water and sediment at various sites in Chesapeake Bay. However, very high apparent sorption coefficients were found in areas where there was high boating activity; this could be due to the presence of tributyltin paint chips in the sediment. Sorption and desorption coefficients were similar, indicating that sorption of tributyltin was reversible. Desorption kinetics indicated initial rapid desorption followed by desorption at a slower rate.

The determination of organotin compounds in bottom sediments is a complex process that requires a number of analytical steps: sample collection, transport, and storage; extraction of analytes from sediment; derivatisation; extract purification; enrichment; and the final chromatographic measurement. Stanbrusska et al. [2] carried out a detailed study of the whole process. It is time and labour consuming and subject to securing a representative sample. In this review, the most frequently encountered problems and the possible analytical solutions are discussed, which encompass the specific steps of the speciation analysis of these toxic compounds.

Organotin compounds occur in the marine environment in various chemical forms which may affect live organisms in different ways, for example toxicity, persistence, or bioavailability effects. Therefore, it is more critical in such cases to determine the content of a specific species (e.g. the most toxic one) than the measurement of the total content of all physicochemical forms present in a given environment solution.

Isolating of butyltin and phenyltin derivatives from the 'primary matrix' – sediments – and transferring them into a 'secondary matrix', for example an appropriate solvent such as methanol, are necessary because of the other analytical steps such as derivatisation.

Various factors influence the concentration of organotin compounds found in bottom sediments. These include factors such as sediment particle size, surface properties, percentage clay, organic matter content and pH, salinity, and temperature of the water column. Various chromatographic techniques have been employed embodying flame photometric detection (LD 0.5–10 pg Sn/g), mass spectrometry (LD1 pg Sn/g), flame ionisation detection (LD pg Sn/g), and electron capture detection (LD pg Sn/g). Liquid chromatography has also been employed to identify and determine organotin compounds in sediments with detection limits in the pg Sn/g region (see Table 10.1).

Ali Sheikh et al. [11] studied the seasonal behaviour of organotin compounds in protected subtropical ecosystems in Okinawa, Japan. Studies were conducted from February to October 2006. Butyltin compounds were frequently detected in all seasons, while phenyltin compounds were found in the winter and early spring. In the Manko Estuary, the total mean concentrations of butyltin and phenyltin compounds were 22.78 ± 30.85 (mean ± SD, $n = 53$) and 0.08 ± 0.27 ng (Sn)/L, respectively. In the Okukubi Estuary, butyltin and phenyltin compounds were 12.58 ± 23.96 and 0.47 ± 1.67 ($n = 55$) ng (Sn)/L, respectively. The Manko sediments can be classified as lightly contaminated, while the Okukubi sediments were uncontaminated with tributyltin. The mean levels of tributyltin shown in the Manko Estuary exceeded the threshold level and represented an ecotoxicological risk to sensitive aquatic life. Generally, this study reports the occurrence and continuous input of organotin compounds in the protected estuaries many years after legal restrictions on tributyltin usage in coastal waters were implemented by the Japanese Environmental Authorities.

Bustamante et al. [12] monitored the levels of polycyclic aromatic hydrocarbons, polychlorinated biphenyls, methylmercury (MeHg$^+$), and butyltin (mono-, di-, and tributyltin) sediments collected from different sampling points of the UNESCO reserve of the biosphere of Urdaibai (Bay of Biscay) from March 2006 to June 2007. Sediment concentrations ranged as follows: total polyaromatic compounds were 856–3495 µg/kg and total polychlorobiphenyls were 58–220 µg/kg. Organometallic species were always below the limits of detection (0.24 µg/kg for MeHg$^+$, 0.6 µg/kg for MBT, 0.48 µg/kg for DBT, and 1.1 ng/kg for TBT). In both sediments and oyster, PAH sources were mostly combustion. In the case of polychlorobiphenyls, 4–6 chlorine-atom congeners were the most abundant ones. Slight differences in the profile of polyaromatic compounds as well as polychlorobiphenyls can be detected when the matrices were compared with each other. Finally, in the case of polyaromatic compounds, sediment and water column played the main role in the accumulation pathway into the organism in all the sampling stations.

Gilmour et al. [13] have developed an extremely sensitive purge and trap method for the determination of methyltin compounds as methylstannanes in marine sediments. Hydride derivatives were prepared with sodium borohydrides in a closed flow-through system consisting of a purge vessel, chromatograph, and mass spectrometer. Borate buffer added to samples generated hydrogen from sodium borohydride, resulting in high purge efficiencies for mono-, di-, and trimethyltin. Selected ion mode monitoring with the mass spectrometer gave detection limits for methyltins of 3–5 pg as tin. The concentration detection limits for 5 g sediment samples were in the sub-microgram per kilogram level, with a standard deviation of 6%–18%, depending on the methyltin species and sample type. The method is both selective

TABLE 10.1

Example of the Application of Liquid Chromatography in the Identification of Measurement of Organotin Compounds in Sediments

Analytes	Chromatographic Mode	Detector	Detection Characteristic		References
			Advantages/Disadvantages	Limit of Detection	
TBT	Ion exchange	ICP-MS	*Advantages*: Excellent sensitivity linear dynamic range, high-speed analysis; ability to perform isotopic analysis 25, 36, 37	40, 20 (pg Sn)	[3]
DBT	Reversed phase	ICP-MS	*Disadvantages*: Clogging the nebuliser and cones by nonvolatile components of mobile phase; plasma stability decreased the increasing percentage of organic solvents; isocratic mode is preferred over the gradient mode because gradient elution changes the solvent load and may cause plasma instability; poor compatibility with most mobile phases	10 (pg Sn)	[4]
TMT TET, TPrT, TpHT, Tbt	Reversed phase	ICP-MS		2.8–16 (pg Sn)	[5]
BTs	Ion pair	ICP-MS		1.8, 2, 5, 1.85 (pg Sn)	[6]
BTs	Reversed phase	ICP-MS		0.7–2 (ng Sn/g)	[7]
PhTs TBT, TPhT, DBT, DPhT	Reversed phase	MS (SIM mode)	*Advantages*: Tolerates mobile phases with high percentage of organic solvents: gradient solvents are acceptable, preferably in RP chromatographic mode 25	1.8, 0.8	[4]

(Continued)

TABLE 10.1 (*Continued*)
Example of the Application of Liquid Chromatography in the Identification of Measurement of Organotin Compounds in Sediments

Analytes	Chromatographic Mode	Detector	Detection Characteristic Advantages/Disadvantages	Limit of Detection	References
TBT	Reversed phase	MS (SIM mode)	*Disadvantages*: Not as sensitive as ICP-MS; salts cause ion suppression and/or contamination of the interface; eluent systems should contain low amounts of dissolved solids and have high volatility 25	20–65 (pg Sn)	[8]
TBT, TPhT	Cation exchange	Fluorimetric	*Advantages*: Sensitive and selective; has been used in (ng Sn, in combination with normal-phase, 200 μL of ion-exchange, or reversed-phase injected sample) chromatographic modes; lower cost than ICP and MS 74 *Disadvantages*: OTs are non-fluorescent compounds; therefore, post-column derivatisation of flavone derivatives is used to allow detection; limited sensitivity for TBT 74	0.9, 0.03	[9,10]

and specific, eliminating most interference while permitting positive identification of individual methyltin species. Sample weights were typically 5 g, and these were treated directly with the borate buffered sodium borohydride reagent.

The recovery of methyltins from sediments was tested by using anoxic, sulfidic clay sediments from a mid-salinity region of Chesapeake Bay. Monomethyltin (11.2 ng) and dimethyltin (11.5 ng) were completely recovered from sediment. However, recovery of 10 ng of trimethyltin chloride from sediment was only about 70%.

Muller [14] described a gas chromatographic method for the determination of tributyltin compounds in sediments. The tributyltin compounds were first converted to tributylmethyltin by reaction with ethyl magnesium bromide and then analysed using capillary gas chromatography with flame photometric detection analyses gas chromatography–mass spectrometry. Tributyltin was found in marine sediments and could be determined in amounts down to less than 1.05 pg/L.

Hattori et al. [15] extracted alkyltin compounds from sediments with methanolic hydrogen chloride, and then following mixture with sodium chloride and water, the mixture was extracted with benzene and converted to hydrides with sodium borohydride and analysed by gas chromatography using an electron capture in amounts down to 0.02 mg/kg with a recovery of 70%–95%.

Labinski et al. [16] speciated organotin compounds in sediments by capillary gas chromatography using helium microwave–induced plasma emission spectrometers as a detector. These researchers used the procedure to determine mono-, di-, tri- and some tetraalkyltin compounds in sediments. The ionic tin compounds were extracted as diethyldithiocarbamates into pentane and then converted to pentyl magnesium bromide derivatives prior to gas chromatography. Detector limits were 0.05 pg tin equivalent to 10–30 ng/kg.

Nemanic et al. [17] evaluated the efficiency of different extraction procedures for the simultaneous determination of butyltin compounds in marine sediments by gas chromatography–mass spectrometry. Three different polar solvents (acetic acid; a mixture of acetic acid with methanol; and a mixture of acetic acid, methanol, and water) and three different extraction approaches (mechanical shaking, ultrasound, and microwave-assisted extraction) were used for the extraction of butyltin compounds from PACS-2 certified marine sediment reference material. Before determination extracted, butyltin species were derivatised with sodium tetraethyl borate and extracted into iso-octane. The results indicated that 30 min of ultrasonic extraction with 100% acetic acid provided satisfactory recoveries for all certified butyltin. The developed analytical method was successfully applied for the determination of butyltin compounds in coastal sediments of the northern Adriatic Sea. The results demonstrated that butyltins were present in all sediments analysed over a wide range of concentrations ranging from 3 to 934 ng Sn/g (monobutyltin), 3 to 434 ng Sn/g (dibutyltin), and 7 to 1215 ng Sn/g tributyltin.

Staniszenska et al. [18] pointed out that the determination of organotin compounds in bottom sediments is a complex process that requires a number of analytical steps: sample collection, transport, and storage; extraction of analytes from sediment; derivatisation; extract purification; enrichment; and the final chromatographic measurement. The whole process is time and labour consuming as well as subject to securing sample representativeness. In this chapter, the most frequently encountered problems and the possible analytical solutions are presented, which encompass the specific steps of speciation analysis of these toxic compounds.

Ali Sheikh et al. [19] investigated the seasonal behaviour of organotin compounds in protected subtropical estuarine ecosystems in Okinawa, Japan.

The freeze-dried sediment was extracted with toluene containing 0.1% tropalone, and acidic ethanol was added. Organotin compounds were determined by gas chromatography, which separated phenyltin and butyltin compounds.

The work revealed the presence of organotin compounds (butyltin and phenyltin) in water and sediments of the Manko and Okukubi protected estuaries. At some locations in the Manko Estuary, the concentrations of tributyltin reported may pose a risk to marine life. The distribution of organotin compounds clearly reflects activities occurring in those areas; notable levels of tributyl and tributyltin were found in the areas of high boating and agricultural activities, respectively. Furthermore, the results showed significant seasonal variations indicating that more monitoring studies of organotin compounds should remain in priority in the preserved estuarine ecosystem. As such, a better understanding of the risk posed by tributyltin pollution in subtropical estuarine ecosystems around the Okinawa Island can be achieved.

Andreae and Byrd [20] have pointed out that methylstannanes produced by the hydridisation of methyltin compounds were both stable and volatile with boiling points ranging from 0°C to 59°C.

Hodge et al. [21] determined nanogram quantities of the halides of methyltin, dimethyltin, trimethyltin, diethyltin, triethyltin, n-butyltin, di-n-butyltin, tri-n-butyltin, phenyltin, and inorganic tin (IV) in marine sediments by a procedure involving reaction with sodium borohydride to convert to tin hydrides, which are then detected by atomic absorption spectrometry. The compounds are separated on the basis of their differing boiling points, which range from 1.4°C (CH_3SnH_3) to 280°C ((n-$C_4H_4)_3SnH$). Detection limits range from 0.4 µg/kg (SnIV) to 2 µg/kg (tri-n-butyltinchloride). Stannane and the organotin hydrides evolve from the hydride trap in such a manner that they can be identified by a 'retention time'. Tin levels in core samples taken in Narragonsett Bay, United States (expressed as total tin) ranged from 1 mg/kg (pre-1900) to 20 mg/kg in present-day samples.

To determine methyltin, butyltin, and inorganic tin in Great Bay Estuary sediments, Randall et al. [22] extracted the freeze-dried sediment with 2.5 mol/L calcium chloride and 2.5 mol/L hydrochloric acid and analysed by hydride generation atomic absorption spectrometry. Detection limits for inorganic tin and tributyltin were 2.2 and 0.6 ng/kg, respectively. Recoveries of methyltin and butyltin species from spiking experiments were greater than 70% ± 10%. Tributyltin was found in all sampled sites, probably originating from tributyltin-based antifouling paints.

Chromatographic methods have been applied with hydridisation. Jackson et al. [23] used a commercial purge and trap apparatus fitted to a packed gas chromatographic column and flame photometric detector to achieve a 0.1 ng detection with sediment samples. Purge and trap procedures followed by boiling point separations and detection by spectrophotometric methods yield detection limits between 0.01 and 1 ng. Detection of SnH emission by flame emission gives the greatest sensitivity.

Sinex et al. [24] determined methyltin compounds in amounts down to 3–5 pg (as Sn absolute), that is, the sub-microgram per kilogram level, in marine sediments by a procedure involving reaction with sodium borohydride to produce tin hydrides, followed by purge and trap analysis and then gas chromatography with mass spectrometric detection.

Chiron et al. [7] determined butyl and phenyltin compounds in sediments by pressurised liquid chromatography coupled with individually coupled plasma mass spectrometry. Rosenberg et al. [8] similarly used high-performance liquid chromatography coupled with mass spectrometry to determine organotin compounds in saline sediments.

As shown in Table 10.2, detection limits available are well within the requirements to be met when considering environmental sediments.

TABLE 10.2
Detection Limit, Organotin Compounds

Determined	Technique	Detection Limit mg/kg unless Otherwise Stated	References
Organotin Compounds			
Mono-, di-, and trimethyltin; mono-, di-, and tri-*n*-butyltin; mono- and di-ethyltin, phenyltin	Reaction with sodium borohydride to form tin hydrides, controlled evaporation and detection by atomic absorption spectrometry	—	[21]
Methyltin, butyltin	Reaction with sodium borohydride to form tin hydrides, controlled evaporation and detection by atomic absorption	0.6×10^{-6}	[22]
Methyltin	Reaction with sodium borohydride to form tin hydrides, purge and trap analysis followed by gas chromatography with mass spectrometric detection	3–5 pg (as Sn) sub µg/kg	[23]
Tributyltin	Conversion to trimethyl–butyltin and determination by gas chromatography with atomic absorption detection	5 pg/L	[14]
Mono-, di-, and trimethyltin	Gas chromatography with helium microwave–induced plasma emission spectrometry	10–30 kg^{-1}	[16]

10.2   ORGANOLEAD COMPOUNDS

Organolead compounds are more toxic than inorganic lead compounds. They can be produced by biological methylation of inorganic and organic lead compounds in the aquatic environment by microorganisms.

Chau et al. [25] described a hexane extraction procedure to extract tetramethyllead, trimethylethyllead, methyltriethyllead, and tetraethyllead from marine sediments. The extracted compounds were analysed in their authentic forms by a gas chromatographic–atomic absorption spectrometric system. Other forms of organic and inorganic lead do not interfere. The detection limit was 0.01 mg/kg as lead.

To digest sediment samples, 5 g of sediment and EDTA solution are extracted with 5 mL of hexane and the resulting solution is gas chromatographed. Concentrations found in a marine sediment ranged from 8.3 mg/kg (tetramethyllead and methyltriethyllead) to 12 mg kg (dimethyldiethyllead and tetraethyllead) and recoveries in spiking experiments were between 81% and 84%. Down to 0.01 mg/kg organolead can be determined by this method.

Chau et al. [26] have described a simple and rapid extraction procedure to extract the five tetraalkyllead compounds (Me_4Pb, Me_3EtPb, Me_2Et_2Pb, $MeEt_3Pb$, and Et_4Pb) from sediment. The extracted compounds are analysed in their authentic form by gas chromatography–atomic absorption spectrometry.

The method is applicable to nonsaline and saline sediments. Other forms of organic or inorganic lead do not interfere. The detection limit for sediment (5 g sample) is 0.01 mg/kg. Recoveries from sediments ranged from 94% for triethyllead to 111% for trimethyllead in the range of 1–20 µg alkyllead spiked to 1g of sediment. An average standard deviation of 4% for trimethyllead and triethyllead and 15% for dialkylead compounds was obtained.

In this method, EDTA is added to the sediments and a hexane extract is examined by gas chromatography.

10.3   ORGANOARSENIC COMPOUNDS

The adsorption of organoarsenic compounds onto sediments under saline conditions varies with the species of the organoarsenic compound involved. It has been shown [27] that inorganic arsenic (III and V), monomethylarsenic, and dimethylarsenic acids are present in natural waters [28,29] and sediments [30,31]. Thus, analytical methods for the separation and measurement of the species are necessary for the study of pathways of accumulation and deposition.

Maher et al. [32] have described a method for the determination of down to 0.01 mg/kg of organoarsenic compounds in marine sediments. In this procedure, the organoarsenic compounds are separated from an extract of the sediment by ion-exchange chromatography, and the isolated organoarsenic compounds are reduced to arsines with sodium borohydride and collected in a cold trap. Controlled evaporation of the arsine fractions and detection by atomic absorption spectrometry completes the analysis.

10.4   ORGANOMERCURY COMPOUNDS

Bartlett et al. [33] observed the unexpected behaviour of methylmercury containing River Mersey sediments during storage. They experienced difficulty in obtaining consistent methylmercury values; supposedly identical samples analysed at intervals of a few days gave markedly different results. They followed the levels of methylmercury in selected sediments. They found that the amounts of methylmercury observed in the stored sediment did not remain constant; initially, there was a rise in the amount of methylmercury observed, and then, after about 10 days, the amount present began to decline to levels which in general only approximate those originally present. They observed this phenomenon in nearly all Mersey sediment samples they examined.

It was noted that sediments sterilised, normally by autoclaving at approximately 120°C, did not produce methylmercury on incubation with inorganic mercury, suggesting a microbiological origin for the methylmercury. A controlled experiment was carried out in which identical samples were collected and homogenised. Some of the samples were sterilised by treatment with an approximate 4% w/w solution of formaldehyde. Several samples of both sterilised and unsterilised sediments were analysed at intervals, and all the samples were stored in ambient room temperature (18°C) in the laboratory. There is a difference in behaviour between the sterilised and unsterilised samples. Some of the samples were separately inoculated into various growth media to test for microbiological activity.

This work suggests that the application of laboratory-derived results directly to natural conditions could, in these cases, be misleading: analytical results for day 10 of the extrapolated directly might lead to the conclusion that natural methylmercury levels and rates of methylation are much greater than they really are. Work in this area with model or laboratory systems needs to be interpreted with particular caution.

Bartlett et al. [33] determined methylmercury compounds by a method in which samples of 2–5 g were extracted with toluene after treatment with copper sulphate and an acidic solution of potassium bromide. Methylmercury was then back-extracted into aqueous sodium thiosulphate. This was then treated with acidic potassium bromide and copper sulphate following which methylmercury was extracted into pesticide-grade benzene containing approximately 100 µg/dL of ethyl mercuric chloride as an internal standard. The extract was analysed by electron capture gas chromatography with Ni[65] detector. The glass column (1 m × 0.4 cm) was packed with 5% neopentyl glycol adipate on Chromosorb G (AW-DMCS). Methylmercury was measured by comparing the peak heights with standards of methylmercuric chloride made up in the methylmercury–benzene solution. The results were calculated as nanograms of methylmercury per gram dry sediment. The detection limit was 1–2 ng/g; 2.2 µg/L means total mercury levels were found in 136 samples of bottom deposits from the Mersey Estuary.

REFERENCES

1. M.A. Unger, W.G. Macintyre, and R.J. Huggett, *Environmental Science and Technology*, 1988, 7, 907.
2. M. Stanbrusska, B. Radke, J. Namiesnik, and J. Bolalek, *International Journal of Environmental Analytical Chemistry*, 2008, 88, 747.

3. J.W. McLaren, K.W.M. Sui, J.W. Lam, S.N. Willie, P.S. Maxwell, A. Palepu, M. Koether, and S.S. Berman, *Fresenius Journal of Analytical Chemistry*, 1990, 337, 721.
4. S. White, T. Catterick, B. Fairman, and K. Webb, *Journal of Chromatography A*, 1998, 794, 211.
5. H.J. Yang, S.J. Yang, and C. Hwang, *Analytica Chimica Acta*, 1995, 312, 141.
6. W.S. Chao and S.J. Jiang, *Journal of Analytical Atomic Spectrometry*, 1998, 13, 1337.
7. S. Chiron, S. Roy, R. Cottier, and R. Jeannot, *Journal of Chromatography A*, 2000, 879, 137.
8. E. Rosenberg, V. Kmetov, and M. Grassorbaur, *Fresenius Journal of Analytical Chemistry*, 2000, 366, 400.
9. E. Grauperer, C. Leal, M. Grandados, M.D. Prat, and R. Compaño, *Journal of Chromatography*, 199, 846, 413.
10. E. Gonzales-Toledo, R. Compaño, M. Granados, and. M.D. Prat, *Journal of Chromatography*, 2000, 878, 69.
11. M. Ali Sheikh, K. Tsuha, Y. Wong, K. Sawano, G.T. Imo, and T. Oomen, *International Journal of Environmental Analytical Chemistry*, 2007, 87, 847.
12. J. Bustamante, A. Albisu, L. Bartolomi, A. Atutza, S. Arrasote, E. Anakate, A. De Diego, A. Usobiaga, and O. Zulnaga, *International Journal of Environmental Analytical Chemistry*, 2010, 90, 722.
13. C.C. Gilmour, J.H. Tuttle, and J.C. Means, *Analytical Chemistry*, 1986, 58, 1848.
14. M.D. Muller, *Fresenius Zeitschrift für Analytica Chemie*, 1984, 317, 32.
15. Y. Hattori, A. Kayabaski, and S. Takemoto, *Journal of Chromatography*, 1984, 315, 341.
16. R. Labinski, W.M. Dirlex, and F.C. Adams, *Analytical Chemistry*, 1992, 64, 159.
17. T.M. Nemanic, R. Milasic, and J. Scanecar, *International Journal of Environmental Analytical Chemistry*, 2007, 87, 615.
18. M. Staniszenska, B. Radka, J. Namiesih, and J. Bolalek, *International Journal of Environmental Analytical Chemistry*, 2008, 88, 747.
19. M. Ali Sheikh, K.R. Suba, X. Wong, K. Sawano, S.T. Imo, and T Oomari, *International Journal of Environmental Analytical Chemistry*, 2007, 87, 4.
20. M.O. Andreae and J.T. Byrd, *Analytica Chimica Acta*, 1984, 156, 147.
21. V.F. Hodge, S.L. Snider, and D. Goldberg, *Analytical Chemistry*, 1979, 91, 1256.
22. L. Randall, J.S. Han, and J.H. Weber, *Environmental Technology Letters*, 1986, 7, 571.
23. J.A. Jackson, W.R. Blair, E.F. Brinkman, and W.P. Iveson, *Environmental Science and Technology*, 1982, 16, 110.
24. S.A. Sinex, A.Y. Centillo, and G.R. Helz, *Analytical Chemistry*, 1980, 52, 2342.
25. Y.K. Chau, P.T.S. Wong, G.A. Bengert, and R.N. Markell, *Water Pollution Central*, 1979, 51, 186.
26. Y.K. Chau, P.T.S. Wong, G.A. Benger, and O. Kramer, *Analytical Chemistry*, 1979, 51, 186.
27. L.W. Jacobs, J.K. Syers, and D.R. Keeney, *Proceedings of the Soil Science of America*, 1970, 34, 750.
28. M.W. Andrea, *Analytical Chemistry*, 1977, 49, 820.
29. J. Persson and K. Irgum, *Analytica Chimica Acta*, 1982, 138, 111.
30. D.L. Johnson and R.S. Braman, *Deep Sea Research*, 1975, 22, 503.
31. D.G. Inverson, A.A. Anderson, T.R. Holm, and R.R. Stanforth, *Environmental Science and Technology*, 1979, 13, 1491.
32. W.A. Maher, *Analytica Chimica Acta*, 1981, 126, 157.
33. P.D. Bartlett, P.J. Craig, and S.F. Moreton, *Nature*, 1977, 267, 606.

Index

9 781032 340098